# Interdisciplinary Approaches to Digital Transformation and Innovation

Rocci Luppicini
*University of Ottawa, Canada*

A volume in the Advances in E-Business Research (AEBR) Book Series

IGI Global
DISSEMINATOR OF KNOWLEDGE

Published in the United States of America by
IGI Global
Business Science Reference (an imprint of IGI Global)
701 E. Chocolate Avenue
Hershey PA, USA 17033
Tel: 717-533-8845
Fax: 717-533-8661
E-mail: cust@igi-global.com
Web site: http://www.igi-global.com



Library of Congress Cataloging-in-Publication Data

Names: Luppicini, Rocci, editor.
Title: Interdisciplinary approaches to digital transformation and innovation / Rocci Luppicini, editor.
Description: Hershey, PA : Business Science Reference, [2020] | Includes bibliographical references and index. | Summary: "This book explores interdisciplinary approaches to e-business technology. It also examines the ethical, legal, social, and cultural issues connected to the current and future developments in the field"-- Provided by publisher.
Identifiers: LCCN 2019034114 (print) | LCCN 2019034115 (ebook) | ISBN 9781799818793 (h/c) | ISBN 9781799818809 (ebook)
Subjects: LCSH: Digital electronics. | Technological innovations.
Classification: LCC TK7868.D5 H323 2020 (print) | LCC TK7868.D5 (ebook) | DDC 658/.05--dc23
LC record available at https://lccn.loc.gov/2019034114
LC ebook record available at https://lccn.loc.gov/2019034115

This book is published in the IGI Global book series Advances in E-Business Research (AEBR) (ISSN: 1935-2700; eISSN: 1935-2719)

British Cataloguing in Publication Data
A Cataloguing in Publication record for this book is available from the British Library.

# Advances in E-Business Research (AEBR) Book Series

ISSN:1935-2700
EISSN:1935-2719

Editor-in-Chief: In Lee, Western Illinois University, USA

## MISSION

Technology has played a vital role in the emergence of e-business and its applications incorporate strategies. These processes have aided in the use of electronic transactions via telecommunications networks for collaborating with business partners, buying and selling of goods and services, and customer service. Research in this field continues to develop into a wide range of topics, including marketing, psychology, information systems, accounting, economics, and computer science.

The **Advances in E-Business Research (AEBR) Book Series** provides multidisciplinary references for researchers and practitioners in this area. Instructors, researchers, and professionals interested in the most up-to-date research on the concepts, issues, applications, and trends in the e-business field will find this collection, or individual books, extremely useful. This collection contains the highest quality academic books that advance understanding of e-business and addresses the challenges faced by researchers and practitioners.

## COVERAGE

- E-procurement methods
- Web Service-Based Business Systems
- E-Marketing
- Web 2.0
- Sustainable E-business
- Economics of e-business
- E-business standardizations
- Electronic Supply Chain Management
- Valuing e-business assets
- Semantic Web



The Advances in E-Business Research (AEBR) Book Series (ISSN 1935-2700) is published by IGI Global, 701 E. Chocolate Avenue, Hershey, PA 17033-1240, USA, www.igi-global.com. This series is composed of titles available for purchase individually; each title is edited to be contextually exclusive from any other title within the series. For pricing and ordering information please visit http://www.igi-global.com/book-series/advances-business-research/37144. Postmaster: Send all address changes to above address. 

## Titles in this Series

***For a list of additional titles in this series, please visit:***
***https://www.igi-global.com/book-series/advances-business-research/37144***

***Implications and Impacts of eSports on Business and Society Emerging Research and Opportunities***
David J. Finch (Mount Royal University, Canada) Norm O'Reilly (University of Guelph, Canada) Gashaw Abeza (Towson University, USA) Brad Clark (Mount Royal University, Canada) and David Legg (Mount Royal University, Canada)
Business Science Reference • © 2020 • 185pp • H/C (ISBN: 9781799815389) • US $165.00

***Tools and Techniques for Implementing International E-Trading Tactics for Competitive Advantage***
Yurdagül Meral (İstanbul Medipol University, Turkey)
Business Science Reference • © 2020 • 395pp • H/C (ISBN: 9781799800354) • US $225.00

***Handbook of Research on Strategic Fit and Design in Business Ecosystems***
Umit Hacioglu (Istanbul Medipol University, Turkey)
Business Science Reference • © 2020 • 775pp • H/C (ISBN: 9781799811251) • US $295.00

***Business Transformations in the Era of Digitalization***
Karim Mezghani (Al Imam Mohammad Ibn Saud Islamic University, Saudi Arabia & University of Sfax, Tunisia) and Wassim Aloulou (Al Imam Mohammad Ibn Saud Islamic University, Saudi Arabia & University of Sfax, Tunisia)
Business Science Reference • © 2019 • 360pp • H/C (ISBN: 9781522572626) • US $215.00

***E-Manufacturing and E-Service Strategies in Contemporary Organizations***
Norman Gwangwava (Botswana International University of Science and Technology, Botswana) and Michael Mutingi (Namibia University of Science and Technology, Namibia)
Business Science Reference • © 2018 • 366pp • H/C (ISBN: 9781522536284) • US $205.00

***Multi-Sided Platforms (MSPs) and Sharing Strategies in the Digital Economy Emerging Research and Opportunities***
Sergey Yablonsky (St. Petersburg University, Russia)
Business Science Reference • © 2018 • 192pp • H/C (ISBN: 9781522554578) • US $165.00

***For an entire list of titles in this series, please visit:***
***https://www.igi-global.com/book-series/advances-business-research/37144***

IGI Global
DISSEMINATOR OF KNOWLEDGE

701 East Chocolate Avenue, Hershey, PA 17033, USA
Tel: 717-533-8845 x100 • Fax: 717-533-8661
E-Mail: cust@igi-global.com • www.igi-global.com

# Editorial Advisory Board

# Table of Contents

# Detailed Table of Contents



Digital transformation is reshaping many areas of work and life within contemporary society. These include healthcare, education, government, politics, law, human rights, and ethical controversies. This chapter addresses the following questions, What is the current conception of digital transformation as an emerging interdisciplinary field of research and study? The objectives of the chapter are twofold: (1) to conceptualize digital transformation as an emerging interdisciplinary field and (2) to identify key research areas that currently constitute digital transformation. The chapter contributes by positing a comprehensive systems definition of digital transformation as an interdisciplinary research field to help guide researchers and other leaders in the field.

Rural economy has been characterized by low incomes and self-consumption production. With the expansion of global markets and the access of customers through internet, the possibility of moving traditional markets to e-commerce increases. This expansion allows the inclusion of rural economy into the e-commerce market. This chapter describes the challenges to be overtaken in order to activate rural economy through e-commerce. The chapter is organized in five sections: The first section focuses on the communication infrastructure available in rural areas. In the second section, the current state and challenges to be addressed for guaranteeing on time delivery are presented. The third section describes payment methods. The fourth section presents schemes of organization required into the communities to guarantee the quality of products. Finally, marketing, advertising, and social networks are discussed.



This chapter explores the concept of trust and issues relating to how the construct is conceptualised and understood in a traditional offline context as well as in online environments. The chapter opens with a mini case study that highlights the complexities of being a privately-owned small firm operating in a dynamic and largely unregulated web environment. The firm is relatively new to the financial sector thus augmenting the challenges that lie in reducing perceived risk in an industry that has a chequered history with customer perceptions of credibility and integrity in the financial services sector. The chapter introduces the theoretical underpinning, which draws from the trust theories and technology adoption at firm level, which is critiqued through the lens of the technology acceptance model. Concepts related to institution-based trust are discussed and managerial implications are considered for pure play firms operating online. Each section of the chapter explores these theoretical perspectives from a FinTech context.

Universities have a meaningful function in motivating young graduates to become entrepreneurs in the technology field. Still, Indonesia has an unemployment issue for people's welfare. Indonesia has many potential markets. The first objective of the research is to investigate the indicators and success factors of business incubators in Indonesian public universities. Second is to examine the critical factors that influence the success of business incubators in the Indonesian public universities The third is to propose and develop successful business incubators in Indonesian public universities, and fourth is to measure the influence of IT and e-commerce assistance factor to the success of the university business incubator. The results indicated that there are four significant success factors (i.e., information technology and e-commerce assistance – abilities of business incubator, mentoring and networking, funding and support, and university regulation) with the moderating factors of age of facility, credit and rewards, and good infrastructure system.



This chapter focuses on predicting web user behaviors. When web users enter a website, every move they make on that website is stored as web log files. Unlike the focus group or questionnaire, the log files reflect real user behavior. It can easily be said that having actual user behavior is a gold value for the organizations. In this chapter, the ways of extracting user patterns (user behavior) from the log files are sought. In this context, the web usage mining process is explained. Some web usage mining techniques are mentioned.



Translational research (TR) is the harnessing of knowledge from basic science and clinical research to advance healthcare. As a sister discipline, translational informatics (TI) concerns the application of informatics theories, methods, and frameworks to TR. This chapter builds upon TR concepts and aims to advance the use of machine learning (ML) and data analytics for improving clinical decision support. A federated machine learning (FML) architecture is described to aggregate multiple ML endpoints, and intermediate data analytic processes and products to output high quality knowledge discovery and decision making. The proposed architecture is evaluated for its operational performance based on three propositions, and a case for clinical decision support in the prediction of adult Sepsis is presented. The chapter illustrates contributions to the advancement of FML and TI.



This study investigates the switching intention and actual behavior of e-wallet users in Greater Jakarta by using the push-pull and mooring model. There were 33 constructs distributed in Greater Jakarta. Using a quantitative technique, this research used non-probability sampling. The samples collected consisted of 357 valid responses. PLS-SEM and bootstrapping were conducted to check the validity, reliability, and hypothesis testing. Based on the results, low efficiency and personal experience had no significant influence towards switching intention. There was a significant influence from switching cost, social image, security, and alternative attractiveness towards switching intention. Lastly, there was a significant influence of switching intention towards actual behavior. The results gained from the study generated managerial and theoretical implications, which later can be used for the e-wallet provider to improve its services.



Thinking in terms of blockchain technology has developed rapidly over the past decade, with the focus primarily being on all the new possibilities created by this essentially technology-driven development. At its core, blockchain technology consists of a combination of reliable communication mechanisms between systems and the performance of joint decision-making processes by these systems. Consensus algorithms enable combinations of hardware, algorithms, and software (i.e., cyber-physical systems) to operate in networks and engage in secure intercommunication and interaction in these networks. Thanks to intercommunication, cyber-physical systems are able to reach consensus on decisions on information transactions to perform. This, in turn, empowers interconnected systems to increase their level of autonomy and adapt their individual and joint operations to changes that emerge from within them or from their environment.

## Section 3
## Trust, Equity, and Security



The main purposes of this research are (1) to reveal the virtual credit card (VCC) awareness of online shoppers and (2) to prepare a to do list for managers and relevant institutions to provide a more secured internet shopping process by enhancing the trust perception of consumers. To this end, a structured questionnaire consists of three sections was designed and conducted during the period April-May 2018 in Turkey. The findings achieved show that online shoppers mostly attach importance to the recognition of the website, which means they seek trust in online shopping process. On the other hand, they mostly use credit and debit cards, which may make consumers encounter serious fraud issues. VCC usage is only 4.9% in general although VCC awareness is 55.1%. This low level of VCC awareness is associated with education level and the information provided by the financial banks.



The influence of social media such as Facebook, LinkedIn, and Twitter plays an increasingly influential role in the daily lives of people. Despite the rise of interest in this topic, the research discussing the ethical concerns of using social media for recruitment purposes remains in exploratory stages. This chapter provides a systematic review of recent research that was published from 2012 to 2018 and focused on ethical issues related to the use of social media for recruitment purposes. The techno-ethical lens, which studies the impact of technology on ethics, was used to explore the social and ethical aspects of how recruiters use social media for recruitment purposes.

# Foreword

It is with great anticipation and pride that I am writing this forward for a new IGI Global Research Handbook on Digital Transformation and Innovation. The editor, Dr. Luppicini has brought together an excellent set of articles that provide a good sampling of current research in this exciting new field.

Dr. Luppicini and I worked together in 2014 to create a new PhD Seminar course for the first cohort of PhD students in an interdisciplinary program between the Telfer School of Management, the Faculty of Arts, and the Faculty of Engineering at uOttawa. Then, as now, the challenge was to bring together students and researchers from diverse backgrounds and bring them up to speed on a wide range of topics from a wide range of perspectives to study a phenomenon that is transforming our world. In both cases, Dr. Luppicini has brought together a wonderful collection of contributions for us to learn from.

The phenomenon of digital transformation and innovation is the third communication revolution to fundamentally transform our world and redefine what it means to be a human society. The first communication revolution was when humans developed the capacity for speech (starting about 1 million years ago[1]) which distinguished them from all other living things. The second communication revolution was when humans developed the capacity for the capacity for the written word, including mathematics and legal documents (3-5000 years ago[2]) which was the basis for civilization well into the 20$^{th}$ century. We are now, since about 1950, in the beginning of a third communication revolution brought about by digital electronic technology. From satellites to computers and telecommunications; and from the Internet to smartphones and ubiquitous sensors; and from online-shopping to global positioning systems and social media, technology is transforming human society. We are literally re-inventing and re-designing the way we are governed, engage in business transactions, and live our lives in community as human beings.

To understand and research this phenomenon, Digital Transformation and Innovation requires multi-disciplinary collaboration between technology, management and humanities researchers. At its heart, the technology enables the collection and communication of huge amounts of data that transforms how business and society work. It also creates a new online environment where the experience of business and social interactions by individuals is being reinvented. Innovation is an important aspect of the program to emphasize the re-invention and creative design of user experiences in business and social interactions

The collection of papers in the Research Handbook are organized into three main groupings. The first set of papers look at the overall impact digital transformation can have in developing economies, and the new opportunities and challenges that individual companies may face. The second set of papers drills into the more technical aspects of the technology and the potential applications that can improve user experience in this new world. While the final set of papers focus on challenges and potential pitfalls in this new world, in terms of trust, equity and security. Throughout the underlying message is that while digital electronic technology offers exciting new opportunities and challenges, human experience in this new world is the critical element that should guide our research and prioritize our focus.

Enjoy.

*Liam Peyton*
*University of Ottawa, Canada*

**Liam Peyton** *is currently Vice-Dean Graduate Studies for the Faculty of Engineering at the University of Ottawa, Canada. He has been a professor there and a principal investigator for the Intelligent Data Warehouse Laboratory since 2002. He had 10 years' experience, before joining uOttawa as a consultant specializing in digital transformation in government, telecommunications, healthcare, and online retail. He and Dr. Dominic Ferrand led the creation of first a diploma (2002), and then a masters (2008) and finally a PhD (2014) at uOttawa in Electronic Business which is now rebranded as Digital Transformation and Innovation (2020). These programs are the only ones in the world that are fully interdisciplinary between Technology, Management and the Humanities.*

## ENDNOTES

1 http://www.historyworld.net/wrldhis/PlainTextHistories.asp?historyid=ab13
2 https://www.ancient.eu/timeline/writing/

# Preface

## OVERVIEW

The increasingly complex, dynamic, and uncertain intertwinement of Internet technologies and business processes within our evolving knowledge society has created new challenges and opportunities that cross disciplinary boundaries and require the expertise and collaboration from experts within multiple fields and disciplines. What organizational changes are influenced by the introduction of new information and communication technologies and computer information systems? How do new information technologies and system architectures manage online business transactions, services, organizations and communities? How does information and telecommunication technology influence society and create innovations in fields such as health, education, law and visual arts? What are the social, cultural, ethical, legal, and political challenges of adopting digital technologies in life and work at the individual, organizational, and public level? Interdisciplinary approaches to Digital Transformation (DT) cross disciplinary boundaries needed solve complex real-world problems at the interface of business, technology and society. *Interdisciplinary Approaches to Digital Transformation and Innovation* provides selective sampling of current research within DT as an emerging interdisciplinary research field. This is the first major research publication that offers such a holistic vision of DT to help leverage knowledge concerning current practices, challenges and future opportunities.

The objective of *Interdisciplinary Approaches to Digital Transformation and Innovation* is to provide a selective sample of coverage of new research in DT as an emerging interdisciplinary field of research. This collection covers contributions that span Business, Engineering, the Social Sciences and Humanities, Health Sciences, and interdisciplinary areas of scholarship focused on the complex interconnections between technology, business, life and society. This edited volume makes a contribution to multiple disciplines by presenting key perspectives on the evolving research field of E-Business Technology. There is a growing demand for such a text to help guide technology designers, information specialists, researchers, engineers, managers, and administrators in areas where emerging new technologies are having an important impact on business and society.

This edited book has a unique interdisciplinary scope which should be of interest to students, instructors, researchers, and managers working in complex technological contexts who need expert knowledge about technology to inform current technology practices and research in multiple technology oriented fields including, Business, Engineering and Computer Science, Health Sciences, Social Sciences, Humanities, and Technology Studies. It will be of particular interest to individuals working closely in technology related areas of research and practice. Readers will gain a thorough understanding of the scope, complexity, and importance of e-business technologies in society from key experts leading the field

## OBJECTIVES AND ORGANIZATION

This edited book pulls together a growing body of work within the emerging interdisciplinary research field of Digital Transformation. This volume contains 12 chapters divided into three sections to highlight a logical flow of writing organized into key thematic areas roughly following Whitehead's (1929) logical description of essential activities for constructing knowledge through human learning cycles of romance, definition, generalization. In the romance phase of learning, *Interdisciplinary Approaches to Digital Transformation and Innovation* gets the reader acquainted with the conceptual roots and boundaries of DT at a global (big picture) and local (organizational) level. Section 1: Transformation and Impact on Developing Economies and Individual Companies focuses on the conceptualization of DT and how it is being applied within developing economies and select organizations. Part of the initiation to DT involves familiarization with where the field came from (roots), what it looks like at a macro level (developing economies), and what it means when applied strategically within particular contexts (individual organizations). This multi-faceted bird's eye view of DT begins to explore the reader to the richness of DT when treated as an interdisciplinary research activity. This is the view from outside.

The next layer of Whitehead's cycle is definition, which in this applied technological oriented field, takes the reader on a journey into the world of DT to meet the DT users interacting with applications in an effort to better define what users want (do not want), need (do not need), and how digital technologies are fulfilling (or not) these interests and needs. To this end, Section 2: Technology and Applications for Improving User Experience, takes a glimpse at the world of the user-technology interface to help define the locus of what matters most in DT. This is the view from insider. The final learning cycle from Whitehead, definition, pulls together the essence of DT and key concerns that remain after the romance is over (or transformed into lasting love) and the sun has set on the known horizon (definition). The generalization phase of the learning cycle leaves the reader with a pervasive list of lessons to take forward. Section 3: Trust, Equity and Security leaves the reader with essential ingredients (trust, equity, security) to DT and other technology-oriented fields where the intertwinement of humans and new technologies create both opportunities and challenges to navigate. With this vision in mind, detailed descriptions of the chapters are as follows:

- Chapter 1, "Digital Transformation and Innovation Explained: A Scoping Review of an Evolving Interdisciplinary Field," by Rocci Luppicini, briefly reviews the recent explosion of Digital Transformation as an interdisciplinary research field to help set the tone for this edited collection. A scoping research is conducted to identify key research areas in the field and advance a comprehensive systems definition of Digital Transformation to help frame the current collection and guide researchers and other leaders in the field.
- Chapter 2, "Rural Economy Activation Through E-commerce: Challenges and Opportunities," by Miguel A. Sánchez-Acevedo, Israel Álvarez-Velásquez, Beatriz A. Sabino-Moxo, José A. Márquez-Domínguez, and Ma. del Rocio G. Morales-Salgado, focuses on the expansion of the global economy and increased user Internet access, particularly with respect to rural economies. This creates a variety of new opportunities and challenges to understand and navigate. This chapter touches upon the role of e-payment methods, quality of products, e-marketing and the role of social networks
- Chapter 3, "In Digital Technology We Trust: A FinTech B2B Context," by Roisin Vize, tackles the challenging issue of trust as it is understood in offline and online context. The practical case covered in the chapter looks at how a small firm operates in this complex new world of trust and business ethics from the perspective of the Technology Acceptance Model.
- Chapter 4, "Final Framework for a Successful Business Incubator for Indonesian Public Universities: The Influence of Information Technology Factor on Business Incubator Success," by Lina Gozali, Maslin Masrom, Teuku Yuri Zagloel, Habibah Haron, Jose Arturo Garza-Reyes, and Benny Tjahjono, investigates key indicators and success factors related to business incubators in Indonesian public universities. Chapter findings uncover multiple success factors and moderating factors that help build success. This practical research study provides a useful case for other similar institutions to learn from and adapt to their own needs.
- Chapter 5, "They Know What You Will Do Next Click," by Serra Çelik, explores Web user behaviour prediction. How can digital technologies be used to improve learner experience? How can web user experience be used

to predict user likes, dislikes and behaviour? Does what you click shape what you like, what you do not like, and maybe even who you are? Can data mining tell users who they are and minimally help them decide what to buy or avoid buying? This chapter opens up the Pandora box to look at what might be looking back at users and learning from their online behaviors.

- Chapter 6, "An Assessment of Federated Machine Learning for Translational Research," by Manoj A Thomas, Diya Suzanne Abraham, and Dapeng Liu, deals with the complexities of Translational informatics (TI) and how the power of machine learning (ML) and data analytics can be harnessed to improve clinical decision support.
- Chapter 7, "Factors that Determine Users' Switching Intention From a Debit/ Credit Card to Adopting E-Wallet as a Payment Method," by Michael Lo and Adhi Setyo Santoso, discusses an empirical study on the switching intention and behaviors of E-wallet users. This chapter helps advance understanding of E-wallet payment methods and other similar applications becoming increasing popular in DT innovation.
- Chapter 8, "The Phenomenon of Blockchain Technology and the Future of Self-Stabilising and Self-Adaptive Systems of Systems," by Ben van Lier, focuses on one of the leading research areas in DT, blockchain technology, and how it is creating new decision-making processes through technologically driven communication mechanisms, increasingly sophisticated algorithms, ever advancing machine autonomy in decision making.
- Chapter 9, "Consumer Protection in Online Shopping: An Investigation on Virtual Credit Card Awareness in Turkey," by Emre Yildirim; Chapter 10, "A Systematic Review of the Use of Social Media as a Recruitment Tool on the Social, Ethical, and Privacy Context," by Ghadah Althawwad; Chapter 11, "Bias and Discrimination in Artificial Intelligence: Emergence and Impact in E-Business," by Jan C. Weyerer and Paul F. Lange; and Chapter 12, "Delivering Effective Cybersecurity Awareness Training to Support the Organizational Information Security Function," by Regner Sabillon, provide a rich take away of new ideas and core considerations for researchers and practitioners to grapple with in the next phase of DT research development.

Taken together, this 12-chapter volume, selectively covers key research areas and issues faced by the exciting new research field of Digital Transformation.

## CONCLUSION

Taken together, the above chapters nicely illustrate the diversity of areas of research opening up to the interdisciplinary research field of Digital Transformation. At the same time, many of the current chapters leave unanswered questions for future researchers to take up. It is expected that scholarly work will continue to flourish within the expanding landscape of new research for years to come.

## REFERENCES

Whitehead, A. N. (1929). *The aims of education and other essays.* New York: MacMillan.

# Section 1

# Transformation and Impact on Developing Economies and Individual Companies

# Chapter 1
# Digital Transformation and Innovation Explained:
## A Scoping Review of an Evolving Interdisciplinary Field

**Rocci Luppicini**
*University of Ottawa, Canada*

## ABSTRACT

*Digital transformation is reshaping many areas of work and life within contemporary society. These include healthcare, education, government, politics, law, human rights, and ethical controversies. This chapter addresses the following questions, What is the current conception of digital transformation as an emerging interdisciplinary field of research and study? The objectives of the chapter are twofold: (1) to conceptualize digital transformation as an emerging interdisciplinary field and (2) to identify key research areas that currently constitute digital transformation. The chapter contributes by positing a comprehensive systems definition of digital transformation as an interdisciplinary research field to help guide researchers and other leaders in the field.*

## INTRODUCTION

*The object of the critic, then, is to seek not the unity of the work but the multiplicity and diversity of its possible meanings, its incompleteness, the omissions which it displays, and above all, its contradictions –C. Belsey, 1980*

DOI: 10.4018/978-1-7998-1879-3.ch001

The above quote illustrates the depth of knowledge and rigor required to capture a comprehensive understanding of the complex and diverse domain of Digital Transformation (henceforce referred to as DT) as an emerging interdisciplinary research field. As an emerging field, critical reflection on the conceptual development and current meaning of the field are required. Conceptual development in DT provides essential theoretical grounding for ongoing academic research and practice. It also provides useful indicators of changes in the field and new directions. Unfortunately, the roots of DT are not easily discerned for multiple reasons. First, the conceptual development of Digital Transformation is complex and intertwined with changing conceptions of E-Business and Technology. Second, the multiplicity of theoretical roots for DT are difficult to access within silos of traditional institutions of Higher Education where DT oriented academic programs and connected research activities are often unknown outside the department or school within which they are created and managed. This chapter focuses primarily on exploring the conceptualization DT as an emerging interdisciplinary research field.

Gleaming back over the short history of DT appears to reveal three primary stages in DT development, namely, E-commerce, E-Business, and DT. Early roots of DT were mainly found under the umbrella of E-Commerce aligning with Internet and computer networking developments in the 1980s and 1990s the. This was followed by a gradual broadening of research areas and professional interests in the 1990s and 2000s under the term E-Business and E-Business Technology. Aghaei, Nematbakhsh, and Farsani (2012) traced this transition to the transition of Web 1.0 to Web 2.0 with the shift from Internet use for online operations to Internet use for online exchange and transactions. E-business continued to flourish into Web 3.0 with the popularization of smart phones which allowed users to have a greater level of online interaction with vendors and online review systems. The rise of DT in the last decade aligns with movement to Web 4.0 (also called Industrial Revolution 4.0) where Internet connectivity continued to expand into a variety of everyday electronic devices and other objects (Erboz, 2017).

This researcher has noticed a recent surge in scholarly research under the umbrella of Digital Transformation (DT) over the last decade, which appears to reflect a conceptual shift to a broader more inclusive field within academia and expansion of research scope across traditional disciplinary boundaries. This raises important questions about current conceptions about this evolving research field and the core research areas that feed into it. What are the main concepts and core research areas that define DT as an interdisciplinary research field? Guided by scoping review procedures, the main chapter objectives are to: (1) conceptualize Digital Transformation as an emerging interdisciplinary field and (2) identify key research areas within Digital Transformation. Recommendations for guiding future research development are offered at the end of the chapter along with a selected reading list.

# BACKGROUND

## Early Roots and Conceptual Developments

What is Digital Technology (DT) and where did it come from? Early roots of DT were mainly found under the umbrella of E-Commerce aligning with Internet and computer networking developments in the 1980s and 1990s. This was followed by a gradual broadening of research areas and interests in the 1990s and 2000s under the term E-Business or E-Business Technology. Parker and Grove (2000) noted that terms e-business, e-commerce, cyber trade, and Internet have been sometimes used interchangeably without paying attention to their scopes, and how they are related to one another. Other scholars, like Corey and Wilson (2009) make a concerted effort to separate terms like e-commerce from e-business, the latter, which they believe is less specific and less restricted to financial and economic transactions. Since this chapter focuses on current work it does not cover earlier work under the umbrella of E-commerce. Rather it begins with a brief review of E-Business Technology and focuses primarily on the conceptualization of DT and the state of current research knowledge.

There are a myriad of definitions of E-business Technology which highlight various elements perceived to be at the core. Many definitions focus exclusively on E-business Technology as a tool (technology) for business. Lowry, Cherrington and Watson (2001) defined E-business broadly as an acceleration of some of the underlying fundamentals of economics and technology while retaining the rules of business. The Collins English Dictionary (2019) defines E-business technologies as "consisting of a variety of software packages, hardware components, mobile solutions, and analytic processes that facilitate business operations via electronic tools". Similarly, Nof, Ceroni, Jeong, and Moghaddam (2015) view Electronic Business Technologies as the application of business process (products, procedures, services, etc.) over the Internet. At the root of the abovementioned definitions is a strong focus on technology and how it can revolutionize how businesses operate. Lopperi and Sengupta (2004) note that at the core of Electronic Business Technologies is the application of Internet and information technology to business processes which allows new ways of conceiving and organizing businesses.

Part of the problem of narrowly defining E-business Technology as a tool (technology) for business is that tools change. Benitez, Chen, Teo, and Ajamieh (2018) observe that what constitutes core-business technologies today will likely seem obsolete a year later. To mitigate risks of becoming obsolete, more recent definitions of E-business Technology provide broader visions of the field that transcend the disciplinary focus on business and business tools. Selected definitions of E-business focus more broadly on Internet use and associated connectivity to advance information sharing between individuals, governments, businesses and financial institutions. This marks the beginning of a conceptual shift in E-business Technology to a broader encompassing more interdisciplinary research domain. A particularly broad encompassing conception of E-Business Technology, and the one highlighted in this chapter comes from the University of Ottawa (2019):

*E-Business Technology is the application of Internet and Information Technology to business processes which leads to remarkable new ways of conceiving and organizing businesses. e-Business Technology includes three main research areas: e-Business, e-Technology and e-Society. E-Business focuses on organizational transformations based on Information Technology such as identification of best practices and trends, and evaluation of performance based on the introduction of new technologies. E-Technology focuses on information technologies and system architectures that are used to create commercial transactions. E-Society is the information and telecommunication technology that drives society and leads to innovation in many fields such as health, education, law and visual arts.*

This last definition captures an explicit openness to multiple knowledge domains and fields of application that align with popular interdisciplinary trends in recent academic research and training.

## Interdisciplinary Expansion of E-Business Technology to Digital Transformation

The abovementioned branching out of E-Business Technology into an interdisciplinary research field aligns well with increasing demands for building multidisciplinary knowledge and cultivating diverse technical and soft skills training needed in an evolving knowledge based economy intertwined with digital technologies and computer networking (Luppicini, 2010). To meet the current and future demands of evolving knowledge based economy, interdisciplinary studies are being looked to as an innovative way to cultivate the diversity of skills and expertise required from Higher Education to help collaborate with public, private and/or not-for-profit sectors to create an innovative, resilient and culturally rich society. According to the *Oxford Handbook of Interdisciplinarity*:

*Interdisciplinary studies is a process of answering a question, solving a problem, or addressing a topic that is too broad or complex to be dealt with adequately by a single discipline, and draws on the disciplines with the goal of integrating their insights to construct a more comprehensive understanding (16).*

Interdisciplinary studies combine two or more academic disciplines or fields of study to bridge multiple bases in knowledge and expertise, which provides a rich base of knowledge and vales to aid in complex decision-making by enabling new models of research capable of dealing with complex problems that require multi-stakeholder interests from public-private partnerships. As stated by Palmer (2001), "The real-world research problems that scientists address rarely arise within orderly disciplinary categories, and neither do their solutions" (p. vii). Instead, a multi-tiered approach that integrates insights derived from interdisciplinary studies provides a richer model to advance knowledge creation in the evolving knowledge economy. The cross-fertilization of interdisciplinary knowledge and expertise represents a potentially powerful set of knowledge building tools to enabling new models of research to emerge with which to create an innovative, resilient and culturally rich society.

Interdisciplinary studies are multi-faceted and impact all subject areas including natural and applied sciences, social sciences, and humanities. The main goal of interdisciplinary work is integration of multiple perspective used to address the challenge of complexity (Frodeman, Klein, Thompson & Mitcham, 2010). Interdisciplinary work provides the opportunity to take theoretical learning and marry it with applied logic to create new and innovate ways to address existing challenges. Interdisciplinary knowledge and methodologies enhance the digital economy by harnessing new computing and digital media tools into new business areas such as social data mining of social, user experience design (UX), and social media marketing. This is largely due to the fact that interdisciplinary research pulls together powerful ideas, people, and tools to focus on shared problems too big and complex to be addressed by a single discipline. In drawing from the National Academy of Sciences, the National Academy of Engineering, and the Institute of Medicine, Frodeman, Klein, Thompson and Mitcham (2010) view interdisciplinary research as:

*a mode of research by teams or individuals that integrates information, data, techniques, tools, perspectives, concepts, and/or theories from two or more disciplines or bodies of specialized knowledge to advance fundamental understanding or to solve problems whose solutions are beyond the scope of a single discipline or area of research practice. (p. 14)*

As becomes evident in chapter sections to follow, the gradual interdisciplinary shift of E-Business Technology into the larger public social sphere has expanded the scope of the field, the base of scholars working within it, and how it is approached. Increasingly, E-Business Technology is characterized by its interdisciplinary nature and vision to advance knowledge and practice from a multidisciplinary standpoint.

This aligns well with the current vision of leading E-Business journals. According to the International Journal of E-Business Research:

*The primary objective of the International Journal of E-Business Research (IJEBR) is to provide an international forum for researchers and practitioners to advance the knowledge and practice of all facets of electronic business. Emerging e-business theories, architectures, and technologies are emphasized to stimulate and disseminate cutting-edge information into research and business communities in a timely fashion. The secondary objective of this journal is to develop a comprehensive framework of e-business by taking a multidisciplinary approach to understanding e-business and its implications on businesses and economies (International Journal of E-Business Research, 2019).*

Continued expansion of E-Business Technology research areas into other disciplines and fields can be taken as a reflection of its increased importance and value in society and selected academic research communities. It also signals an important expansion in the field and the need for a conceptual framework that accommodates changing views and research interests within it. One of the most interesting results of this expansion is that it has paved the way for an alternative conceptualization of the research field under the umbrella of DT and revealed concerted effort to define DT. For instance, i-SCOOP.eu (2019), describes DT as, "the profound transformation of business and organizational activities, processes, competencies and models to fully leverage the changes and opportunities of a mix of digital technologies and their accelerating impact across society in a strategic and prioritized way, with present and future shifts in mind". Similarly, Westerman et al. (2019) conceptualized DT as a replacement for a function or process already performed (substitution), as a means of improving the functionality of a process or product (Extension) or as a means of fundamentally redefining as process or product (transformation). Although multiple definitions have been published, more continue to be appear in the research literature. Pramanik, Kirtania, and Pani, (2019) note, "While most operate on some common premise related with the construct of digital transformation there is no established definition yet". This chapter attempts to fill this gap by providing a comprehensive conceptualization of DT based on a holistic synthesis of current definitions within existing research literature.

The rise of this alternative conceptualization can be attributed to at least two factors. First, at a general societal level, there is a greater demand for digital technology, not only in the workplace but also in other areas of society (education, healthcare, politics, law, etc). Second, the growing trend in Higher Education towards greater interdisciplinarity in technology oriented academic programs across traditional university departments, schools, and faculties can be captured under the umbrella of DT as a multi-faceted approach to digital technologies within society in need of multiple types of knowledge and expertise. This alternative conceptual development in the field appears promising in terms of broadening the scope of real world applications and leveraging interdisciplinary collaboration among separate units within academic institutions.

Of particular interest in this chapter is how DT as an interdisciplinary field is reflected in current DT definitions and core research areas. While conceptual papers and small scale research reviews have been published in recent years (Pihir et al., 2019; Reis et al., 2018), there is need for broader scoping scholarly reviews on DT as an interdisciplinary field that draws on a greater diversity of multidisciplinary databases to better capture the expanding scope of DT as an interdisciplinary research field.

With the objective of exploring the conceptualization and key research areas that constitute Digital Transformation as an emerging interdisciplinary field, a scoping research review was conducted using leading multi-disciplinary databases. From this, the following analyses were made. First, there was analysis of digital transformation conceptualization derived from peer reviewed conceptual papers, research reviews, conference papers, and selected documentation from professional associations and corporate leaders. Based on the criteria of publication citation success, content quality (interdisciplinary focus), and publication date (2015-2019), ten publications were synthesized to identify a comprehensive definition of DT as an interdisciplinary research field. Second, there was a review of the most published research areas in DT to highlight current research trends, address research gaps, and inform future research directions. The next sections discuss the methodological approach, followed by the findings and discussion.

## METHOD

To explore Digital Transformation, a scoping review was selected as a suitable approach to capture the recent developments and change in this evolving research field. Following Grant and Booth (2009), scoping reviews are useful when interested in a "preliminary assessment of potential size and scope of available research literature. Scoping reviews are well suited to identify the roots and extent of research evidence produced" (Grant & Booth, 2009). In terms of search strategies, a number of databases were used with a focus on multidisciplinary database searches, namely Directory of Open Access Journals, SCOPUS, ABI/INFORM Global, Web of Science, and 1findr. These databases were selected for their rigor in indexing references and citations as well as their multidisciplinary coverage and broad accessibility. The research was conducted on October 1, 2019, based on keyword set "Digital Transformation" appearing in the title, abstract and, subject headings restricted to peer-reviewed journal and conference articles (books, book chapters, and reviews excluded) with full-text access from the online University of Ottawa Library Database System. Follow-up searches of selected documents from professional associations and corporate leaders identified in article bibliographies were also included to provide more comprehensive coverage of DT. Coverage included publications between 2015-2019 to capture key concepts and leading research areas defining the current research field.

## MAIN FOCUS OF THE CHAPTER

Following the applied parameters, eliminating duplicate articles, and eliminating articles unrelated to the topic focus, a total of 1678 documents were retained. A preliminary review of the 1678 documents uncovered DT to be the most popular research theme followed respectively by Industry 4.0, Design & Entrepreneurship, Social Media, Healthcare, Business, Management, Blockchain, Economics, Sustainability: Smart societies, Internet of Things, Computer Science, Social Science, Information Systems, Artificial Intelligence, Education, Cloud computing, Big Data, Privacy Preserving Data Mining, Engineering, Cybersecurity, Supply Chain, Competitiveness, Trade and Marketing digitization, Ethical issues, and Digital Leadership. Because the main chapter focus is on conceptualizations of DT, only documents related to the main theme were retained for analysis.

Conceptualizations of DT were identified in 189 documents. One of the major findings (and challenges) revealed that there was no common definition of DT employed in documents reviewed. Instead, a myriad of definitions were revealed from multiple literatures with different elements highlighted. A representative sample of 10 typical definitions capturing the diversity of DT was selected for further analysis. Following Reis, et al. (2019), DT definitions were first classified according to their core orientation (s) or elements highlighted. In this chapter, 10 representative definitions of DT were selected from the 189 for further analysis based on key aspects highlighted, namely, technological (T), organizational (O), social (S) and/or system (SYS). These appear in Table 1.

Each of the classified definitions in Table 1 effectively highlight DT's technological, organizational, and social orientation. For instance, the organizational orientation aligns DT with traditional business transformation (Wright, Worlf, Heneghan, & Clamp, 2017), transformation of business activities, processes, competencies, and models (Demirkan, Spohrer, & Welser, 2016), industrial processes, internal and cross-company processes in production and logistics, products and services offered to customers (Ivanov, Dolgui, Das, & Sokolov, 2019), providing novel ways to interact and engage with customers, digitization of a firm's products and services, tactical or strategic business moves by data-driven insights, and the launch of digital business

*Table 1. A Systems Breakdown of Digital Transformation*

| Orientation | Definitions of Digital Transformation |
|---|---|
| O and SYS | Digital transformation is a natural progression from traditional business transformation, one more suited to the modern world. The critical difference is the opportunities for innovation, scalability and agility that are possible in any transformation process, now that digital is part of the mix. Rather than setting long-term goals and moving steadily towards them, sometimes over a number of years, digital transformation has the ability to deal with uncertainty and respond quickly to change. What's more, digital transformation is not a one-off program. Once it has begun, it opens up an organization to continuous and much more effective evolution (Wright, Worlf, Heneghan, & Clamp, 2017). |
| S and SYS | Digital transformation is the profound transformation of business and organizational activities, processes, competencies and models to fully leverage the changes and opportunities of a mix of digital technologies and their accelerating impact across society in a strategic and prioritized way, with present and future shifts in mind. Digital transformation is a journey with multiple connected intermediary goals, in the end, striving towards continuous optimization across processes, divisions and the business ecosystem of a hyper-connected age where building the right bridges (between front end and back office, data from 'things' and decisions, people, teams, technologies, various players in ecosystems etc.) in function of that journey is key to succeed (I-Scoop, n.d.) |
| T and SYS | Digital transformation has been defined as the use of new digital technologies, such as mobile, artificial intelligence, cloud, blockchain, and the Internet ofthings (IoT) technologies, to enable major business improvements to augment customer experience, streamline operations, or create new business models (Warner & Wager, 2018). |
| T and SYS | In the digital category of the fourth industrial revolution, there is an emphasis in cloud computing, IoT, Artificial Intelligence and Big Data, which are part of the so-called Digital Transformation, which is a perspective of the use of information and communication technology (ICT) where it comes to act as a preponderant element in the transformation and reconfiguration of organizational elements, such as: strategy, processes, culture and structures (Mendonça & Andrade, 2018, p.23) |
| O and SYS | Digital transformation is the profound and accelerating transformation of business activities, processes, competencies, and models to fully leverage the changes and opportunities brought by digital technologies and their impact across society in a strategic and prioritized way (Demirkan, Spohrer, & Welser, 2016) |
| O and SYS | The fusion of the digital world with industrial processes is the so-called digital transformation. In addition to internal and cross-company processes in production and logistics, this also applies to the products and services offered to customers that need to be refined through the use of digital technologies (Ivanov, Dolgui, Das, & Sokolov, 2019) |
| S and SYS | In recent years, an important phenomenon in companies research strategic called Digital Transformation (DT) has appeared. DT encompasses the changes taking place in society and industries through the use of digital technologies. Recent research has shown that technology is only part of a complex puzzle that must be solved for organizations to remain competitive in a digital world (Angelopoulos, Kontakou, & Pollalis, 2019) |
| O and SYS | Digital transformation encompasses the digitization of sales and communication channels, which provide novel ways to interact and engage with customers, and the digitization of a firm's offerings (products and services), which replace or augment physical offerings. Digital transformation also describes the triggering of tactical or strategic business moves by data-driven insights and the launch of digital business models that allow new ways to capture value (Haffke, Kalgovas, & Benlian, 2016) |
| T and SYS | Digital transformation highlights the impact of IT on organizational structure, routines, information flow, and organizational capabilities to accommodate and adapt to IT. In this sense, digital transformation emphasizes more the technological root of IT and the alignment between IT and businesses (Li, Su, Zhang, & Mao, 2018) |
| O and SYS | Digital transformation is a complex and demanding process that requires the engagement of all enterprise resources - human, technological, physical, organizational and financial. It marks the digitalization and its implementation throughout the organization with a special focus on people and business processes. DT is the transformation of an organization driven by digital technology, and it directly affects the change of organization's core, that is, the change of its business model (Kutnjak, Pihiri, & Furjan, 2019) |

models (Haffke, Kalgovas & Benlian, 2016). Similarly, technological and social orientations align with key activities captured in the table above. This, however, did not capture the variety of activities aligned with DR as an interdisciplinary concept.

A unique inter-categorical orientation that emerged from the definitions above is a systems orientation which captures how digital transformation deals with uncertainty, responds quickly to change, and accommodates organizational evolution (Wright, Worlf, Heneghan, & Clamp, 2017). A systems orientation aligns with the conception of DT as a fusion of the digital world with internal and cross-company processes in production and logistics, and services (Ivanov, Dolgui, Das, & Sokolov, 2019). Most importantly, a systems orientation acknowledges DT as a complex and demanding process that requires the engagement of all enterprise resources--human, technological, physical, organizational and financial (Kutnjak, Pihiri, & Furjan, 2019) that provides new ways to capture value (Haffke, Kalgovas, & Benlian, 2016). The systems orientation expands on the technological, social, and organization orientation identified in recent research (Reis, et al. 2019).

A detailed examination of the main research themes within DT is outside the chapter scope. It is noteworthy that following DT as the main research theme, the three most popular research themes published in descending order were namely, Industry 4.0, Design and Entrepreneurship, and Social Media. Industry 4.0 (also referred to as Fourth Industrial Revolution) describes disruptive technologies (digital and physical) that offer innovative values and services to customers and organizations (Pacchini, Lucato, Facchini, & Mummolo, 2019). This appears to connect well with DT as an interdisciplinary field since it touches upon technological, organization, social, and systems orientations. Researchers recognize that Industry 4.0 is not limited to industrial production but is intended to apply to all aspects of society, including technology, production, consumption and business (Li et al., 2017).

Design and Entrepreneurship represents a third main research that surfaced during preliminary research review. As noted by Tschimmel (2012), “Design Thinking is a way of thinking which leads to transformation, evolution and innovation, to a new form of living and to new ways of managing business”. Design thinking integrates human, business, and technological factors in defining a problem and finding and designing a solution for it. There is a growing recognition in the literature that this approach integrates expertise from design, social sciences, engineering, and business (Plattner, Meinel, & Leifer, 2010). This also appears to connect well with DT as an interdisciplinary field in addressing technological, organization, social, and systems orientations. Finally, Social media was found to be the forth leading research theme. Social media was typically used to describe as part of the technology orientation of DT as a tool to enable major business improvements by improving customer experience, streamlining operations and creating new business models (Vogelsang, Liere-Netheler, Packmohr, & Hoppe, 2019). Similarly, Nwankpa and Roumani

(2016) view social media Changes and transformations that are driven and built on a foundation of digital technologies. Within an enterprise, digital transformation is defined as an organizational shift to big data, analytics, cloud, mobile and social media platform. Whereas organizations are constantly transforming and evolving in response to changing business landscape, digital transformation are the changes built on the foundation of digital technologies, ushering in unique changes in business operations, business processes and value creation. Based on scoping review findings, the next section posits a comprehensive definition of DT as an interdisciplinary research field from a technological, social, and organization, and systems standpoint.

## Advanced Analysis: DT Explained

The main chapter findings revealed technological, social, and organization, and system orientations within current conceptualizations of DT. These findings support and expand on prior research (Reis et al., 2019) by uncovering an inter-categorical systems dimension within published DT research literature. A synthesis of scoping research results yields a unique conceptualization of DT as an interdisciplinary research field.

DT is an interdisciplinary research field concerned with the complex intertwinement of humans, organizations, digital technologies (egs., Internet, big data, computer networks, wearable technologies, AI, cloud computing, blockchain, Internet of things (IoT), cybersecurity) technologies and systems within society. A systems definition of DT is characterized as a goal oriented problem-solving approach utilizing tools, techniques, theories, and methods from multiple knowledge domains to guide change and the transformation of sociotechnical systems in society. A systems definition of DT includes the following four main research orientations: organizational, technological, social, and systems.

1. The organizational orientation of DT focuses on organizational transformation, organizational activities, processes, competencies, models, logistics, products and services.
2. The technological orientation of DT focuses on the design, development, and implementation of digital technologies (egs., Internet, big data, computer networks, wearable technologies, AI, cloud computing, blockchain, Internet of things (IoT), cybersecurity) in areas of activity that rely on technology.
3. The social orientation of DT focuses on human values and practices connected to the transformation of key areas of society being reshaped by technological innovation such as science, health, education, law, politics, and visual arts.

4. The systems orientation of DT focuses on the interaction of humans, organizations, and digital technologies within complex, uncertain, and changing socio-technical systems driving digital transformation and innovation.

A systems orientation provides DT an overarching framework to critically examine and evaluate organizational, technological, and social aspects of DT to help leverage holistic knowledge creation from multiple domains and guide complex problem solving and decision making for advancing human needs and interests in all areas of society that rely on technology. This unique systems orientation of DT is appealing for multiple reasons. First, it provides an overarching conceptual framework for DT's technological, social, and organization orientation identified in recent research (Reis et al., 2019). Second, it helps distinguish DT from related conceptualizations of E-Commerce and E-Business Technology by providing unique conceptual grounding. Third, the systems orientation aligns well with the conception of DT as an interdisciplinary research field as an alternative approach to deal with change and complexity in real world problems that requires a multi-pronged integration of data, techniques, tools, perspectives, concepts, and/or theories from multiple disciplines and bodies of specialized knowledge (Frodeman, Klein, Thompson & Mitcham, 2010).

Systems orientation provides the missing relational orientation which captures the broad reaching scope of DT as an interdisciplinary research field that reaches through and across disciplinary boundaries and looks at technology in a completely different way. From a systems perspective, technology is not merely a tool or instrument. Rather it is a relational concept and an integral part of an evolving system of human values and social activity in society. As such, technology is value laden and entrenched within a larger system co-shaped by social, cultural, political, and environmental influences. McGinn (1978) pointed out a number of ways technology is value-laden: (1) The value of a technique reflects the values of who make it and use it, (2) Technology is optimistic in assigning value to "technological progress", (3) Technology is value laded insofar as use of resources for advance may preclude their use in other work that may improve life, (4) The institutionalization of modern technology allows the direction of technology to be influenced, externally by companies rather than by practitioners, and (5) Products of technology are expressions of individual and cultural values of designers. This much broader systems oriented view of technology aligns well with current conceptualizations of technology and with DT as an interdisciplinary research field.

## SOLUTIONS AND RECOMMENDATIONS

The systems perspective on DT as an interdisciplinary research field offers a viable conceptual grounding for DT by providing a relational interconnection (or glue) that captures the broad reaching scope of DT as an interdisciplinary research field that reaches through and across disciplinary boundaries and looks at technology in a completely different way. This requires a change in mindset on how it is that technology is viewed. From a systems perspective, technology is not merely a tool or instrument. Rather it is a relational concept and an integral part of an evolving system of human values and social activity in society. As such, technology is value laden and entrenched within a larger system co-shaped by social, cultural, political, and environmental influences. McGinn (1978) pointed out a number of ways technology is value-laden: (1) The value of a technique reflects the values of who make it and use it, (2) Technology is optimistic in assigning value to "technological progress", (3) Technology is value laded insofar as use of resources for advance may preclude their use in other work that may improve life, (4) The institutionalization of modern technology allows the direction of technology to be influenced, externally by companies rather than by practitioners, and (5) Products of technology are expressions of individual and cultural values of designers. This systems view of technology aligns well with DT as an interdisciplinary research field. It is suggested that research scholars embrace the deeper level of meaning that technology, as a social-system perspective offers in continued efforts to advance DT research scholarship.

## FUTURE RESEARCH DIRECTIONS

It is often the case in scoping reviews that more questions arise as others are answered. This chapter has contributed to research on the conceptualization on DT but more detailed research is required to address many remaining questions. Fortunately, there is a small but promising research trend focused on the question of what characterizes and defines DT. Some scholars have already begun pushing this line of inquiry forward that spurs new ideas, debates, and questions. Why does DT have no established definition yet for scholars and researchers to rally around and is this a problem (Pramanik, Kirtania, & Pani, 2019)? Is DT old wine in new bottles or is it a radical change of doing business in the digital age (Hrustek, Tomicic Furjan, & Pihir, 2019)? How can DT leverage productivity and value within organizations (Ustinova, Ivanova, & Aletdinova, 2019)? Should DT be treated as an ecosystem based on cloud computing (Al-Ruithe, Benkhelifa, & Hameed, 2018)? Is DT an effective way to attain competitive advantages and company differentiation (Ferreira,

Fernandes, & Ferreira, 2019)? Why do some view DT as the new corporate mandate for companies with pre-digital success (Al-Ali & Phaal, 2019)? Is DT the most important element of the fourth Industrial Revolution that is changing the way of doing business (Orbik & Zozul'aková, 2019)? Should new technological models for marketing be created to adapt to the growth of e-commerce and the changing landscape of customer loyalty (García, Lizcano, Ramos, & Matos, 2019)? Does it make sense to view DT as a path to value creation (Galindo-Martín, Castaño-Martínez, & Méndez-Picazo, 2019) or value destruction? Is DT a path to growth and prosperity (Samerski & Müller, 2019) or the road to hell? Can DT have an impact at a national and organization level (Pousttchi, Gleiss, Buzzi, & Kohlhagen, 2019)? Will DT improve the flow in governing data processes and operations in different organizations (Koilada, 2019)? Continued work in these areas to address the questions above would help further ground DT and help guide future researchers and other leaders in the field.

A second area of emerging scholarly interest revolves around User Experience (UX) and visual aesthetics in DT. Chapter scoping research review results of 1678 documents uncovered the following popular research themes in DT, namely, DT, Industry 4.0, Design & Entrepreneurship, Social Media, Healthcare, Business, Management, Blockchain, Economics, Sustainability: Smart societies, Internet of Things, Computer Science, Social Science, Information Systems, Artificial Intelligence, Education, Cloud computing, Big Data, Privacy Preserving Data Mining, Engineering, Cybersecurity, Supply Chain, Competitiveness, Trade and Marketing digitization, Ethical issues, and Digital Leadership. It was found that User Experience (UX) and visual aesthetics in DT were underrepresented in this researcher's view. More and more attention in DT innovation is being placed on the user and this is a promising avenue for future research direction in multiple domains, including: Visual aesthetics and web design (Michailidou, Harper, & Bechhofer, 2008; Strebe, 2016), techno-aesthetics (Luppicini, 2019), aesthetic experience in industrial design (Sathikh, 2017) and social robot design (Van Wynsberghe, 2013).

## CONCLUSION

This chapter has come full circle and ends where it began. So what is DT anyway and what is an appropriate conceptual framework to help guide future research within this emerging field. Based on a scoping research review of DT, this chapter made the case that DT is an interdisciplinary research field concerned with the complex intertwinement of humans, organizations, and digital technologies (egs., Internet, big data, computer networks, wearable technologies, AI, cloud computing, blockchain, Internet of things (IoT), cybersecurity) technologies within society. In terms of what conceptual framework of DT to adapt, this chapter proposed a systems definition of DT characterized as a goal oriented problem-solving approach utilizing tools, techniques, theories, and methods from multiple knowledge domains to guide change and the transformation of sociotechnical systems in society.

## REFERENCES


Aghaei, S., Nematbakhsh, M. A., & Farsani, H. K. (2012). Evolution of the world wide web: From web 1.0 to web 4.0. *International Journal of Web & Semantic Technology*, *3*(1), 1–10. doi:10.5121/ijwest.2012.3101

Al-Ali, A. G., & Phaal, R. (2019). Design Sprints for Roadmapping an Agile Digital Transformation. Proceedings - 2019 IEEE International Conference on Engineering, Technology and Innovation, ICE/ITMC 2019. doi:10.1109/ICE.2019.8792597

Angelopoulos, M., Kontakou, C., & Pollalis, Y. (2019). *Digital Transformation and Lean Management. Challenges in the Energy Industry of Utilities. A Review.* Academic Press.

Aronica, V., Abbiati, E., Bonfiglio, F., Idone, G., & Ruffino, C. (2018). Digital Transformation. Retrieved October 31, 2019, from https://www.eng.it/resources/whitepaper/doc/digital-transformation/engineering-digital-transformation-whitepaper-en.pdf

Belsey, C. (1980). *Critical Practice.* London: Methuen; doi:10.4324/9780203328439.

Benitez, J., Chen, Y., Teo, T. S., & Ajamieh, A. (2018). Evolution of the impact of e-business technology on operational competence and firm profitability: A panel data investigation. *Information & Management*, *55*(1), 120–130. doi:10.1016/j.im.2017.08.002

Chaffey, D. (2011). E-Business and E-Commerce Management - Strategy, Implementation and Practice. Financial Times Prentice Hall.

Consulting, C. (2011). Digital transformation: A roadmap for billion-dollar organizations. Retrieved October 31, 2019, from MIT Center for Digital Business and Capgemini Consulting website: https://www.capgemini.com/wp-content/uploads/2017/07/Digital_Transformation__A_Road-Map_for_Billion-Dollar_Organizations.pdf

Corey, K. E., & Wilson, M. I. (2009). e-Business and e-Commerce. International Encyclopedia of Human Geography. doi:10.1016/B978-008044910-4.00148-6

Demirkan, H., Spohrer, J. C., & Welser, J. J. (2016). Digital innovation and strategic transformation. *IT Professional, 18*(6), 14–18. doi:10.1109/MITP.2016.115

Dictionary, C. E. (2019). *E-business technologies*. Author.

Ferreira, J. J. M., Fernandes, C. I., & Ferreira, F. A. F. (2019). To be or not to be digital, that is the question: Firm innovation and performance. *Journal of Business Research, 101*, 583–590. doi:10.1016/j.jbusres.2018.11.013

Frodeman, R., Klein, J. T., & Mitcham, C. (Eds.). (2010). The Oxford Handbook of Interdisciplinarity. Retrieved from http://www.universitystudy.ca/search-programs/?k=Interdisciplinary%20Studies&description_language=ENGLISH&cipcode_description=Multidisciplinary/interdisciplinary%20studies,%20other

Galindo-Martín, M.-Á., Castaño-Martínez, M.-S., & Méndez-Picazo, M.-T. (2019). Digital transformation, digital dividends and entrepreneurship: A quantitative analysis. *Journal of Business Research, 101*, 522–527. doi:10.1016/j.jbusres.2018.12.014

García, J. J. L., Lizcano, D., Ramos, C. M. Q., & Matos, N. (2019). Digital marketing actions that achieve a better attraction and loyalty of users: An analytical study. *Future Internet, 11*(6). doi:10.3390/fi11060130

Gimpel, D. H., & Röglinger, D. M. (2015). Digital transformation: Changes and chances. Retrieved October 31, 2019, from https://www.fim-rc.de/wp-content/uploads/Fraunhofer-Studie_Digitale-Transformation.pdf

Grant, M. J., & Booth, A. (2009). A typology of reviews: An analysis of 14 review types and associated methodologies. *Health Information and Libraries Journal, 26*(2), 91–108. doi:10.1111/j.1471-1842.2009.00848.x PubMed

Haffke, I., Kalgovas, B. J., & Benlian, A. (2016). *The Role of the CIO and the CDO in an Organization's Digital Transformation*. Academic Press.

Hasibovic, A. C., & Tanovic, A. (2019). *PRINCE2 vs Scrum in digital business transformation. In 2019 42nd International Convention on Information and Communication Technology, Electronics and Microelectronics, MIPRO 2019 - Proceedings* (pp. 1514–1518). Academic Press; doi:10.23919/MIPRO.2019.8756716.

Hrustek, L., Tomicic Furjan, M., & Pihir, I. (2019). *Influence of digital transformation drivers on business model creation. In 2019 42nd International Convention on Information and Communication Technology, Electronics and Microelectronics, MIPRO 2019 - Proceedings* (pp. 1304–1308). Academic Press; doi:10.23919/MIPRO.2019.8756666.

I-Scoop. (n.d.). Digital transformation: online guide to digital business transformation. Retrieved October 31, 2019, from https://www.i-scoop.eu/digital-transformation/

Ivanov, D., Dolgui, A., Das, A., & Sokolov, B. (2019). Digital Supply Chain Twins: Managing the Ripple Effect, Resilience, and Disruption Risks by Data-Driven Optimization, Simulation, and Visibility. In *Handbook of Ripple Effects in the Supply Chain* (pp. 309–332). Cham: Springer; doi:10.1007/978-3-030-14302-2_15.

Koilada, D. K. (2019). Value-based digital transformation: Innovating customer experiences. *2019 IEEE Technology and Engineering Management Conference, TEMSCON 2019*. 10.1109/TEMSCON.2019.8813559

Kutnjak, A., Pihiri, I., & Furjan, M. T. (2019, May). Digital Transformation Case Studies Across Industries–Literature Review. In 2019 42nd International Convention on Information and Communication Technology, Electronics and Microelectronics (MIPRO) (pp. 1293-1298). IEEE. doi:10.23919/MIPRO.2019.8756911

Leonard, K. (2018). Retrieved November 15, from, https://smallbusiness.chron.com/benefits-e- business-technology-4048.html

Li, L., Su, F., Zhang, W., & Mao, J. Y. (2017). Digital transformation by SME entrepreneurs: A capability perspective. *Information Systems Journal*, ▪▪▪, 1–29.

Lopperi, K., & Sengupta, S. (2004). Are we ready? The state of wireless e-business in the USA. *Information Systems and e-Business Management*, *2*(4), 293–307. doi:10.1007/s10257-004-0044-4

Lowry, P. B., Cherrington, J. O., & Watson, R. R. (Eds.). (2001). *The E-Business Handbook*. CRC Press.

McGinn, R. (1978). What is technology. *Research in Philosophy and Technology*, *1*, 179–197.

Mendonça, C. M. C. D., & Andrade, A. M. V. D. (2018). Dynamic Capabilities and Their Relations with Elements of Digital Transformation in Portugal. *Journal of Information Systems Engineering & Management*, *3*(3), 23.

Michailidou, E., Harper, S., & Bechhofer, S. (2008). Visual complexity and aesthetic perception of web pages. In Proceedings of the 26th annual acm international conference on design of communication (pp. 215-224). doi:10.1145/1456536.1456581

Nof, S. Y., Ceroni, J., Jeong, W., & Moghaddam, M. (2015). Revolutionizing Collaboration through e-Work, e-Business, and e-Service. Springer. doi:10.1007/978-3-662-45777-1

Nwankpa, J. K., & Roumani, Y. (2016). *IT capability and digital transformation: A firm performance perspective*. Academic Press.

Orbik, Z., & Zozul'aková, V. (2019). Corporate Social and Digital Responsibility. Management Systems in Production Engineering, 27(2), 79–83. doi:10.1515/mspe-2019-0013

Pacchini, A. P. T., Lucato, W. C., Facchini, F., & Mummolo, G. (2019). The degree of readiness for the implementation of Industry 4.0. *Computers in Industry*, *113*, 103125. doi:10.1016/j.compind.2019.103125

Pihir, I., Oreski, D., & Zestic, M. (2017). Cost and benefits of webshop implementation for sme's in first three years. In Economic and Social Development: Book of Proceedings (pp. 685-694). Retrieved December 1 from, https://search.proquest.com/docview/2070392460?accountid=14701

Plattner, H., Meinel, C., & Leifer, L. (2010). *Design thinking: Understand–improve–apply*. Springer Science & Business Media.

Pousttchi, K., Gleiss, A., Buzzi, B., & Kohlhagen, M. (2019). Technology impact types for digital transformation. In *Proceedings - 21st IEEE Conference on Business Informatics* (Vol. 1, pp. 487–494). CBI; doi:10.1109/CBI.2019.00063.

Pramanik, H. S., Kirtania, M., & Pani, A. K. (2019). Essence of digital transformation—Manifestations at large financial institutions from North America. *Future Generation Computer Systems*, *95*, 323–343. doi:10.1016/j.future.2018.12.003

Pramanik, H. S., Kirtania, M., & Pani, A. K. (2019). Essence of digital transformation—Manifestations at large financial institutions from North America. *Future Generation Computer Systems*, *95*, 323–343. doi:10.1016/j.future.2018.12.003

Samerski, S., & Müller, H. (2019). Digital health literacy in Germany – requested, but not supported? Results of the empirical study TK-DiSK [Digitale Gesundheitskompetenz in Deutschland - gefordert, aber nicht gefördert? Ergebnisse der empirischen Studie TK-DiSK]. *Zeitschrift für Evidenz, Fortbildung und Qualität im Gesundheitswesen*, *144–145*, 42–51. doi:10.1016/j.zefq.2019.05.006 PubMed

Sathikh, P. (2017). *Aesthetics of experience: Industrial design in the era of design thinking and user experience*. Advances in Transdisciplinary Engineering; doi:10.3233/978-1-61499-779-5-338

Strebe, R. (2016). Aesthetics on the web: Effects on approach and avoidance behaviour. *Behaviour & Information Technology*, *35*(1), 4–20. doi:10.1080/0144929X.2015.1070202

Thorlacius, L. (2007). The role of aesthetics in web design. *Nordicom Review*, *28*(1), 63–76.

Thüring, M., & Mahlke, S. (2007). Usability, aesthetics and emotions in human–technology interaction. *International Journal of Psychology*, *42*(4), 253–264. doi:10.1080/00207590701396674

Tschimmel, K. (2012). Design Thinking as an effective Toolkit for Innovation. In ISPIM Conference Proceedings. The International Society for Professional Innovation Management (ISPIM).

University of Ottawa. (2019). E-Business Technology (2019). Retrieved October 31, 2019, from https://catalogue.uottawa.ca/en/graduate/doctorate-philosophy-electronic-business/

Ustinova, L., Ivanova, E., & Aletdinova, A. (2019). Innovative processes of digital transformation in radio electronics. *IOP Conference Series. Materials Science and Engineering*, *497*(1). doi:10.1088/1757-899X/497/1/012098

vAl-Ruithe, M., Benkhelifa, E., & Hameed, K. (2018). Key Issues for Embracing the Cloud Computing to Adopt a Digital Transformation: A study of Saudi Public Sector. *Procedia Computer Science*, *130*, 1037–1043. doi:10.1016/j.procs.2018.04.145

Van Wynsberghe, A. (2013). Designing robots for care: Care centered value-sensitive design. *Science and Engineering Ethics*, *19*(2), 407–433. doi:10.1007/s11948-011-9343-6 PubMed

Vogelsang, K., Liere-Netheler, K., Packmohr, S., & Hoppe, U. (2019). Success factors for fostering a digital transformation in manufacturing companies. Journal of Enterprise Transformation, 1-22.

Wade, M. (2015). *Digital Business Transformation*. Global Center for Digital Business Transformation.

Warner, K. S. R., & Wager, M. (2018). *Building dynamic capabilities for digital transformation: An ongoing process of strategic renewal*. Academic Press.

Westerman, G., Bonnet, D., & McAfee, A. (2014). The nine elements of digital transformation. *MIT Sloan Management Review*, *55*(3), 1–6.

Wright, R., Worlf, D., Heneghan, L., & Clamp, A. (2017). Destination (un) known. Retrieved October 31, 2019, from https://home.kpmg/content/dam/kpmg/uk/pdf/2017/09/digital_transformation_guide_2017.pdf

Zimmerman, P., Gilbert, T., & Salvatore, F. (2019). Digital engineering transformation across the Department of Defense. *Journal of Defense Modeling and Simulation, 16*(4), 325–338. doi:10.1177/1548512917747050

## ADDITIONAL READING

Andriole, S. J. (2017). Five myths about digital transformation. *MIT Sloan Management Review, 58*(3), 20–22.

Capurro, R. (2017). Digitization as an ethical challenge. *AI & Society, 32*(2), 277–283. doi:10.1007/s00146-016-0686-z

Dobrick, F. M., Fischer, J., & Hagen, L. M. (2018). Research Ethics in the Digital Age. Springer. doi:10.1007/978-3-658-12909-5

Duernberger, C. (2018). Digitization in the stable-Ethical perspectives on a trend of the future. *TIERAERZTLICHE UMSCHAU, 73*(11), 391–394.

Hess, T., Matt, C., Benlian, A., & Wiesboeck, F. (2016). Options for formulating a digital transformation strategy. *MIS Quarterly Executive, 15*(2), 123–139.

Nambisan, S., Wright, M., & Feldman, M. (2019). The digital transformation of innovation and entrepreneurship: Progress, challenges and key themes. *Research Policy, 48*(8), 103773. doi:10.1016/j.respol.2019.03.018

Royakkers, L., Timmer, J., Kool, L., & van Est, R. (2018). Societal and ethical issues of digitization. *Ethics and Information Technology, 20*(2), 127–142. doi:10.1007/s10676-018-9452-x

## KEY TERMS AND DEFINITIONS

**Artificial Intelligence:** Artificial intelligence (AI) refers to the creation of computer programs and devices for simulating brain functions and activity. It also refers to the research program aimed at designing and building intelligent artifacts.

**Industry 4.0:** Is a collection of technologies (disruptive digital and physical) that offer new value to users, groups, and organizations.

**Systems Approach:** A logical process or complex strategy to identify needs, analyze problems and possible solutions, or design procedures to improve systems operations and eliminate needs.

**Technology:** (Process) construction uses and the organization of knowledge for the achievement of practical purposes in intellectual and social contexts; (tool): material construction and operation of physical systems based on systematic knowledge of how to design artifacts; and (system): a self-producing and operationally-closed system situated within and differentiated from the environment.

**Techno-Aesthetics:** Deals with study of human aesthetic values, experiences, and judgements of beauty connected to technology. Techno-aesthetics questions the origin and nature of aesthetic values and human experience concerning technology and its role in human life and society.

**User Experience (UX):** Refers to a recent multi-disciplinary domain of research and practice that deals with new ways of understanding and studying the quality-in use of interactive products. US draws on related design concepts, including: Usability testing, user interface, interaction design; web design; visual aesthetics, and experiential aspects of technology use.

# Chapter 2
# Rural Economy Activation Through E-Commerce:
## Challenges and Opportunities

**Miguel A. Sánchez-Acevedo**
https://orcid.org/0000-0002-0996-0038
*Universidad de la Cañada, Mexico*

**Israel Álvarez-Velásquez**
*Universidad de la Cañada, Mexico*

**Beatriz A. Sabino-Moxo**
*Universidad de la Cañada, Mexico*

**José A. Márquez-Domínguez**
https://orcid.org/0000-0003-2552-2289
*Universidad de la Cañada, Mexico*

**Ma. del Rocio G. Morales-Salgado**
*Universidad Popular Autónoma del Estado de Puebla, Mexico*

## ABSTRACT

*Rural economy has been characterized by low incomes and self-consumption production. With the expansion of global markets and the access of customers through internet, the possibility of moving traditional markets to e-commerce increases. This expansion allows the inclusion of rural economy into the e-commerce market. This chapter describes the challenges to be overtaken in order to activate rural economy through e-commerce. The chapter is organized in five sections: The first section focuses on the communication infrastructure available in rural areas. In the*

DOI: 10.4018/978-1-7998-1879-3.ch002

*second section, the current state and challenges to be addressed for guaranteeing on time delivery are presented. The third section describes payment methods. The fourth section presents schemes of organization required into the communities to guarantee the quality of products. Finally, marketing, advertising, and social networks are discussed.*

## INTRODUCTION

With the Internet revolution, many sectors of society have changed the way some activities are carried out. The business community is a sector that has improved the way of carrying out its activities through the use of Internet, giving rise to the electronic business or e-business. E-business involves the use of Internet to improve business process, commerce, communication and collaboration within company, customers, providers and stakeholders (Combe, 2006). An e-business activity which has gained popularity is the electronic commerce or e-commerce, which use information technology to increase sales, wide the market and provide insights for new products and services (Isoraite & Miniotiene, 2018). E-commerce involves the purchase or transfer of data, services or goods through Internet and is constituted by three major activities: fulfillment of orders, payment, and delivery (Turban, King, Lee, Liang, & Turban, 2015). All processes, since the customer performs a purchase until the product is delivery, need to be followed to guarantee customer satisfaction. These activities can be performed by retailers or third-party companies. The advantages of relying on third-party companies are reduction of transportation cost, reduction on delivery time, and secure payment options. Another important concept developed in e-business is coopetition, which induces to cooperation among competing companies (Waskowsky, 2017). Through coopetition, the market is expanded, production costs are reduced, and better customer service can be provided. Once retailers have established a point of sale online, the necessity of maintaining the loyalty of current customers and attract new ones arises; social networks have been adopted as a medium to establish brand communities (Shabbir, Zhang, Pitsaphol, & Hassan, 2015).

E-commerce has gained market worldwide by 10.2% of the global retail sales and there is an expecting increase of 17.5% in 2021 (Statista, 2018). In 2017, China, South Korea, UK, Germany, Indonesia, India, United States, Taiwan, Poland, and Thailand were the ten leading countries in online shopping (Statista, 2017b). Although the items more acquired online are books, movies, music, video games, consumer electronics, computers, clothing and footwear (Statista, 2017a), there is a tendency of expanding e-commerce to new markets like the agri-food sector (Zeng,

Jia, Wan, & Guo, 2017). This expansion allows the inclusion of rural economy into the e-commerce market.

Rural economy has been characterized by lack of decent jobs, low incomes, informality, under-investment and lack of social protection and services (Rivera, Seebacher, & Diaz Puente, 2015). E-commerce provides a window to offer goods and services produced in rural areas, widening the market and generating more revenues; however, several challenges need to be addressed to make of e-commerce a real option to improve rural economy. Rural economy maintains six major categories for income generation: crop production, livestock production, agricultural employment, non-agricultural employment, self-employment, and remittances (Carletto, Covarrubias, Davis, Krausova, Stamoulis, & Winters, 2007). An important characteristic of rural communities is their union; this cooperation can be strengthened to constitute productive cooperatives. Once cooperatives are constituted, channels for distribution are required; those channels can be generated through e-commerce. According to the OECD (2018), there are six mega-trends which will influence rural areas to be an important part of economy in the countries: "population ageing and migration, urbanization, global shift of productions, climate change and environmental pressures, rise of emerging economies, and technological breakthroughs". Natural resources, infrastructure, and force o labor available in the rural communities are essential for taking advantage of the new global economy. The participation of local governments is crucial to improve the cooperation among communities and maintain a connection with urban centers. The areas which need to be strengthened for the inclusion of e-commerce in rural communities are communication infrastructure, logistic, payment methods, quality assurance, and marketing and advertising.

The chapter is organized in five sections; the first section focuses on the communication infrastructure available in rural areas; in the second section rural logistic is described; third section presents payment methods; four section proposes schemes of organization required into the communities to guarantee the quality of products; finally, marketing and advertising, and social networks are discussed.

## COMMUNICATION INFRASTRUCTURE IN RURAL AREAS

In 2016, 22% of inhabitants in Africa, 39% in Arab States, Asia and Pacific, 60% in America, and 76% in Europe, had access to Internet (Internet Society, 2016). The percentage of rural population in 2017 was of 47.18% in Africa, 49.91% in Arab States, Asia and Pacific, 26% in America and 24.5% in Europe (World Bank, 2017). Two limitations for accessing Internet services in rural areas are the lack of infrastructure and the cost of services. By extending the Internet coverage the number of individuals with access to Internet can be increased (Fig. 1); however, low costs must be offered in order to ensure service adoption.

Currently there are six technologies for providing broadband Internet access: mobile broadband, fixed wireless, cable, DSL, fiber, and satellite. Among the most extended technologies are mobile broadband, cable and DSL. Since rural areas are populated with people with low incomes, most of ISPs have not invested to extend their coverage due to low revenues. There are areas where only one provider is available; therefore, in the absence of competition, the quality of the service decreases (Connected Nation, 2018). Some governments have started programs to provide Internet access to rural areas but most of the countries have not invested in communication infrastructure; however, there are other possibilities to communicate producers with distributors in rural areas.

Despite Internet access is not available in most rural areas, there are technologies developed to establish communication among specific group of persons like those involved in security and rescue operations. Among systems of mobile land radio communication, which are in currently improvement, are: Terrestrial Trunked Radio System (TETRA), Project 25/Project 34 (P25/P34), Digital Integrated Mobile Radio System (DIMRS), TETRAPOL System, Enhanced Digital Access Communication System (EDACS), Public Access Mobile Radio (PAMR), TETRA Enhanced Data Service (TEDS), Global Open Trunking Architecture (GoTa), Next Generation Digital Network System (NXDN), and Broadband Trunking Communication (B-TrunC)

*Figure 1. Individuals without Internet compared to percentage of rural population*

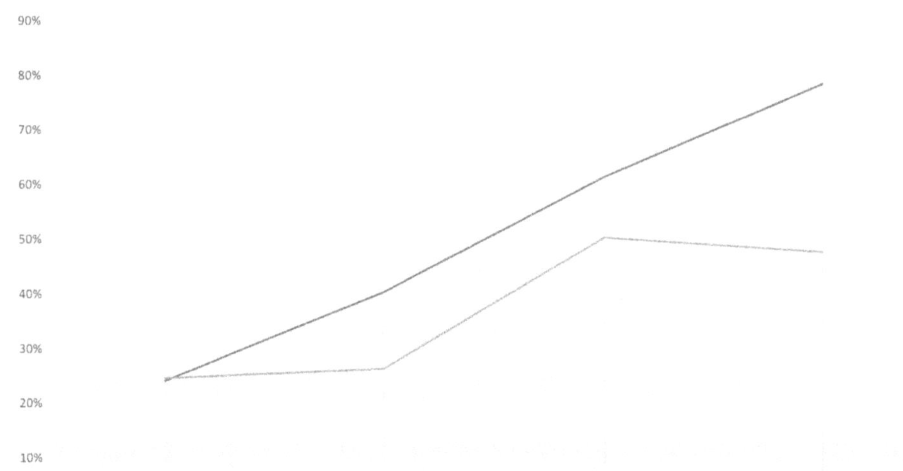

(ITU-R, 2017). The radio communication technology has a coverage around 50 kms; this area can be reduced by hills depending on the standards in which rely. TETRA is a standard defined by the European Telecommunications Standards Institute (ETSI) which supports voice and data transmission with a transfer rate up to 28.8 Kbit/s (ETSI, 2019). Project 25 supports six voice calls and three dedicated data channels; the standard is specified by the Telecommunications Industry Association (TIA) (Motorola, 2014); in the revision of 2009, a minimum data speed of 9600 bps was established (APCO, 2010). DIMRS, as defined in the technical specification of 2013 of ETSI, establishes 2 channels with a transfer rate up to 9600 bps (ETSI, 2013).

By placing distribution centers in communities where Internet connection is available, the communication with production centers, located in rural areas, can be achieved through mobile radio systems (Figure 2). This communication can be limited to the requisition of specific products. The distribution center will process the orders received from customers. Customers can order products through the web page of the company or through the different channels established for the acquisition of products. If the products are not available in the distribution center, the center will request the orders through a mobile radio system to the production centers. Production centers are in charge off the manufacturing of specific products. When a production order is received, division of labor among the community is performed during the manufacturing process to reduce the production time. After products are finished, they must be sent to distribution centers.

Unfortunately, the main characteristic of rural communities is the lack of resources to invest in mobile radio systems; also, people living in rural communities are not subject of credits from bank companies. In their initial stage, the cooperatives constituted will require the support of the government to define the communication strategy among nearby towns where production centers are located. The establishment of cooperatives and the support of governments to them for exporting their products out of the community will change the rural economy, not only by increasing the incomes of families, but improving the communication infrastructure, which can be used in emergency situations also.

## RURAL LOGISTICS

Once the digital communication has been solved, the terrestrial communication among communities must be evaluated for transporting the products from production centers to distribution centers. Roads must be improved for reducing the transportation time. There are communities where transport is available once a week only. After products are manufactured in the rural communities, they need to be transported to the distribution center for packaging and delivery to final customers or for placing

*Figure 2. Communication infrastructure among production centers and distribution center*

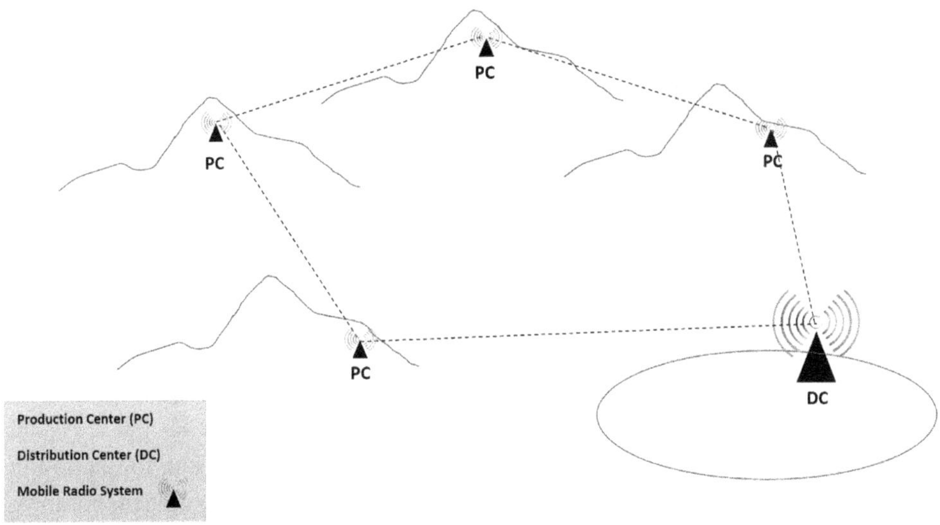


them in the warehouse. The operations related to transportation, distribution, storage, material handling, and packaging of goods in rural areas are the main concern of rural logistics (Asian Development Bank, 2017).

Rural logistics oversees the establishment of an efficient way for transporting goods from production centers to the distribution center and the delivery of the products on-time to final customers. Rural communities are characterized by the far distances and the hilly roads to be traveled for arriving to them. Mobility from rural areas to urban centers is achieved, in most cases, through public transportation. The availability of public transportation in the area depends on the distance to be traveled; farthest communities are less communicated. There are places where people have to walk long distances to reach an area communicated by public transport; however, the communities are well adapted to the conditions of the area where they live; so, most of them have established days when people travel to most populated communities for offering their products.

The adoption of strategies for transporting goods among communities in an efficient and cost-effective manner is of vital relevance to overcome the low availability of transportation among communities. A schedule for production and recollection of goods must be proposed taking into consideration the distance from the distribution center, capacity of production of the community, capacity of transportation, expiration of the goods produced, conditions of the roads, and demand of products. Historical records of sales are of high value to forecast the demand for products. With the resulting forecast, schedule for production and recollection can be improved. Based on the demand of the products, orders to production centers can be processed in advance. The kind of products which are manufactured in rural areas have the characteristic of having the raw material as a resource of the community, which avoids transportation of raw material to the community. The manufacturing process of products with short expiration dates must be performed in communities nearest to the distribution center to avoid decomposition due to long transportation or storage times.

After products are transported from the production centers to the distribution center, they must be packaged to fulfill the customer orders. Packaging is not only concerned with the protection of the products for transportation, but also, it is of great relevance to attract customers; packaging is used as a marketing tool for promoting products and incentivizing the buyer to acquire the product (Mahmoodi & Khataee, 2016). The allocation of human resources to the packaging process is important to reduce the time required to have the products ready for delivery and for ensuring the quality of the products before delivery. Resources optimization in the logistic process can be performed through lean strategies (Ramachandran & Neelakrishnan, 2017).

The last step in the logistic process is the delivery of orders to final customers. In this step, the services of third-party companies with regional, country, or world-wide coverage can be acquired. Most of the parcel companies offer the service of picking the package at the distribution center of the seller and deliver the product at the postal address of the customer. Customers in the global market are more demanding; so, there may be occasions where products could be returned to sellers. The service of transporting from client to the distribution center also could be offered by the same third-party company, which is responsible of the initial delivery, by an extra cost. Those costs must be absorbed and included in the cost of products to offer a better service for final customers; this way, loyalty of customers can be increased.

## E-PAYMENT METHODS

The payment methods used in e-commerce allow customers to give a direct electronic order to their bank for making a payment to the seller. When purchasing a good or service, the customer performs an electronic transfer to seller corresponding to the total cost of the products. As it is a direct transfer of money between customer and seller, there is no compensation between banks. Currently, most payment methods are based on credit card, payments by email, payment service providers, mobile payment systems, among others. The proper characteristics of an electronic payment system are the following: easy to use, low cost per transaction, unlimited amount to transfer, high level of security, payment guarantee, privacy in the purchase, and proof of purchase. Nowadays, the most used types of payment in e-commerce are PayPal, on-line payments through card, cash on delivery, bank transfers, mobile payment systems, and online banks (Monreal, 2012; Rodríguez, 2015).

## PayPal

At the beginning, it was a platform to guarantee transactions at auctions and returns when the product did not meet the expectations of the buyer; then, it was extended to become one of the most used payment systems for e-commerce. PayPal is an intermediary between the buyer and seller. The seller only receives an email from the notification of the deposit of the item sold. The seller does not know the personal data of the buyer nor his bank account number. The security and simplicity of the transaction increases the conversion rates at checkout and the recovery of lost sales with card payment. The main characteristics of this service are: PayPal access data is required to buy, the money can be transferred to PayPal through card, transfer, or another PayPal account, commissions for the seller depends on the number of sales, and if there is currency exchange, commissions are charged to buyers.

## Online Payments Through Cards

The cards are issued by a bank that authorizes to use them as a method of payment. The process begins when a buyer places an order, chooses the type of card payment and their data is requested; then, the payment gateway sends the information to the card issuer, who accepts or rejects the transaction and reports the status of the card transaction. Finally, the on-line store notifies the customer of the acceptance or rejection of their purchase. In the use of credit or debit card, a verification code is requested to avoid possible fraud. On-line payment platforms must have essential attributes such as: security, ease of use, flexibility, and personalization.

*Security:* on-line payment platforms protect the data of users by encrypting information when traveling from the user computer to the bank server. Buyers do not provide their credit card information on the website where they buy, they only make it once in the payment platform. With this, the information is completely secure; so, it will not be necessary to capture the information again whether another purchase is performed. *Ease of use:* on-line platforms allow payments to be made immediately. The Internet buyer is a person who looks for speed and its use should be simple. *Flexibility:* the buyer can choose the most convenient form of payment from several options. The seller can offer to customers the option of paying with credit cards without having to purchase a point-of-sale terminal. *Personalization:* with on-line payment platforms the money can be transferred to a private and personalized account; this creates a history of the movements that are made through the platform and can be consulted at any time.

## Bank Transfers

The bank transfer is very little used, the on-line stores provide the customer with the data of a bank account so that they can proceed with the payment of the order, which will be managed once the payment is confirmed. Their main characteristics are: the money is received in advance, low cost for trade because it does not require technical developments in the on-line store, and there are no commissions for the seller; however, the delay in the purchase process is increased because until the seller receives the transfer, the order is delivered. There exists risk that the seller does not send the order and commissions are charged for the buyer according to the bank conditions.

## On-line Payments Through Mobile

Among the latest trends in e-commerce is the payment through mobile devices. Mobile applications have been developed to work as electronic wallets that store the personal and banking data of the users in order to make transactions in a faster way. Their main advantages are the reduction of long lines at the points of sale, the elimination of the problem of change and lack of cash, and security and fast control of expenses, income and transactions.

## On-line Banks

It consists of a mobile platform in which all operations are concentrated (payments, movements, balance, hiring, etc.), so that the user does not have to move to a physical place or to another associated web page. Their main characteristics are easy mobile payments, security and control of money, centralization of operations, and free account without conditions.

## Google Wallet

This system is like PayPal, acts as a virtual wallet and charges a percentage of the money. It is an excellent way to buy in virtual stores without the need to provide personal data. In addition, Google Wallet receives payment for purchases of Google Play applications. This system allows to acquire products and services through smartphones, enhancing the mobile technology in the e-commerce.

## Financing

The seller offers the possibility of dividing the payment of the purchase in periodical partialities. This method is used in purchases where the price is very high thus facilitated the sale of the product.

The payment methods are changing constantly according to the progress of technology and the needs of companies, vendors and customers, with the aim of making a safe and reliable purchase. The seller can choose those technologies which their customers are more familiar; however, if more payment methods are offered, there are more opportunities of closing purchases. In the proposed cluster infrastructure for improving rural economy, distribution centers will collect the customer payments through a banking account; then, the payment to production centers will be performed in cash, since there are no banks in the rural communities (Figure 3).

# QUALITY ASSURANCE

As previously mentioned, e-commerce is a method of buying and selling products or services through the Internet, presenting the following advantages: availability 24 hours a day, 7 days a week, competitive advantage with respect to traditional

*Figure 3. E-commerce payment architecture*

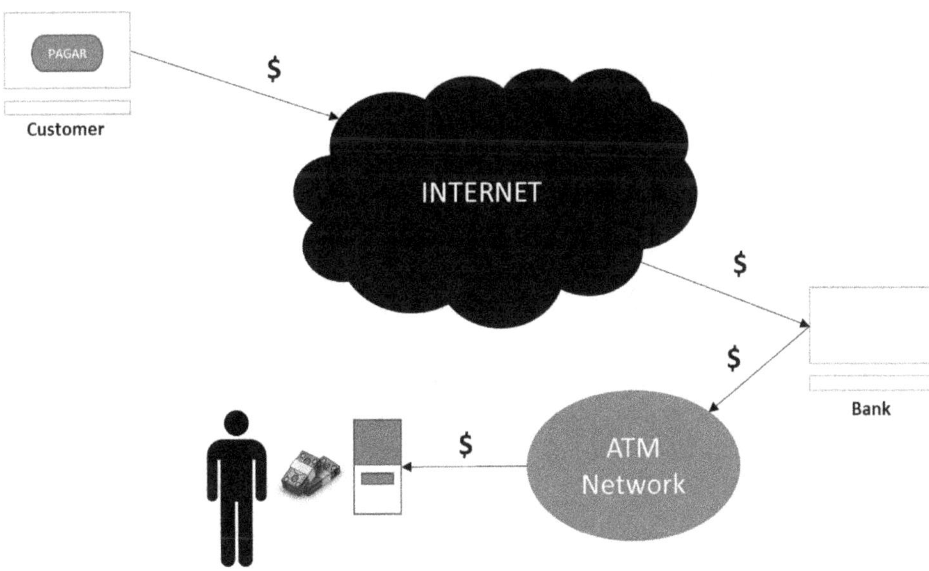

commerce, extend the reach of products and services to other customers, equality between customers regardless of their nationality or place of residence, the customer can find different offers of a product, and the purchase process is relatively easy. In this environment, quality control is important since when a customer makes different purchases online, whether it's clothing, footwear, beauty products, electronics, among others, what he wants is to find the best product at a reasonable price.

Products manufactured in rural communities are developed made the most of the creativity of their creators; however, when those products are manufactured in large quantities, the quality of the product is lost. In a local market, the cost is most valuated than quality of the product, but when a product is offered to a global market, the beauty of the product will attract more to buyers although the product is not cheap. To ensure quality products, more effort from producers is required, but better incomes can be obtained. Another characteristic of products developed in rural communities is that the products are always maintained in the same line of design. Variations on designs are scarce. This problem can be overcome through the interchange of ideas among producers, which could help to innovate or create new products by mixing the existing ones. Producers with more experience can be responsible of the implementation of new ideas, and then, train to young producers to reproduce the manufacturing process.

Through training and supervision, the quality of the products can be improved. Products with designs inspired by creativity, which guarantee their difference among other products, are more appreciated by customers. Design and creativity are skills available in rural producers; so, they must be directed to reach specific objectives. Furthermore, a good election of raw material must be performed in order to obtain quality products. On the other hand, feedback received from customers is also important to improve production process. Quality assurance guarantee the customer satisfaction. A customer satisfied with the product is more likely to buy again and pass on his experience with the products to other persons. While more persons acquire the products offered, more revenues are obtained, improving the economy of the community.

## MARKETING AND ADVERTISING

Marketing is the set of methods and tools used to reach, seduce, attract, satisfy, as well as keep the customer loyal (Martínez, 2016). This process within companies has undergone a constant evolution; communication is the most relevant element, as it is the link to establish a solid relationship with customers and achieve a profitable exchange in both directions (Kotler & Armstrong, 2012). Marketing and communication pursue the same objectives as always: to know the market better, to understand their needs and to be able to respond to them in an efficient and satisfactory way; however, the ways to achieve them have changed radically. The new online tools put in crisis the previous paradigm of company-consumer relationship and force a new exchange model that requires new skills, new knowledge and new approaches.

With the arrival of digital marketing and socialization platforms applied to business, the strategies developed by companies have been resized to improve their presence on the Internet, thus leading to the development of spaces in social media that allow interaction with the consumer public. In order to know the consumer main needs and be able to provide content of interest, companies interact with the consumer public through social media, taking advantage of the promotion of the product or service within the web. In this sense, the evolution of this system of business activities has gone through three stages of development: product orientation, sales orientation, and market orientation. The latter begins to emerge in the mid-forties with the end of the World War II, where many companies recognized that they should apply their work capacity for making available to consumers what they wanted to buy, instead of what the companies wanted to sell (Stanton, Etzel & Walker, 2007).

Nowadays, thanks to technological advances and their incursion into marketing, communication in the market orientation stage has seen better results displacing traditional media, giving way to the use of the Internet. Social networks (Facebook, WhatsApp, Instagram, etc.), streaming platforms (Netflix, Blim, YouTube, etc.) and e-commerce platforms (Amazon, E-Bay, Alibaba, Mercado Libre, etc.) are the most visited and used by people (Kotler, Kartajaya & Setiawan, 2012). By 2019, the growth of social networks as a means of communication also led to an increase in Internet users reaching 4, 388 million of users, which means 57% of the total world population. This exponential growth has not been invisible to companies, making of these media a fundamental part of their marketing processes, achieving not only the promotion and publicity of their products or services, but also generating a deeper and more direct relationship with consumers.

This scenario of evolution is proposed to be carried out through the inclusion of rural areas into e-commerce to rescue the rural economy, as established in the 2030 Agenda for Sustainable Development (United Nations, 2015). The mechanism of application is the one proposed by Porter (1998) through the formation of clusters for the generation of competitive advantages, under the premises of cooperation and collaboration between producers. The cooperation among groups of independent companies that work together, with the intention of directing the flow of goods and services in the value chain to the client, could lead to the formation of Competitive Productive Systems (Johnston & Lawrence, 1988). These groups, united under the same objective, can grow and strengthen, giving the possibility of improving the rural economy and, with it, improving the quality of life of their members. With the integration of digital marketing and the use of social networking platforms in the process of forming Competitive Productive Systems, it is proposed to promote the growth and development of micro and small producers, which will generate an improvement in productive sectors based on the formulation of common development objectives.

As we have seen, marketing has turned out to be a powerful tool within the industry, demonstrating that from a solid and close relationship between producer and consumer could have influence on the increase in sales for products, this being an important indicator of growing. With the arrival of new communication and information technologies, and the exponential growth of platforms for socialization, this relationship is increasingly profitable; also, it is easy to establish a communication channel suitable between customers and sellers. With the generation of production clusters in countries, strategies for growth and productive development can be established, strengthening national economies based on improvement of the primary sector, which is composed of rural economies. This has generated the incursion of national economies into foreign markets, managing to offer production under rigorous quality standards.

## DISCUSSION

Enterprises have been concentrated in urban areas for long time due to high availability of resources, causing problems like population concentration in small areas, resource over-exploitation, economic inequality due to high salaries for small groups of professionals and low salaries for big groups of workers, among others. E-business introduces the possibility to democratize the enterprise activities through technology, by reducing production costs, increasing salaries equitably, hiring workers in distant regions without concentrating them in a specific area, replacing competition by coopetition, and opening new markets; moreover, rural communities have the opportunity to be part of global markets, generating a more sustainable economy. However, to achieve this goal it's required to continue improving applications for logistics, payments, and security by exploiting technologies like cloud computing, big data, artificial intelligence, block-chains, crypt-currency, Internet of Things, wireless networks, principally. Also research have to be intensified to diminish cost and time to collect, compute, analyze, and present information such that decision makers can do their work efficiently to reduce risks and costs.

## CONCLUSION

E-commerce has changed the way consumers acquire products around the world. Its penetration has made possible that small companies compete shoulder to shoulder against big ones. The quality of the products plays an important role to gain clients, keep their loyalty, and expand the market. This new paradigm opens a window to rural communities to offer their products to a global market. However, the geographic and infrastructure characteristics of rural communities impede the access to this market in an individual manner; nevertheless, the cooperation among members of the community is a strength which could be used to constitute clusters of production. Through this integration and with the government support, rural economy can be boosted.

Through the chapter, challenges and opportunities to address the inherent problems to rural communities with respect to offer their products to a global market have been presented. Alternatives of communication, logistic strategies, payment methods, quality and innovation required to succeed, marketing strategies, and the constitution of clusters of production have been addressed. As the communities and the governments work together to activate the rural economy, better life conditions can be provided to rural areas. There is a long way to walk in this process; however, through the documentation of this efforts, other communities can be encouraged to take a step further in the inclusion to the global economy through e-commerce.

## REFERENCES

APCO. (2010). *APCO Project 25: Statement of requirements*. APCO.

Asian Development Bank. (2017). *Promoting Logistics Development in Rural Areas*. ADB.

Carletto, G., Covarrubias, K., Davis, B., Krausova, M., Stamoulis, K., & Winters, P. (2007). Rural income generating activities in developing countries: Re-assessing the evidence. *Electronic Journal of Agricultural and Development Economics*, *4*(1), 146–193.

Combe, C. (2006). *E-business management and strategy*. Elsevier Ltd.

Connected Nation. (2018). *Rural Broadband: A Texas Tour*. Connected Nation.

ETSI. (2013). *Electromagnetic compatibility and Radio spectrum Matters (ERM); Digital Mobile Radio (DMR) Systems; Part 1: DMR Air Interface (AI) protocol*. ETSI.

ETSI. (2019). *Terrestrial Trunked Radio (TETRA); Voice plus Data (V+D); Part 2: Air Interface (AI)*. ETSI.

Internet Society. (2016). *Global Internet Report 2016*. Internet Society.

Isoraite, M., & Miniotiene, N. (2018). Electronic commerce: Theory and practice. *Integrated Journal of Business and Economics*, *2*(2), 73–79. doi:10.33019/ijbe.v2i2.78

ITU-R. Radiocommunication Sector of ITU. (2017). *Digital land mobile systems for dispatch traffic*. Report ITU-R M.2014-3. Geneva: ITU.

Johnston, R., & Lawrence, P. R. (1988). Beyond vertical integration: The rise of the value-adding partnership. *Harvard Business Review*, *66*(4).

Kotler, P., & Armstrong, G. (2012). *Marketing* (14th ed.). Pearson Education.

Kotler, P., Kartajaya, H., & Setiawan, I. (2012). *Marketing 3.0* (3rd ed.). LID.

Mahmoodi, A., & Khataee, S. (2016). Observing the role of packaging handicrafts from the aspect of mobile outdoor advertising on branding and cultural exports. *International Research Journal of Applied and Basic Sciences*, *10*(3), 190–199.

Martínez, D. (2016). Key factors in marketing focused services businesses. *Scientific E-Journal of Human Sciences, 12*(34), 42-58.

Monreal, M. (2012). *Sistemas de pago para comercio electrónico*. Tesis de Licenciatura. CIMAT.

Motorola. (2014). *Project 25 and interoperability in land mobile radio*. White paper. Motorola.

OECD. (2018). *Rural 3.0. A framework for rural development*. Organization for Economic Co-operation and Development.

Porter, M. (1998). *Competitive Strategy: Techniques for Analyzing Industries and Competitors*. Free Press.

Ramachandran, G. M., & Neelakrishnan, S. (n.d.). An approach to improving customer on-time delivery against the original promise date. *South African Journal of Industrial Engineering*, *28*(4), 109–119.

Rivera, M., Seebacher, A., & Diaz Puente, J. M. (2015). Rural economy: a geographical distribution of scientific discourse. In A. Raupelienė (Ed.), *Proceedings of the 7th International Scientific Conference Rural Development 2015*. (pp. 1-6). Aleksandras Stulginskis University. 10.15544/RD.2015.083

Rodríguez, M. (2015). *Nuevos Métodos de Pago Online, Seguridad y Confiabilidad.* Tesis de Licenciatura. Universidad de Cantabria.

Shabbir, R., Zhang, J., Pitsaphol, C., & Hassan, W. (2015). The dynamics of value creation in online brand communities: Strategy map approach. *Information Management and Business Review*, *7*(1), 48–58.

Stanton, W., Etzel, M., & Walker, B. (2007). *Fundamentals of Marketing* (14th ed.). McGraw-Hill. doi:10.4324/9780203030783

Statista. (2017a, February). Online or in-store shopping preference for selected product categories by consumers worldwide as of 2017. *Statista - The Statistics Portal*. Retrieved from https://www.statista.com/statistics/692871/global-online-shopping-preference-by-category/

Statista. (2017b, October). Global markets with the highest online shopping penetration rate as of 2nd quarter 2017. *Statista - The Statistics Portal*. Retrieved from https://www.statista.com/statistics/274251/retail-site-penetration-across-markets/

Statista. (2018, March). E-commerce share of total global retail sales from 2015 to 2021. *Statista - The Statistics Portal*. Retrieved from https://www.statista.com/statistics/534123/e-commerce-share-of-retail-sales-worldwide/

Turban, E., King, D., Lee, K. Y., Liang, T.-P., & Turban, D. C. (2015). *Electronic commerce: a managerial and social networks perspective* (8th ed.). Springer. doi:10.1007/978-3-319-10091-3

United Nations. (2015). *Transforming our world: the 2030 Agenda for Sustainable Development*. UN.

Waskowsky, Z. (2017). The strategy of coopetition. Value-creating networks of partnership relations: the case of sport markets. In *Proceedings of the Management International Conference*. (pp. 13-21). Monastier di Treviso.

World Bank, World Development Indicators. (2017). Rural population (% of total population) [Data file]. Retrieved from https://data.worldbank.org/indicator/sp.rur.totl.zs

Zeng, Y., Jia, F., Wan, L., & Guo, H. (2017). E-commerce in agri-food sector: A systematic literature review. *The International Food and Agribusiness Management Review*, *20*(4), 439–459. doi:10.22434/IFAMR2016.0156

## ADDITIONAL READING

Congressional Research Service. (2019). *Broadband Internet Access and the Digital Divide: Federal Assistance Programs*. CRS.

Laudon, K. C., & Traver, C. G. (2018). *E-commerce 2018* (14th ed.). Person.

Odrowsky, J. (2018). *Digital Business Ecosystem Patterns: Architecture Blueprints for Applying IoT, Blockchain, and Data Analytics to Achieve Business Outcomes*. ComponentWave Press.

Schau, H. J., Muñiz, A. M. Jr, & Arnould, E. J. (2009). How brand community practices create value. *Journal of Marketing*, *73*(1), 30–51. doi:10.1509/jmkg.73.5.30

Sick, D. (Ed.). (2014). *Rural Livelihoods, Regional Economies, and Processes of Change*. Routledge. doi:10.4324/9780203798416

World Economic Forum. (2018). *Delivering the goods: e-commerce logistics transformation*. Author.

## KEY TERMS AND DEFINITIONS

**Agri-Food Chain:** Term used to refer all process involved in the production of food, from planting, harvesting, processing, packaging, and selling.

**Coopetition:** Cooperation among competing companies to expand their markets, share customers, and reduce operational costs.

**E-Commerce:** Constitutes the operations of buying, selling, or interchange of goods, services, or information through internet.

**Rural Economy:** Describes the economic activities through which persons living in rural areas obtain incomes and the well being generated with those incomes.

# Chapter 3
# In Digital Technology We Trust:
## A FinTech B2B Context

**Roisin Vize**
*Technological University Dublin, Ireland*

**Tara Rooney**
*Technological University Dublin, Ireland*

**Lesley E. Murphy**
*Technological University Dublin, Ireland*

## ABSTRACT

*This chapter explores the concept of trust and issues relating to how the construct is conceptualised and understood in a traditional offline context as well as in online environments. The chapter opens with a mini case study that highlights the complexities of being a privately-owned small firm operating in a dynamic and largely unregulated web environment. The firm is relatively new to the financial sector thus augmenting the challenges that lie in reducing perceived risk in an industry that has a chequered history with customer perceptions of credibility and integrity in the financial services sector. The chapter introduces the theoretical underpinning, which draws from the trust theories and technology adoption at firm level, which is critiqued through the lens of the technology acceptance model. Concepts related to institution-based trust are discussed and managerial implications are considered for pure play firms operating online. Each section of the chapter explores these theoretical perspectives from a FinTech context.*

DOI: 10.4018/978-1-7998-1879-3.ch003

## INTRODUCTION

This chapter critically discusses trust in an online B2B context. The chapter opens with a B2B company that operates in the financial lending sector of the FinTech industry. The case company provides a backdrop to the context and as such, case illustrates the importance of digital businesses developing and maintaining trust in an environment that is disruptive, highly fragmented, dynamic and largely unregulated. Issues relating to increased uncertainty, risk perceptions, and lack of knowledge regarding the firms' solution services create challenges for pure play FinTech firms that are looking to develop long-term relationships with clients. Empirical research in B2B services acknowledge that trust plays a pivotal role in relationships operating in an online context as it shifts the focus to future conditions and is considered a necessary ingredient for long-term orientation (Briggs & Douglas, 2010; Brun et al., 2014; Harris & Goode, 2004). This chapter explores issues relating to digital trust - how it is defined, the process to trust creation online and institutional trust mechanisms that need to be considered when operating in an intangible environment that is perceived to be high risk and afflicted with reduced trust. The managerial implications indicate that it is prudent for firms to develop trust building procedures, and systems that are designed to generate trust and alleviate firms risk perceptions. The chapter introduces the Technology Acceptance Model (TAM) by way of gaining conceptual insight into research that has examined factors influencing web adoption at firm level. This theoretical lens specifically explores the concept of digital trust and the key factors that influence it. To develop this agenda, the theory of trust is explored, followed by an assessment of the dimensions of institution-based trust and the trust building mechanisms used to facilitate secure online transactions and privacy protection. The main objectives of this chapter are to interrogate the literature on trust and digital trust & understand the process of the creation of trust in web based business environments.xplore the theoretical underpinning of web adoption at the firm level through the lens of the Technology Acceptance Model. This model continues to be the most widely applied theoretical framework in Information Systems research.understand key institution-based trust dimensions and trust building mechanisms to facilitate secure online transactions and privacy protection as well as to highlight how management may implement web design strategies to reduce risk perceptions and build trust online.

# BACKGROUND

## Opening Case: Invoicefair.com – A Revolution in SME Finance

InvoiceFair was established in 2014 by an experienced team of industry professionals to answer the urgent need for an alternative to traditional funding within the SME sector in Ireland. The founders, Helen Cahill, Peter Brady, and Ivan Fox have backgrounds in the financial services industry and collectively have a combined 75+ years' experience in lending, treasury, investments business ownership and entrepreneurship, Their vision was to combine their experience and expertise with the power of technology to create a Platform that provides SME's with an alternative way to fund their business. 2015 sees the launch of the first version of the InvoiceFair online platform. In just one year they reach a milestone €100 million traded mark. Having gained a reputation as a 'trusted online platform' the company secured a €600k funding round in 2017 and appointed a CTO to drive technological innovation as it entered a period of further growth. By 2019 InvoiceFair have successfully funded over €350 million multi-currency receivables via their online platform and have grown to become a team of 20 employees.

The swift growth of InvoiceFair is reflected in the rapidly growing invoice trading industry. Top level data from Nesta/ Berkley and Cambridge (2013) show the top sectors using invoice trading are Construction, Technology, Manufacturing and Engineering. The average value per transaction is £57,094 with greater than 5,000 businesses using Invoice Trading. Online financial marketplaces are growing in more established and larger markets such as, the USA, the UK and Australia and are providing an alternative source of finance for companies.

### InvoiceFair Corporate Strategy and Business Model

InvoiceFair is an emerging business model that is part of a new movement globally who aim to provide alternative forms of finance to companies through the use of web based digital technology. InvoiceFair is an Irish-based, online platform that connects SMEs looking to raise working capital finance, with institutional investors seeking attractive returns form this specific asset class. It is a form of 'alternative finance' and falls into the category of a 'FinTech' brand, a growing sector in Financial Services in Ireland and internationally.

Invoicefair.com is considered a digital pure player, that is, its business model is primarily delivered online between buyers and sellers. It provides an online platform to facilitate trading of receivables (invoices, purchase orders or recurring revenue) between sellers and buyers. It enables sellers to raise short-term finance/working capital, and buyers to make a low risk, high return gain in the short term. Fees are deducted as a percentage of the trade – the Seller pays a processing fee of between 0.5% &1% of the receivable amount being sold and the Buyer is charged 16.26 plus VAT % of the gain the make. The model is 'pay as you go' with no lock-ins, whereas most competitor models have significant fixed overheads and exit fees. It is fast, flexible and requires no collateral or guarantees, with funds available within 24 hours of the initial trade request.

### *How the Platform Works:*

The process is relatively straightforward as depicted in figures one and two but rigorous nonetheless in approval and verification procedures as shown in appendix one. The seller goes through a due diligence process and once approved the seller can offer 90% of an invoice (the Receivable) for sale. Trading takes place on the platform on a daily basis and buyers actively deploy or auto-deploy funds to purchase a receivable. Buyers can purchase 100% of the receivable or select to purchase tranches of the receivable. Once the trading closes the amount of the winning bid is transferred to the Seller less the processing fee. SME's receive their working capital within 24 hours. When the invoice is settled by the debtor on the due date, the unsold portion (10%) is transferred to the seller and remainder to the buyer less a transaction fee equivalent to 16.26% plus vat of the buyers gain. All funds flow through a client segregated trust account.

### *Benefits & Challenges*

According to Co-founder Peter Brady, there are a significant amount of SMEs in Ireland and the UK seeking fast online access to working capital solutions. Many of these companies do not want to hand over their entire debtor book to traditional banks or discounters, which reduces their control, is time consuming, incurs significant hidden costs and often is conditional on the provision of personal guarantees. This financial marketplace provides a single invoice solution without excessive bureaucracy and with a significantly higher level of flexibility. The fact that there is not one but a large pool of funders across the capital markets purchasing the receivables eliminates the requirement for imposing concentration or geographic limits on the SME. Such limits are a common feature of traditional lending models. The *Receivables Trading* model enables SMEs to leverage on their customer relationships resulting in companies being able to accelerate their working capital cycle, release cash and continue to focus on growing their business.

Under this innovative model, the pricing of 'Trade Receivables' is via a market based price discovery. This is an appealing proposition to both Sellers and Buyers. SMEs quickly see the benefits of a positive track record of invoices settling on time translating into a lower cost of credit. Buyers are able to employ classic portfolio

*Figure 1. The buyer registration process at www.invoicefair.com*

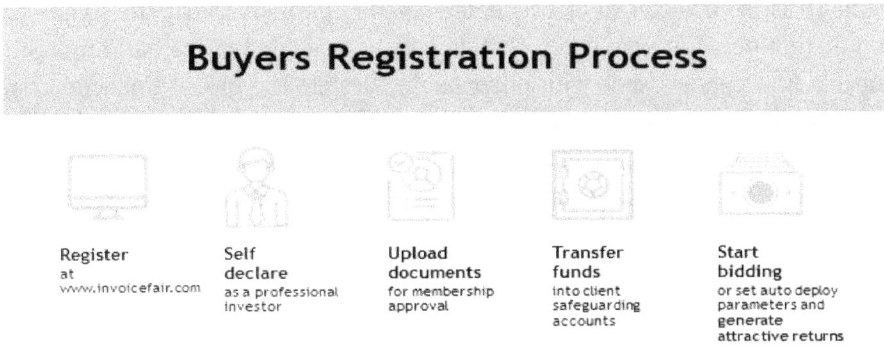


management techniques to assemble an array of assets that reflect their risk appetite and liquidity/duration requirements in an asset class that is un-correlated to the broader, more traditional investment market.

As mentioned above there is great potential to grow business as SMEs are frustrated with existing sources of corporate finance, interest rates, and the restrictive

*Figure 2. Receivable buying process at www.invoicefair.com*

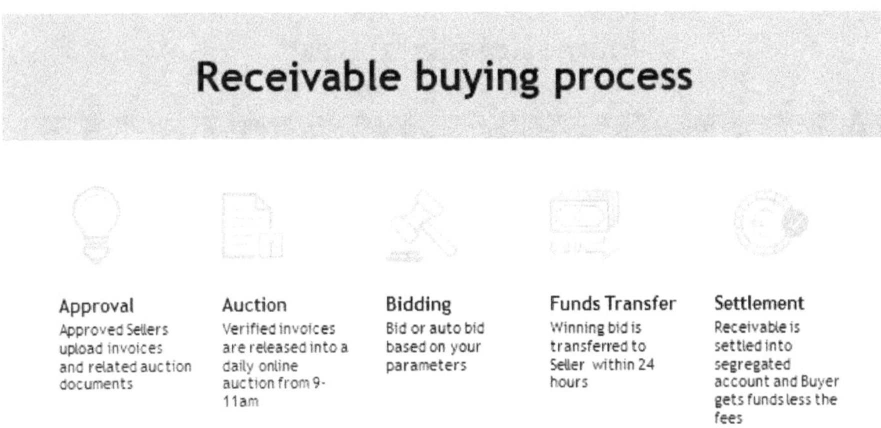

operational model of traditional banks. With trust in online platforms now greatly exceeding trust in Financial Services (Edelman, 2017), the market is ripe for significant digital disruption. However, a key challenge is to evaluate how to build trust in a sector that has legacy issues with pillar banks, despite the global financial crisis and difficulties that arise when transferring trust in an emerging online financial platform (Edelman, 2019). In their continuing effort to maintain their reputation as a trusted online platform particularly in a largely unregulated environment they need to ensure they can not only build trust, but maintain it as illustrated in figure three, which how Invoice Fair consider trust throughout the customer journey. Moreover, Invoice Fair's rigorous approval process and verification procedures are illustrated in the appendix.

## What Does the Future Hold for Invoice Fair?

By 2025, InvoiceFair aim to have significantly altered the funding landscape for SME's in Ireland and beyond. The goal is for the brand to be widely known and understood in the business and finance professional communities in Ireland and the UK. It is anticipated that by then, SME's will understand that their customers and the associated receivables are assets, which they can leverage in order to release cash and drive their own growth and profitability. SME's will recognise this as a potential for competitive advantage, as they are liberated from the normal restrictions

*Figure 3. The journey of digital trust*

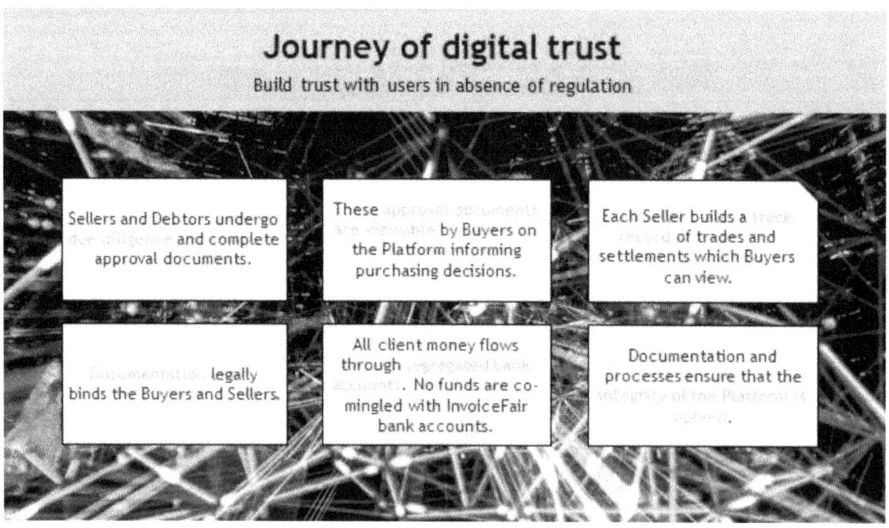

associated to their size and scale of their customers. Across the industry financial professionals, Accountants, Commercial Solicitors and enterprise bodies will consider Receivables Trading as a bedrock and highly appealing working capital funding product. Referrals will be standard and frequent. Looking at the working capital funding market overall, Business owners and Finance Directors will have a range of funding options, with 60%+ of all working capital funding through Receivables Trading. A new channel, enabled by InvoiceFair will be in place - an MNC Procurement policy will optimise the financial strength of their supply chain by proactively embedding the model into Terms of Business. InvoiceFair will have scaled significantly with a depth of business functions, increased automation, and consistency of messaging, procedure and approach across the business. The company will have traded €1billion multi-currency receivables across Ireland, UK and globally. In summary, InvoiceFair want alternative finance to be less alternative and far more mainstream in the marketplace of Ireland and the UK.

## CONCEPTUAL BACKGROUND

The following section explores the literature on technology adoption. Specifically the Technology Acceptance Model (Davis et al., 1989) is discussed in order to get a deeper understanding of people's adoption of 'new' technologies. This model provides insight into key factors that influence consumer's acceptance or rejection of new technologies.

## TECHNOLOGY ACCEPTANCE MODEL: A CONCEPTUAL OVERVIEW

The theoretical origins of the Technology Acceptance Model (Davies et al., 1989) (referred to as TAM henceforth) lie in The Theory of Reasoned Action (TRA) (Fishbein & Ajzen, 1975) and Theory of Planned Behaviour (TPB) (Ajzen, 1991). These theories assume that people are rational and systematic in how they use information to inform decision making and actions. Broadly speaking the theory was developed to better understand and predict behaviour and attitudes. TAM is proposed by Davies et al., to explain why users' accept or reject new technologies. TAM has been given substantial attention in business research to better understand acceptance, adoption and usage intention of new technologies since its inception in 1989. TAM has been used in many contexts from the adoption of communication and office-based systems such as e-mail and fax machines (Adams et al., 1992; Lee et al., 2009), to more recent digital and web based technologies for example, e-commerce

(Pavlou, 2003), e-banking (Lee et al., 2009), mobile services (Lopez-Nicolas et al., 2008) and wearable technologies (Canhoto & Arp, 2017). Its extensive use within Information Systems (IS) research is unsurprising given the model was developed specifically for technology systems use; it has excellent measurement properties, is parsimonious and empirically sound (Pavlou, 2003).

The fundamental assumption of TAM is that technology acceptance behaviour or attitude towards technology is determined by the user's perception of the usefulness (PU) and ease of use (PEOU) of the technology. PU is described as a user's subjective probability that using a specific information system will increase performance and PEOU refers to the degree to which the user expects the system to be free of effort (Davies et al., 1989; Pavlou, 2003). The model as per figure four proposes three overarching relationships that impact behavioural intentions (BI), (1) PEOU leading to PU, (2) PEOU leading to BI, and (3) PU leading to BI and BI leads to actual usage (U). These three links have revealed a high consistency and validity in explaining the acceptance of several technological products and services in many contexts. Research suggests that to fully appreciate technology acceptance, there is a need to consider the features and utility of the technology, the context of usage and the user of the technology (Canhoto & Arp, 2017; Miltgen et al., 2013). Furthermore the technology needs to be perceived as superseding alternative or previous technologies.

## TAM: A FinTech Context

With rapid advances in technology and widespread uptake of digital technology at organisational level has resulted in the transformation of business models in practically all sectors from retail, tourism, healthcare, and financial services. Recent research considers the Financial Technology industry (FinTech) as a new paradigm in information technology and is considered a driving force in innovation in the financial services industry (Lee & Shin, 2018). The digitization of the financial services industry has revolutionized the process of how business customers access capital to facilitate business transactions. The impact of digital technology has been particularly obvious in the banking sector (Featherman & Pavlou, 2003). Technological advancements in this sector have led to the development of electronic finance (e-finance). Lee and Shin (2018: 36) define e-finance as allowing 'individuals or businesses to access accounts, transact business, and obtain information on financial products and services without being in physical contact with financial firms'. FinTech and e-finance has facilitated the growth of online financial marketplaces and platforms, online banking and mobile payment apps. The widespread adoption of these digital innovations is primarily due to the advantages they offer financial services entities in terms of comfort and efficiencies when performing client transactions, as well as increasing

*Figure 4. Technology acceptance model*

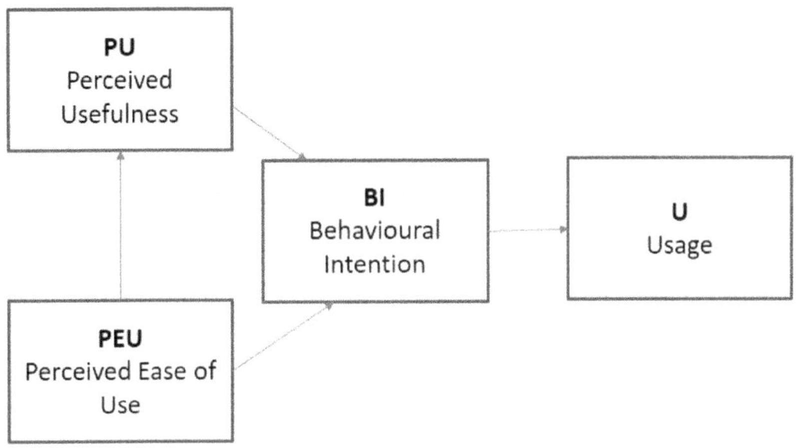


Technology Acceptance Model – Davis et al., (1989)

market coverage and improved service quality (Laukkanen, 2007). Digital banking services provide more functions and functionalities in a more cost effective way by comparison to traditional banking.

However, several studies argue that comfort and ease with using the technology cannot fully explain the increasing popularity and adoption of new digital technologies particularly within the financial services industry (see; Aldas-Manzano et al., 2009; Featherman & Pavlou, 2003; Li, 2013). These studies adopted TAM to investigate the phenomenon and extended the model by incorporating measures of perceived risk, and perceived trust to capture an individual's' security and privacy concerns about online financial transactions in the banking sector. Lack of trust and perceived risk are generally the main reasons why individuals may be slow to adopt online banking or opt to switch financial institutions (Lee, 2009; Manari & Manari, 2007). Risk perceptions increase when an individual perceives an environment to be uncertain or complex and research shows trust to be of high significance in web based uncertain environments (Pavlou, 2003). Lee (2009) found perceived risk factors to supersede perceived benefits when customers intend to adopt online banking inferring that online banking companies need to search for risk-reducing strategies that induce confidence and trust in potential customers. McKnight and

Chervney (2001) suggest that management should consider institutional trust building strategies when designing websites. For example, building secure firewalls to avoid intrusion, developing methods for strengthening encryption, and authenticating websites in order to prevent fraud and identity theft, statements of guarantee and long-term customer service are all measures that could be taken in online financial services sector. In conclusion, it is reasonable to assume the increasing adoption of FinTech appears to be much more complex and multifaceted where an individual's intention to use a digital platform for financial transactions, whether it is online banking, mobile payments, or online lending platforms (e.g. crowdfunding) goes beyond perceptions of ease of use and usefulness.

## FROM TRUST TO DIGITAL TRUST: UNDERSTANDING THE ROLE OF TRUST IN THE DEVELOPMENT OF DIGITAL TRUST

Trust has been widely acknowledged as an important antecedent of any relational exchange (Dwyer *et al.,* 1987; Donaldson & O'Toole, 2007; Morgan & Hunt, 1994). Rotter (1967, p 651) defines trust "an expectancy held by an individual … that the word, promise, verbal or written statement of another individual … can be relied upon." Many of our conceptualisations of trust have stemmed from Rotter's definition, which infers two prerequisite for trust to be present; a customer's expectancy of a provider and a customer's intention to rely on a provider based on that expectancy. While there has been a sustained interest in Trust Theory over the last three decades (Kantsperger & Kunz, 2010), there remains limited understanding regarding how it is operationalised in a consumer context (Young, 2006), with some suggesting that our understanding of trust is somewhat incomplete (Brown, Crosno & Tong, 2019). Conversely, the benefits of trust in consumer interactions are well documented and include; increased co-operation (Lewicki et al., 1998), reduced levels of conflict, risk reduction and decreased transactional costs (Rousseau et al., 1998).

From a theoretical perspective, models of trusting behaviour are quite very varied. Consequentially, this can lead to confusion in understanding how trust functions and how it can be created in a consumer context. Part of this stems from the fact that trust can move from a very basic and almost innate level, to an emotionally significant complex level. At a transactional level, when the outcome of an experience is uncertain or unimportant, trust can simply be 'implicit' (Egan, 2008, p. 106). However, where trust is broken or promises not fulfilled, distrust can manifest.

## Types of Trust

While trust is an indicator of a healthy relationship, most research in business views it as a static construct, failing to acknowledge that it can oscillate as a relationship evolves. Rousseau *et al.* (1998) refer to such change as the "ebb and flow of relationships" (p. 396). Because trust is a multidisciplinary concept, a broad array of criteria has been identified as central to determining the presence of trust. Moorman et al. (1992) consider trust to be a function of consumer's perceptions of capability and dependability. Similarly, Zhou and Tian (2010) identify transferability, calculative assessments, capability, and intentionality as important trust constructs. Other well cited factors include honesty (Kumra & Mittal, 2004), integrity, ethics (Coulter & Coulter, 2002) and most commonly reliability (Rotter, 1967), credibility and benevolence (Doney & Cannon, 1997; Garbarino & Johnson, 1999). The latter three factors are significant in developing trust. Reliability in particular is a crucial determinant to building trust as it relates to a partners perceived intention to stand by their word (Kantsperger & Kunz, 2010). Central to this is the fulfilment of promises (Schurr & Ozanne, 1985) and perceived levels of competence in the provider. Consequentially, it is reinforced through perceptions of dependability, quality and consistency and subject to high levels of predictability by the consumer (Egan, 2008). These facets are nourished by perceptions of reputation, service guarantees and branding. Credibility is linked to reliability as it concerned with the customers belief that the provider has the expertise required to perform the role in a reliable manner. Conversely, Benevolence is somewhat more intangible, focusing on the customer's belief that the provider will act in their best interests. The insinuation here is that the provider will not act opportunistically but remain motivated to behave in a way that promotes the customers prosperity.

Table 1 presents an overview of the types of trust that can be present in customer interactions.

**Calculative Trust (CT)** is a basic level of trust that presents itself early in a customer exchange (Lewicki & Bunker 1995, 1996). As a person seeks to satisfy their self-interests they make a simple economic assessment of the costs associated with behaving opportunistically (Doney & Cannon, 1997; Rao & Bergen, 1992). Thus, CT it is a rationally motivated choice based on the simple principles of exchange (Rousseau *et al.,* 1998, p, 399). As you would expect, it is therefore given on a conditional basis and is a prerequisite for exchange to take place. Conditionality is key here as it gives the consumer the contemplative space to strategically evaluate the progress of the interaction, acting as a self-preservation buffer to give protection against negative outcomes or consequential loss (Wooten & Reed II, 2004). Similarly to CT, **Deterrence-Based Trust (DBT**) is based on utilitarian factors where the perceived costs of switching deter a customer from behaving opportunistically

*Table 1. Types of relational trust based on existing literature*

| Type of Trust | Characteristics | Evaluation Criteria |
|---|---|---|
| **Calculative** | Opportunism of both parties.<br>Rationalisations | Credibility and Reliability |
| **Deterrence** | Controls and Bonding Mechanisms | Perceived levels of dependence |
| **Knowledge-Based** | Familiarity and Interaction<br>Prior Knowledge | Predictability |
| **Capability** | Loyalty and Satisfaction | Benevolence |
| **Transference** | Acquired | Social Dimension |

(Rousseau et al., 1998 p, 398). So, DBT can be calculative in nature. It is most commonly associated with exchanges that have shallow or superficial levels of dependence, which then stimulate 'deterrence' based motivations to mitigate behaving opportunistically (Rousseau *et al.,* 1998). Unlike CT, this level of trust is based on the perceived risks of making a switch (Sheppard & Sherman, 1998 p, 424) and not on egocentricity factors.

More meaningfully, **Knowledge Based Trust (KBT)** (Lewicki & Bunker, 1995, 1996) evolves when familiar situational cues permit individuals to evaluate and compare interactions. It is heavily influenced by past-experiences and prior knowledge, which allows individuals to predict expectations (Morgan & Hunt, 1994) and includes concepts such as perceived competence, benevolence and integrity (Zhou, 2011). Given its reliance on past-experiences, it is related to **Capability Trust**, sometimes referred to as competence trust. This type of trust is linked to trust processes such as dependability and reliability as it is based on the assessment of another party's ability to fulfil their obligations (Doney & Canon, 1997: McAllister, 1995). As trust is multidimensional, Capability based trust can co-exist with CT or DBT, so types of trust are not mutually exclusive. For instance, in order to have a sufficient level of calculative trust, an individual must also have perceived adequate levels of capability trust.

Of particular relevance is **Transference Based Trust (TBT)** which is a function of word of mouth and represents trust building by external influences (Doney & Canon, 1997). This advocates a strong social component to trust (Selnes, 1998). TBT can minimise risk as individuals seek reassurances from others before engaging in an exchange. While these external influences can be a positive influence, they are largely uncontrollable by an organisation and so can produce negative consequences by facilitating the transference of distrust. Distrust, how it manifests and how it is transmitted, remains poorly researched in the extant literature.

## DIGITAL TRUST

Moving on from theory, the interplay of trust and the digital environment is complex and poorly understood. It has been acknowledged that trust is a necessity for any online businesses (Hoffman, Lutz & Meckel, 2014). Arguably, trust is of greater importance in digital exchanges given the relational dynamics of online buyers and sellers (Rice, 2012). For example, in a digital exchange a buyer has to pay for a good or service before receiving it. They must trust that the provider will honour their promise to deliver that product/service. Unfortunately, the theory of trust has failed to keep pace with the rate of our digital evolution. Subsequently a chasm has developed between trust theory and digital trust practice. Indeed, the question remains, is existing trust theory robust enough and appropriate in digital context? Most researchers take a granular view of trust by applying existing theory to the digital environment. Arguably, this is rather flawed and myopic as many additional factors come in to play when building digital trust such as privacy, data security, access to personal information, system functionality and system hacks/failures. Chakravorti, Bhalla and Chaturvedi (2018) refer to the presence of 'trust friction' between digital consumers and digital providers, concluding that how we progress as digital users will be predicated on our ability to successfully build digital trust. They argue that trust friction has many causes citing; infrastructural issues such as design and functionality; systemic issues, such as regulatory/legal requirements and data security measures. Others factors can relate to levels of uncertainty between parties in the transaction. Uncertainty is arguably more of an issue in a digital space as interactions are not face-to-face but rather computer mediated forms of communication. However, research has also shown that, when compared to face-to-face communication, technologically mediated exchanges can actually enhance trust (Burgoon et al 2003). To further complicate the situation, user characteristics have been shown to influence the development of digital trust. In particular, gender differences have been attributed to digital behavioural adaptability. Males for instance will enhance digital security measures as a basic online trust action (Sheehan, 1999), while females are more likely to seek third party recommendations to build trust (Awad & Ragowsky, 2008). The former is arguably a calculative approach to trust and the latter a transference based approach. This suggests that digital consumers should be treated as heterogeneous rather than homogenous, which is often the default strategy for digital providers. The privacy paradox also makes assessing the importance of privacy and trust difficult. Individuals routinely and willingly disclose personal data online for little return yet appear to value privacy above all other things) This is a complex issue that highlights just how difficult it is to conceptualise digital trust while demonstrating how divergent it can be from extant theory concepts. One suggestion is that research should evaluate the different types of personal information being divulged in tandem with behavioural studies rather than self-reported studies (Kokolakis, 2017), which can be subject to social bias and dissonance.

Like its offline counterpart, digital trust is also multidimensional requiring different approaches to measurement (Chakravorti, Bhalla & Chaturvedi, 2018). Digital trust and trust theory overlap significantly. For instance, providers must still deliver on promises, be credible and reliable with consumer's interests at heart. Divergence occurs on an operational level whereby the former issues will be dependent on perceptions of governance, levels of data ownership and control. Zhou (2011) argues that trust has significance given higher perceptions of risk and uncertainty inherent in online exchanges. These issues will become more significant as data collection extends into biometrics, the internet of things, AI and cyber technologies.

As previously highlighted, the literature suggests that our perceptions of trust are based on assessments of benevolence, reliability, credibility and dependability. Consequentially, how consumers assess trustworthiness in a digital environment will be based on their subjective understanding of these concepts and a systems ability to convey these dimensions effectively. Digital trust is also highly dynamic. From a technology perspective, trust in a digital environment is often viewed as something businesses can manage and build into their systems in a way that communicates trustworthiness. However, as trust is a subjective assessment, based on prior cumulative experiences and interactions, it is arguably more difficult to assess in the digital space. According to Chakravorti, Bhalla and Chaturvedi (2018) digital trust-building is costly and resources need to be effectively allocated to guarantee trust. They further argue that businesses need to assess where is best to prioritize trust building spend.

## Trust in FinTech Services

The nature of financial services require the establishment of high levels of trust as involvement and risk are more significant, and it is the case that customers place a higher levels of trust in financial services than other categories (Alexander and Colgate 1998). This is largely because we deposit our money with banks and therefore have to trust they will not lose that money or act opportunistically. As such, we have always held financial institutions to higher standards than other service providers. We have regulators, ombudsmen, government overseers and global financial watchdogs. The applicable trust and digital trust concepts align with those of financial trust. For instance, Van Esterik-Plasmeijer and van Raaij (2017) identify integrity as one of the most important determinants of trust in banking. They also identify key concepts such as transparency, customer orientation, and competence as having significance along with systems trust. Zhou (2011a) argues that trust in this context involves two levels; initial and continuance trust (p 528). To secure the latter a firm must enable the former. Initial trust is based on perceptions of website quality and appeal. Other antecedent factors include information quality, perceived security and perceived privacy. Zhou (2011b) further posits that TAM related concepts such as perceived ease of use and perceived usefulness also have influence along with customer characteristics such as an individual's propensity to trust. Both if these have a moderating effect on the establishment of initial trust. Zhou (2011b) also suggests that online vendor's reputational signals will further determine the establishment of trust in an inline environment. The financial crisis in 2008 witnessed the mass erosion of consumer trust in financial institutions (Gounaris & Prout, 2009) which has slowly been recovering. However pre-crisis, most banking interactions were delivered by people. Now they are increasingly being delivered by mediated technologies and in some instances in the absence of any bricks and mortar capacity. The challenge has become not just about restoring trust in financial institutions but reconceptualising how trust operates in a remote impersonal financial context.

Platform based financial lenders have both an advantage and a disadvantage in this regard. Arguably, individuals who are willing to engage with a financial lenders operating purely online already have a predisposition to trusting technology and therefore are likely to have a level of trust to begin with. The challenge however is to grow and convince other customers to trust in platform based forms of finance. This necessitates FinTech businesses to disrupt the traditional view of financial lending and engender sufficient levels of trust in traditional banking consumers to convince them to switch sides. As demonstrated by Zhou, Tsiga, Li, Zheng and Jiang (2018), who identify trust as a significant moderator on the relationship between satisfaction and continuance intention in e-finance situations, if this is to be achieved, trust will be a central element of success., Key to establishing successful trust building mechanisms is the concept of Institutional Based Trust.

## INSTITUTION BASED TRUST AND TRUST BUILDING MECHANISMS

It has been acknowledged that the growth of the digital eco system and the convergence of technology have meant that businesses are constantly having to re-evaluate their business processes (Deloitte). In order to navigate this ever changing ecosystem, trust can be seen as a currency for interdependent actors within an ecosystem. Businesses need to understand how to build trust in such a constantly changing environment. Trust is important in an exchange system because such relationships are characterised by uncertainty and risk (McKnight et al., 2017). Uncertainty and risk are key features of the online ecosystem because business partners have less interpersonal contact, the fact that new technology continues to be developed, increased competition in the global online market place and the unknown (McKnight et al., 2017). The presence of trust helps to alleviate exchange partner's uncertainty, increase perceptions of the exchange system and also the relationship itself (Morgan & Hunt, 1994; McKnight et al., 200). This section defines what is termed as "institutional based trust" and explores trust related concepts such as web based institutional trust mechanisms.

What is institutional trust? The role of industry trust is becoming increasingly important in online contexts and even more so in emerging industries such as online FinTech and ebanking. The underlying principle of institutional-based trust stems from ones trust in traditional corporate and political institutions such as the banking institutes, government departments, and the legal system. However, following Bachman (2011), we concur that the concept of institutional based trust refers to the process by which individuals or organisations develop trust in the face of specific institutional mechanisms in the digital eco system. Institution-based trust has been defined as "a buyer's perception that effective third-party institutional mechanisms are in place to facilitate transaction success" (Pavlou & Gefen, 2004, pg. 37) such as online security measures, legal structures, recourse, and encryption technologies for the secure transfer of information and financial transactions online.

It is of utmost importance for online companies to be able to demonstrate institutional mechanisms associated with institutional trust. Furthermore, a positive corporate reputation has also been considered a key factor for reducing uncertainty and instilling trust as it indicates that the online firm has met its obligations towards others in the past (Pavlou & Gefen, 2004; Vize et al., 2013). This is particularly pertinent to the banking industry, which witnessed a rapid erosion of trust in recent years. It is therefore reasonable to assert that companies operating within the FinTech sector need to embrace and promote a positive corporate image to minimise risk perception online. This can be achieved via online service reviews, third party accreditation bodies and product / service awards. . Furthermore Vance et al. (2008) underline that institutional trust is related to the belief that the digital ecosystem has appropriate safeguards and protections. The last decade has witnessed an exponential growth in academia and practice around trust and legitimacy in the FinTech industry as a result of the 2008 global financial crisis. Institution based trust is helpful for businesses operating in a digital eco system as it can be a sense making strategy for the parties concerned who may be treading in unchartered territory (Fugslong, 2015). McKnight et al. (2004) argue that institutional trust comprise of two dimensions namely: *situational normality* and *structural assurances*. In another research study, McKnight et al. (2002) propose a trust building model to explain how initial trust boosts consumer's behavioural intention. The following briefly highlights how these dimensions are defined and applied in a digital context.

**Situational Normality** has been defined as the degree to which the work setting appears customary, with everything in proper order (Baer et al., 2018). Normality is based on a perception of how things are meant to be and is the fulcrum in understanding what is normal and what is not normal (Misztal, 2001, Taylor et al., 2013) Situation normality has been defined as the perception of security in a trust relationship that the expected outcomes and the remedies for violations are guaranteed by the structural assurances and situational normality built into the relationship (McKinght et al., 1998). This facilitates for example, website visitors with feelings of certainty and familiarity (Misztal, 2001). Previous studies have found that situational normality has effects on trust (e.g Baer et al., 2017). Web users who regard a situation as having high levels of normality perceive the environment to be an appropriate and orderly place to conduct business (McKnight et al., 2002). In this instance trust is a protective mechanism, which prevents chaos and promotes order by providing users with feelings of safety, certainty and familiarity (Misztal, 2001).

**Structural Assurances** can be seen as the institutional guarantee structures built into the relationship such as formal contracts, guarantees, and legal recourse that are available to the trusting party to ensure success of a transactional exchange. They provide assurance that protects parties from financial loss or privacy exposure by creating protective and technological structures (McKnight et al., 2002; Huang et al., 2005). In a similar vein, Fuglsang (2015) calls this the regulative dimension, which includes legal requirements such as litigation and arbitration and congruence with laws or rules. Structural assurances include: escrows seals of approval, vendor-specific guarantees, protections from credit card fraud and transaction protection may have an effect on trusting intentions. This regulative dimension can increase a party's level of trust and lower perceive risk (McKnight et al., 2002; Huang, 2005; Sha, 2009).

## INSTITUTIONAL TRUST APPROACHES ONLINE: MANAGEMENT IMPLICATIONS

Trust has traditionally been a key component in the area of financial institutions and historically there has been an underlying expectation that financial institutions follow various financial and control mechanisms in order to minimise risk and maximise public trust (Dhar, 2017). As Grazel (2016) highlights, "the battle is about 'trust' not 'tech'. Although there have been a number of trust gaps which have been closed, there is still an opportunity for fintech organisations to improve. Evidently, the 2008 financial crisis was a watershed where serious questions were asked about governance and compliance in the financial services sector. The onus in the future rests on the digital players in the eco system to build trust in ideas, platforms and individuals. This will be the engine driving growth for the financial services sector (Pratini, 2018).

Structural assurance provides assurance that the online business is operating in a secure and safe environment that protects consumers from financial losses or privacy exposure by establishing protective and technological structures (Hoffman & Novak & Peralta, 1999; McKnight & Choudhury& Kacmar, 2002). Structures such as escrows, up to date data encryption software, secure shopping carts, password authentication and so on. These can be displayed to consumers via recognisable logos, symbols, and easy to understand contracts of agreement. At a micro level having these reassurances built into systems enables basic levels of trust in consumers and businesses operating in the digital ecosystem. From a Macro perspective, they are essential for markets to build a strong, safe and trustworthy digital ecosystem. Given that businesses now operate in a low trust environment where brands can no longer demand trust but must earn it (Edelman, 2019), the role of institution based trust has become a prerequisite for building affective levels of trust which are vital to promoting advocacy, purpose and community.

Thus, external signals, such as the assurance by independent public parties may be good sources of information for this purpose. Doney and Cannon (1997) suggests that vendors may cultivate trust transferred from trusted third party, such as friends, certified institutions, government offices, etc. Moreover, structural assurance implies that a web site has protective legal (e.g. privacy warranty) or technological structures (e.g. SSL) based on the institution-based trust perspective (McKnight & Choudhury & Kacmar, 2002; Shapiro, 1987; Zucker, 1986). Yet, merely providing privacy or other company policies is insufficient to guarantee a customer their online transaction is completely safe and free from compromise (Kaplan & Nieschwietz, 2003). Consequently, a vendor may employ web-based structural assurances by displaying certified web seals issued by trusted third parties, for example, BBBOnline, TRUSTe, and WebTrust to enhance the creditability and integrity on it services as well as honouring promises of safe and secure transactions online (Kaplan & Nieschwietz, 2003; McKnight & Cummings & Chervany, 1998). The greater the degree of risk perception the greater the urgency to implement such strategies, especially for transactions that may have financial consequences.

Other challenges facing fintech companies relates to the leveraging of digital platforms that replace conventional banking or payment systems, which may give rise to security breaches or compromise data protection. For fintech applications, sensitive information may be stored on mobile devices that have a greater chance of being lost or stolen (Lee & Shin, 2018). Furthernmore, security of mobile devices can also be compromised through digital payment wallets such as Google Wallet, Apple Pay, and Samsung Pay. With increased adoption and use of social media and other web based platforms consumers can easily file complaints related to data security and privacy breeches to regularity agencies and governing bodies. Therefore fintech companies need to implement appropriate measures to protect sensitive data from unauthorised access and reassure their customers that security mechanisms are in place to prevent unscrupulous behaviour (Chakravorti, & Chaturvedi, 2017). As trust plays an increasingly important role in the adoption of new technologies it is in the company's interest to maintain privacy and security as a top priority. This issue is compounded by the prospects of an ever increasingly fractured financial services sector where traditional banks are significantly disrupted by 'neobanks' and fintech operators. The dissolution and replacement of traditional bricks and mortar banking systems with financial platforms and financial mobile apps will force governments to protect customer's privacy and prevent financial disruptors amassing and misusing customer data. The future beckons the requirement for safety systems that protect individual's assets and data, which will include vetting new entrants to ensure they are adequately capitalised while balancing this with innovation responsive regulation.

## FUTURE DIRECTIONS

Trust as a key driver of consumer relationships has been widely established. However, as relationships with consumers move into digital and virtual environments, how concepts such as trust operate have to be reflected upon. While it may be practical to transpose existing theories into more digital spaces, we argue that research has to ask if existing theories and concepts, which were developed for offline interactions, are appropriate for the more nuanced digital ecosystem. Taking the view that digital environments and connected technologies are omnipresent, multi-channelled, multifaceted (Verhoef, Stephen, Kannan, Luo, Abhishek, Andrews, Bart, Datta, Fong, Hoffman, Hu, Novak, Rand & Zhang, 2017) and heterogeneous (Saarikko, Westergren, Ulrika, Blomquist & Tomas, 2017), we suggest that trust as a moderator on business to customer interactions in this context requires considerable attention. In response to this era of 'technological infusion' where consumers will begin to engage with robots, increased automation and technology objects (Van Doorn, Mende, Noble, Hulland, Ostrom, Grewal & Petersen, 2017), how trust can be embedded into these systems and experiences will become increasingly important particularly in the absence of human to human engagements. Reflecting this, Falkenreck, Christine, Wagner, Ralf (2017) identify trust as core to online consumer relationship success, arguing that firm credibility is essential online as it significantly influences consumer perceptions of how useful a technology is. Furthermore, we argue that technology empowers consumers (Laing, Angus, Hogg & Gillian, 2008), such that trust can arguably become a differentiation factor, assisting consumers in the decision making process, reducing perceived risk and helping them make service provider assessments in an increasingly crowed and complex online environment. While recent research has witnessed an increase in the application of TAM as a lens for customer experience in online environments (Balaji. Roy, Kumar, 2017), the role that trust plays in co-creating experience and value in online relationships requires further exploration. Consequentially, we propose that future research needs to focus on the role of privacy and reputation on consumer's functional interactions with providers, of which trust plays a key role (MSI Research Priorities 2018-2020). Finally, there have been calls for research to address the darker side of online engagements and interactions (Verhoef et al., 2017). While some studies have shed light on how technologies can influence management and firm misbehaviour (De Cremer, Nguyen & Lyndon, 2017; Lee & Shin, 2018), there remains a distinct research deficit in this area. In particular, we suggest that research exploring concepts such as distrust and consumer misbehaviour are also lacking.

## CONCLUSION

In conclusion, services whose attributes are predominantly credence-based are difficult for buyers to verify and evaluate (Darby & Karni, 1973) and can often lead to customer uncertainty and heightened risk perceptions. Such services lack physical evidence of the service process, so customers often do not have clear expectations of what to expect from them resulting in customer uncertainty about the outcome of the value of the service (Vize et al., 2013). Information asymmetry increases the scope for opportunism and conflict to occur. If the customer is at an information disadvantage vis-à-vis the service provider it makes it difficult to ascertain if their provider's recommendations are benevolent or an opportunistic attempt to just 'sell' another service. To manage uncomfortable levels of uncertainty and risk customers often tend to place additional importance on the relational component between the parties (Storbacka, 2011). This approach can, be thwarted by the lack of personal interactions within a digitally focused industry (Obal & Lancioni, 2013) and even more so in organisations that operate in industries perceived as high risk. Relatively speaking as the web industry is still in the early stages of development, there are a complex range of financial solution services available, and a general lack of structural assurances, safeguards, guarantees and lines of legal recourse for customers when things go wrong (Chien *et al.,* 2012; Mysen *et al.,* 2011). However, recent changes in data protection laws and advances in security and encryption software as well as transparent online service review websites may enable management to implement institutional mechanisms online thus mitigate against malware or other harmful activities giving every opportunity to encourage customer interactions and trusted transactions online. The trust relationship is a dynamic one especially a relationship underpinned by changing technologies, which as business and society becomes more digitally evolved, the expectations for instilling trust increases. Therefore, companies that act as 'trust guarantors' must ensure that their digital services are offered efficiently, speedily and in a convenient manner, reassuring the client that any confidential and sensitive information is kept private and secure.

## ACKNOWLEDGMENT

This research was supported by The Hothouse Discovery project, TU Dublin.

## REFERENCES

Adams, D. A., Nelson, R. R., & Todd, P. A. (1992). Perceived Usefulness, Ease of Use, and Usage of Information Technology: A Replication. *Management Information Systems Quarterly*, *16*(2), 227–247. doi:10.2307/249577

Ajzen, I. (1991). The Theory of Planned Behavior. *Organizational Behavior and Human Decision Processes*, *50*(2), 179–211. doi:10.1016/0749-5978(91)90020-T

Aldás-Manzano, J., Lassala-Navarré, C., Ruiz-Mafé, C., & Sanz-Blas, S. (2009). The role of consumer innovativeness and perceived risk in online banking usage. *International Journal of Bank Marketing*, *27*(1), 53–75. doi:10.1108/02652320910928245

Alexander, N., & Colgate, M. (2000). Retail financial services: Transaction to relationship marketing. *European Journal of Marketing*, *34*(8), 938–953. doi:10.1108/03090560010331432

Awad, N. F., & Ragowsky, A. (2008). Establishing trust in electronic commerce through online word of mouth: An examination across genders. *Journal of Management Information Systems*, *24*(4), 101–121. doi:10.2753/MIS0742-1222240404

Baer, M. D., Van Der Werff, L., Colquitt, J. A., Rodell, J. B., Zipay, K. P., & Buckly, F. (2018). Trusting the 'look and feel': Situational normality, situational aesthetics, and the perceived trustworthiness of organisations. *Academy of Management Journal*, *61*(5), 1718–1740. doi:10.5465/amj.2016.0248

Balaji, M. S., & Roy, S. K. (2017). Value co-creation with Internet of things technology in the retail industry. *Journal of Marketing Management*, *33*(1-2), 7–31. doi:10.1080/0267257X.2016.1217914

Briggs, E., & Douglas, G. (2010). Service Performance—Loyalty Intentions Link in a Business to-Business Context: The Role of Relational Exchange Outcomes and Customer Characteristics. *Journal of Service Research, 13*(1), 37–51. doi:10.1177/1094670509345683

Brown, J. R., Crosno, J. L., & Tong, P. Y. (2019). Is the theory of trust and commitment in marketing relationships incomplete? *Industrial Marketing Management, 77*, 155–169. doi:10.1016/j.indmarman.2018.10.005

Brun, I., Rajaobelina, L., & Ricard, L. (2014). Online relationship quality: Scale development and initial testing. *International Journal of Bank Marketing, 32*(1), 5–27. doi:10.1108/IJBM-02-2013-0022

Burgoon, J. K., Stoner, G. A., Bonito, J. A., & Dunbar, N. E. (2003, January). Trust and deception in mediated communication. *36th Annual Hawaii International Conference on System Sciences.*

Canhoto, I. A., & Arp, S. (2017). Exploring the factors that support adoption and sustained use of health and fitness wearables. *Journal of Marketing Management, 33*(1-2), 1–2, 32–60. doi:10.1080/0267257X.2016.1234505

Chakravorti, B., Bhalla, A., & Chaturvedi, R. S. (2018). The 4 Dimensions of Digital Trust, Charted Across 42 Countries. *Harvard Business Review Digital Articles*, 2-8.

Chakravorti, B., & Chaturvedi, R. S. (2017). *Digital Planet 2017; How competitiveness and trust in digital economies vary across the world.* Retrieved from https://sites.tufts.edu/digitalplanet/files/2017/05/Digital_Planet_2017_FINAL.pdf

Chien, S.-H., Chen, Y.-H., & Hsu, C.-Y. (2012). Exploring the impact of trust and relational embeddedness in e-marketplaces: An empirical study in Taiwan. *Industrial Marketing Management, 41*(3), 460–468. doi:10.1016/j.indmarman.2011.05.001

Coulter, K. S., & Coulter, R. A. (2002). Determinants of trust in a service provider: The moderating role of length of relationship. *Journal of Services Marketing, 16*(1), 35–50. doi:10.1108/08876040210419406

Darby, M. R., & Karni, E. (1973). Free competition and the optimal amount of fraud. *The Journal of Law & Economics, 16*(1), 67–88. doi:10.1086/466756

Davis, F. D., Bagozzi, R. P., & Warshaw, P. R. (1989). User acceptance of computer technology: A comparison of two theoretical models. *Management Science, 35*(8), 982–1003. doi:10.1287/mnsc.35.8.982

De Cremer, D., Nguyen, B., & Simkin, L. (2017). The integrity challenge of the Internet-of-Things (IoT): On understanding its dark side. *Journal of Marketing Management*, *33*(1-2), 145–158. doi:10.1080/0267257X.2016.1247517

Deloitte. (2017). *Digital ecosystems in telecom, media and entertainment*. Author.

Dhar, V., & Stein, R. (2017, October). Fintech platforms and strategy. *Communications of the ACM*, *60*(10), 32–35. doi:10.1145/3132726

Donaldson, B., & O'toole, T. (2007). *Strategic market relationships: from strategy to implementation*. John Wiley & Sons.

Doney, P. M., & Cannon, J. P. (1997). An examination of the nature of trust in buyer–seller relationships. *Journal of Marketing*, *61*(2), 35–51.

Dwyer, F. R., Schurr, P. H., & Oh, S. (1987). Developing buyer-seller relationships. *Journal of Marketing*, *51*(2), 11–27. doi:10.1177/002224298705100202

Edelman. (2017). *Edelman Trust Barometer*. Retrieved from https://www.edelman.com/trust2017/

Edelman. (2019). *Edelman Trust Barometer*. Retrieved from https://www.edelman.com/sites/g/files/aatuss191/files/201904/2019_Edelman_Trust_Barometer_Technology_Report_0.pdf

Egan, J. (2008). *Relationship marketing: Exploring relational strategies in marketing*. Pearson education.

Falkenreck, C., & Wagner, R. (2017). The Internet of Things–Chance and challenge in industrial business relationships. *Industrial Marketing Management*, *66*, 181–195. doi:10.1016/j.indmarman.2017.08.007

Featherman, M. S., & Pavlou, P. A. (2003). Predicting e-services adoption: A perceived risk facets perspective. *International Journal of Human-Computer Studies*, *59*(4), 451–474. doi:10.1016/S1071-5819(03)00111-3

Fishbein, M., & Ajzen, I. (1975). *Belief, attitude, intention and behavior: An introduction to theory and research*. Reading, MA: Addison–Wesley.

Fuglsang, L., & Jagd, S. (2015). Making sense of institutional trust in organizations: Bridging institutional context and trust. *Organization*, *22*(1), 23–39. doi:10.1177/1350508413496577

Garbarino, E., & Johnson, M. S. (1999). The different roles of satisfaction, trust, and commitment in customer relationships. *Journal of Marketing*, *63*(2), 70–87. doi:10.1177/002224299906300205

Gounaris, K. M., & Prout, M. F. (2009). Repairing relationships and restoring trust: Behavioral finance and the economic crisis. *Journal of Financial Service Professionals*, *63*(4).

Graze, J. (2017). *World FinTech Report 2017*. Available online: https://www.capgemini.com/service/introducing-the-world-fintech-report-2017/

Harris, C. L., & Goode, M. M. H. (2004). The four levels of loyalty and the pivotal role of trust: A study of online service dynamics. *Journal of Retailing*, *80*(2), 139–158. doi:10.1016/j.jretai.2004.04.002

Hoffmann, C. P., Lutz, C., & Meckel, M. (2015). Content creation on the Internet: A social cognitive perspective on the participation divide. *Information Communication and Society*, *18*(6), 696–716. doi:10.1080/1369118X.2014.991343

Kantsperger, R., & Kunz, W. H. (2010). Consumer trust in service companies: A multiple mediating analysis. *Managing Service Quality: An International Journal*, *20*(1), 4–25. doi:10.1108/09604521011011603

Kokolakis, S. (2017). Privacy attitudes and privacy behaviour: A review of current research on the privacy paradox phenomenon. *Computers & Security*, *64*, 122–134. doi:10.1016/j.cose.2015.07.002

Kumra, R., & Mittal, R. K. (2004). Trust and its Determinants in Internet Banking: A Study of Private Sector Banks in India. *Decision, 31*(1).

Laing, A., & Hogg, G. (2008). Re-conceptualising the professional service encounter: Information empowered consumers and service relationships. *Journal of Customer Behaviour*, *7*(4), 333–346. doi:10.1362/147539208X386860

Laukkanen, T. (2007). Internet vs mobile banking: Comparing customer value perceptions. *Business Process Management Journal*, *13*(6), 788–797. doi:10.1108/14637150710834550

Lee, I., & Shin, Y. J. (2018). Fintech: Ecosystem, business models, investment decisions, and challenges. *Business Horizons*, *61*(1), 35–46. doi:10.1016/j.bushor.2017.09.003

Lee, M.-C. (2009). Factors influencing the adoption of internet banking: An integration of TAM and TPB with perceived risk and perceived benefit. *Electronic Commerce Research and Applications*, *8*(3), 130–141. doi:10.1016/j.elerap.2008.11.006

Lewicki, R. J., & Bunker, B. B. (1995). Trust in relationships. *Administrative Science Quarterly*, *5*(1), 583–601.

Lewicki, R. J., & Bunker, B. B. (1996). Developing and maintaining trust in work relationships. *Trust in organizations. Frontiers of Theory and Research, 114*, 139.

Lewicki, R. J., McAllister, D. J., & Bies, R. J. (1998). Trust and distrust: New relationships and realities. *Academy of Management Review, 23*(3), 438–458. doi:10.5465/amr.1998.926620

Lundahl, N., Vegholm, F., & Silver, L. (2009). Technical and functional determinants of customer satisfaction in the bank-SME relationship. *Managing Service Quality: An International Journal, 19*(5), 581–594. doi:10.1108/09604520910984382

Maier, E. (2016). Supply and demand on crowdlending platforms: Connecting small and medium-sized enterprise borrowers and consumer investors. *Journal of Retailing and Consumer Services, 33*, 143–15. doi:10.1016/j.jretconser.2016.08.004

Manrai, L. A., & Manrai, A. K. (2007). A field study of customers' switching behavior for bank services. *Journal of Retailing and Consumer Services, 14*(3), 208–215. doi:10.1016/j.jretconser.2006.09.005

Marketing Science Institute. (n.d.). *2018-2020 Research Priorities: Marketers' Strategic Imperatives*. Retrieved from https://www.msi.org/articles/marketers-top-challenges-2018-2020-research-priorities/

McAllister, D. J. (1995). Affect and cognition-based trust as foundations for interpersonal cooperation in organizations. *Academy of Management Journal, 38*, 24–59.

McKnight, D. H., Choudhury, V., & Kacmar, C. (2002). Developing and Validating Trust Measures for e-Commerce: An Integrative Typology. *Information Systems Research, 13*(3), 334–359. doi:10.1287/isre.13.3.334.81

Mcknight, D. H., Cummings, L. L., & Chervany, N. L. (1998). Initial trust formation in new organizational relationships. *Academy of Management Review, 23*(3), 473–490. doi:10.5465/amr.1998.926622

McKnight, D. H., Kacmar, C. J., & Choudhury, V. (2004). Shifting Factors and the Ineffectiveness of Third Party Assurance Seals: A Two-stage Model of Initial Trust in a Web Business. *Electronic Markets, 14*(3), 252–265. doi:10.1080/10196780420000245263

McKnight, H., & Chervany, N. (2001). What Trust Means in ECommerce Customer Relationships: An Interdisciplinary Conceptual Typology. *International Journal of Electronic Commerce, 6*(2), 35–59. doi:10.1080/10864415.2001.11044235

Misztal, B. A. (2001). Normality and Trust in Goffman's Theory of Interaction Order. *Sociological Theory, 19*(3), 312–324. doi:10.1111/0735-2751.00143

Moorman, C., Zaltman, G., & Deshpande, R. (1992). Relationships between providers and users of market research: The dynamics of trust within and between organizations. *JMR, Journal of Marketing Research, 29*(3), 314–328. doi:10.1177/002224379202900303

Morgan, R. M., & Hunt, S. D. (1994). The commitment-trust theory of relationship marketing. *Journal of Marketing, 58*(3), 20–38. doi:10.1177/002224299405800302

Mysen, T., Svensson, G., & Payan, G. M. (2011). The key role of opportunism in business relationships. *Marketing Intelligence & Planning, 29*, 426-449.

Nesta / Berkley and Cambridge. (2013). Retrieved at https://www.nesta.org.uk/news/almost-1bn-raised-by-alternative-finance-in-2013/

Obal, M., & Lancioni, R. A. (2013). Maximizing buyer–supplier relationships in the Digital Era: Concept and research agenda. *Industrial Marketing Management, 42*(6), 851–854. doi:10.1016/j.indmarman.2013.06.002

Pavlou, P. A. (2003). Consumer Acceptance of Electronic Commerce: Integrating Trust and Risk with the Technology Acceptance Model. *International Journal of Electronic Commerce, 7*(3), 101–134. doi:10.1080/10864415.2003.11044275

Pavlou, P. A., & Gefen, D. (2004). Building Effective Online Marketplaces with Institution-Based Trust. *Information Systems Research, 15*(1), 37–59. doi:10.1287/isre.1040.0015

Pratini, N. (2018). *How Fintech is (and isn't) leading the way in digital trust*. Available online https://fin.plaid.com/fintech-leading-the-way-in-digital-trust/

Rao, A. R., & Bergen, M. E. (1992). Price premium variations as a consequence of buyers' lack of information. *The Journal of Consumer Research, 19*(3), 412–423. doi:10.1086/209311

Rice, S. C. (2012). Reputation and uncertainty in online markets: An experimental study. *Information Systems Research, 23*(2), 436–452. doi:10.1287/isre.1110.0362

Rotter, J. B. (1967). A new scale for the measurement of interpersonal trust 1. *Journal of Personality, 35*(4), 651–665. doi:10.1111/j.1467-6494.1967.tb01454.x PMID:4865583

Rousseau, D. M., Sitkin, S. B., Burt, R. S., & Camerer, C. (1998). Not so different after all: A cross-discipline view of trust. *Academy of Management Review*, *23*(3), 393–404. doi:10.5465/amr.1998.926617

Saarikko, T., Westergren, U. H., & Blomquist, T. (2017). The Internet of Things: Are you ready for what's coming? *Business Horizons*, *60*(5), 667–676. doi:10.1016/j.bushor.2017.05.010

Schurr, P. H., & Ozanne, J. L. (1985). Influences on exchange processes: Buyers' preconceptions of a seller's trustworthiness and bargaining toughness. *The Journal of Consumer Research*, *11*(4), 939–953. doi:10.1086/209028

Selnes, F. (1998). Antecedents and consequences of trust and satisfaction in buyerseller relationships. *European Journal of Marketing*, *32*(3/4), 305–322. doi:10.1108/03090569810204580

Sha, W. (2009). Types of structural assurance and their relationships with trusting intentions in business-to-consumer e-commerce. *Electronic Markets*, *19*(1), 43–54. doi:10.100712525-008-0001-z

Sheehan, K. (1999). An investigation of gender differences in on-line privacy concerns and resultant behavior. In Internet Marketing. McGraw Hill.

Sheppard, B. H., & Sherman, D. M. (1998). The grammars of trust: A model and general implications. *Academy of Management Review*, *23*(3), 422–437. doi:10.5465/amr.1998.926619

Storbacka, K. (2011). A solution business model: Capabilities and management practices for integrated solutions. *Industrial Marketing Management*, *40*(5), 699–711. doi:10.1016/j.indmarman.2011.05.003

Taylor, N. J., Dennis, A. R., & Cummings, J. W. (2013). Situation normality and the shape of search: The effects of time delays and information presentation on search behavior. *Journal of the American Society for Information Science and Technology*, *64*(5), 909–928. doi:10.1002/asi.22782

Van Doorn, J., Mende, M., Noble, S. M., Hulland, J., Ostrom, A. L., Grewal, D., & Petersen, J. A. (2017). Domo arigato Mr. Roboto: Emergence of automated social presence in organizational frontlines and customers' service experiences. *Journal of Service Research*, *20*(1), 43–58. doi:10.1177/1094670516679272

Van Esterik-Plasmeijer, P. W., & van Raaij, W. F. (2017). Banking system trust, bank trust, and bank loyalty. *International Journal of Bank Marketing*, *35*(1), 97–111. doi:10.1108/IJBM-12-2015-0195

Vance, A., Elie-Dit-Cosaque, C., & Straub, D. W. (2008). Examining trust in information technology artifacts: The effects of system quality and culture. *Journal of Management Information Systems, 24*(4), 73–100. doi:10.2753/MIS0742-1222240403

Vize, R., Coughlan, J., Kennedy, A., & Ellis-Chadwick, F. (2013). Technology readiness in a B2B online retail context: An examination of antecedents and outcomes. *Industrial Marketing Management, 42*(6), 909–918. doi:10.1016/j.indmarman.2013.05.020

Wang, Y.-S., Wang, Y.-M., Lin, H.-H., & Tang, T.-I. (2003). Determinants of user acceptance of Internet banking: An empirical study. *International Journal of Service Industry Management, 14*(5), 501–519. doi:10.1108/09564230310500192

Wooten, D. B., & Reed, A. II. (2004). Playing it safe: Susceptibility to normative influence and protective self-presentation. *The Journal of Consumer Research, 31*(3), 551–556. doi:10.1086/425089

Young, L. (2006). Trust: Looking forward and back. *Journal of Business and Industrial Marketing, 21*(7), 439–445. doi:10.1108/08858620610708920

Zhou, M., & Tian, D. (2010). *An Integrated Model of Influential Antecedents of Online Shopping Initial Trust: Empirical Evidence in a Low-Trust Environment.* Academic Press.

Zhou, T. (2011a). An empirical examination of initial trust in mobile banking. *Internet Research, 21*(5), 527–540. doi:10.1108/10662241111176353

Zhou, T. (2011b). The effect of initial trust on user adoption of mobile payment. *Information Development, 27*(4), 290–300. doi:10.1177/0266666911424075

Zhou, W., Tsiga, Z., Li, B., Zheng, S., & Jiang, S. (2018). What influence users'e-finance continuance intention? The moderating role of trust. *Industrial Management & Data Systems, 118*(8), 1647–1670. doi:10.1108/IMDS-12-2017-0602

# APPENDIX

## Invoice Fair's Rigorous Process of Approval and Verification Procedures and Sample Testimonials

*Figure 5. Legal framework*

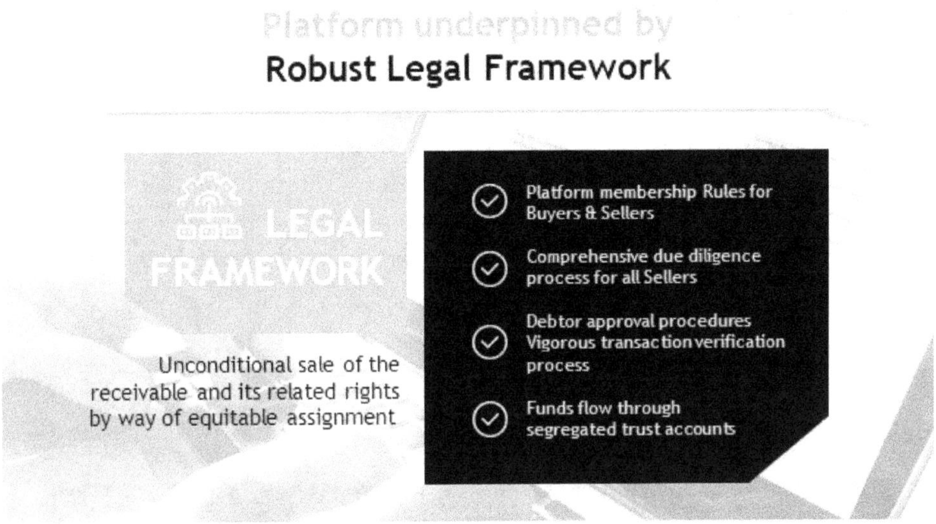


*Figure 6. Seller approval procedures*

Figure 7. Debtor approval procedures

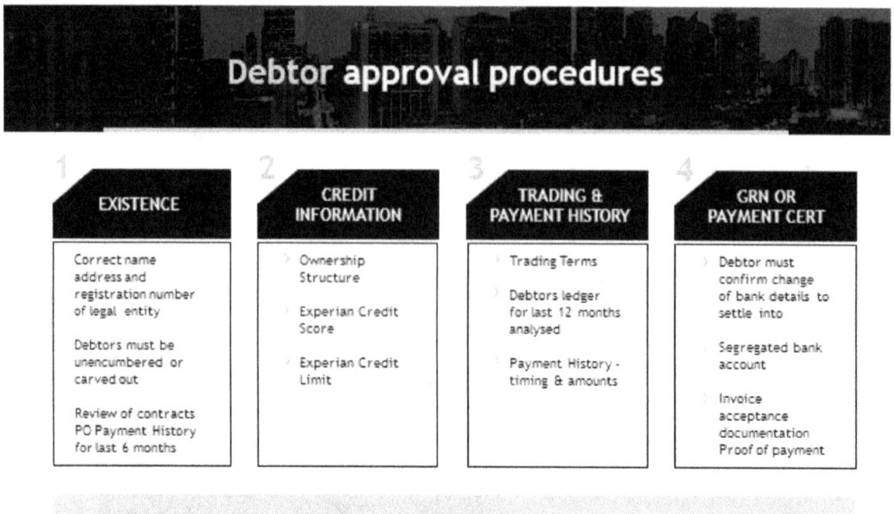


Figure 8. Transaction verification procedures

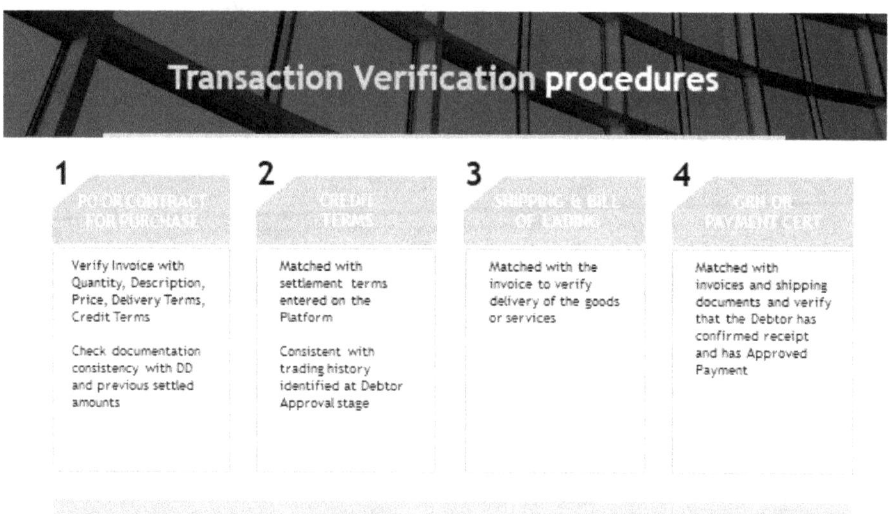

*Figure 9. Confidential & secure*

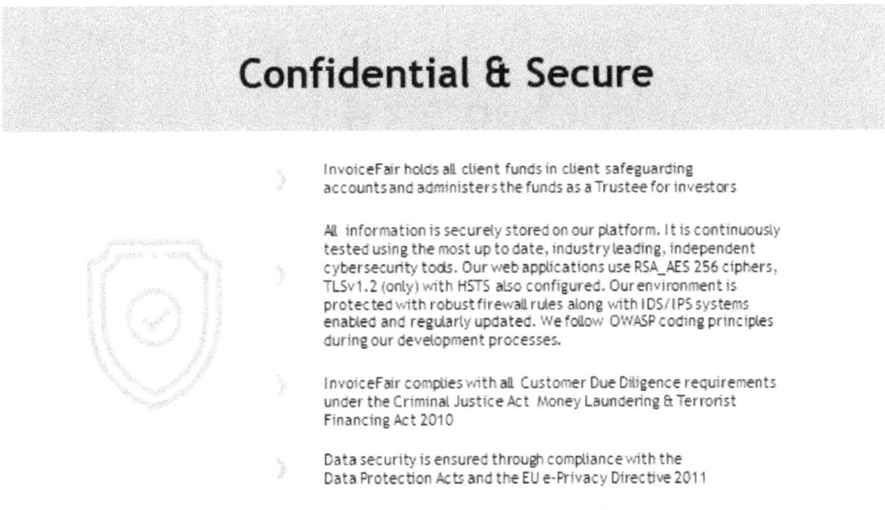


*Figure 10. Testimonials*

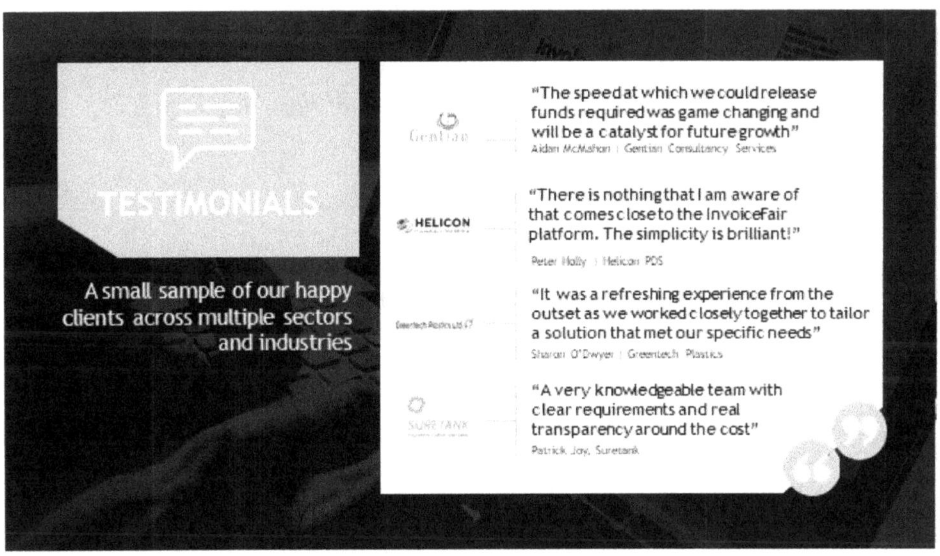

# Chapter 4
# Final Framework for a Successful Business Incubator for Indonesian Public Universities:
## The Influence of Information Technology on Business Incubator Success

**Lina Gozali**
https://orcid.org/0000-0002-3071-0762
*Universitas Tarumanagara, Indonesia*

**Maslin Masrom**
*Universiti Teknologi Malaysia, Malaysia*

**Teuku Yuri Zagloel**
*University of Indonesia, Indonesia*

**Habibah Norehan Haron**
https://orcid.org/0000-0003-3079-9119
*Universiti Teknologi Malaysia, Malaysia*

**Jose Arturo Garza-Reyes**
*Derby University, UK*

**Benny Tjahjono**
*Centre for Business in Society, Coventry University, UK*

## ABSTRACT

*Universities have a meaningful function in motivating young graduates to become entrepreneurs in the technology field. Still, Indonesia has an unemployment issue*

DOI: 10.4018/978-1-7998-1879-3.ch004

*for people's welfare. Indonesia has many potential markets. The first objective of the research is to investigate the indicators and success factors of business incubators in Indonesian public universities. Second is to examine the critical factors that influence the success of business incubators in the Indonesian public universities The third is to propose and develop successful business incubators in Indonesian public universities, and fourth is to measure the influence of IT and e-commerce assistance factor to the success of the university business incubator. The results indicated that there are four significant success factors (i.e., information technology and e-commerce assistance – abilities of business incubator, mentoring and networking, funding and support, and university regulation) with the moderating factors of age of facility, credit and rewards, and good infrastructure system.*

## 1. INTRODUCTION

Business incubators perform as an active mover to help the structure of new enterprises, and deliver them the assistance and support their need to raise (Rice & Matthews, 1995). The role of business incubation is based on its ability to create jobs at a generally cheap public expense that remains in the environment and leads to economic development (Molnar *et al.,* 1997). The most significant component of incubators in their early phase is tangible incubator services. Tangible incubator services include: (i) Physical resources that an organization own that can be inspected and measured; (ii) Financial revenues such as fresh money, and the borrowing treasure; (iii) Organizational assets, for example the formal reporting frame; (iv) Physical building such as construction, machinery, materials, and productive room; (v) Technological resources: center of technology, intellectual property, and patents (Utami & Lantu, 2014). The incubator actually wants to serve the entrepreneur to exceed the obstacles and to help the accomplishment of promising business concept through specific works (Dietrich, 2010).

The university business incubator (UBI) is an ingenious system designed to helps businessmen, particularly businessmen in technology, and to build of new companies. By giving a variety of services and help the start-ups and newly firms, the incubator conduct to accelerate endowment, technology, resources, and knowledge effectively to leverage their talent, to link the growing of new firms, and thus rapidly to commercialize the technology (Smilor, Gibson, & Dietrich, 1990).

Faculty further give an effort to elevated emphasis on the trade of institutional culture and academic opinion in acknowledgement of the invention idea or innovation and the synergy of varied point of views and abilities are key supporting factors in the improvement of research (Babbar, 2017). Researchers, regulation creator, business-incubation managers and stakeholders have deficient of a systematic approach for managing and controlling the achievement of business incubators across industrial area and geographic zone. Despite increasing emphasis from both faculties and practitioners to assess the accomplishment and influence of business incubators, many studies continue to suffer from methodological, theoretical and empirical limitations (Lewis, 2001; Cornelius & Bhabra-Remedios, 2003; Cheng & Schaeffer, 2011; Motoyama & Wiens, 2015).

Indonesia has seen an increasing figure of new ventures or entrepreneurs. Entrepreneurship policies applied at the micro-level still focusing on expansion of ability, future chance, and encouragement. Abilities include the organisational, commercial and technological skills; encouragement relates coaching, mentoring, leadership, and exposure; while future chance relates advertising, channel of selling and channel of funding. At the macro level, government support stress on funding, business courses, entrepreneurship spirits, business infrastructure systems, and workshop for trainers (Mirzanti, 2015).

The objective of this research is to develop and propose a successful business incubator framework for Indonesian public universities.

## 2. CONCEPTUAL FRAMEWORK AND HYPOTHESES

The initial framework by Campbell, Kendrick and Samuelson (1985) emerged with the simple business incubator services and facilities. The business incubator framework started with entry criteria, selection process, funding, and mentoring-networking for tenant business growth.

Smilor (1987) introduced a non-profit business incubator framework whose model implicates the tenant business mission, such as economic development, successful product, tenant's profit, technology diversification, and job creation. Smilor's framework involved support system (i.e. administration, facilities, and business expertise), universities and the government. His work was probably the most extensive in ascertaining and elaborating the different elements of an incubation system.

Campbell (1989) introduced the new incubation process model consisting of pre-incubation process, entry criteria and selection process, monitoring and controlling processes. The previous models by Campbell et al. (1985) and Smilor (1987) had not introduced the processes and activities from the pre-incubation and incubation processes until successful outcomes were achieved.

Mian (1997) gave more detailed processes, criteria, policies, and programmes; and had sought the involvement of universities, communities and other stakeholders in developing a theoretical model for the evaluation and management of the university-based technology business incubators' (UTBI) performance. Mian for the first time introduced the university involvement and developed performance criteria for technology business incubators in the public and private sectors.

The business incubator model is categorised as pre-incubation activities or input (entry criteria), incubation or process, and graduation or output (exit criteria) (Costa-David, Malan, & Lalkaka, 2002). Costa-David et al. for the first time gave the detailed skill requirement (such as management, finance, business advice, networking and training for start-ups until their graduation).

Verma's (2004) framework introduced more detailed success factors (i.e. shared services, facilities and location, financing and support, control of incubators, mentoring-networking, entry criteria, exit criteria), as well as moderating factors (i.e. age and quality of facilities for successful business incubator framework). Voisey (2006) introduced the concepts of hard (profitability, sales turnover, etc.) and soft (business skill improvement, cost saving, etc.) performance measurement of business incubator practice achievement.

All the indicators, successful factors and moderating factors of successful framework of business incubator for Indonesian public universities have been explored in (Gozali, 2015) and (Gozali, 2016). The hypotheses for final framework such as:

## 2.1 Successful Factors

### 2.1.1 The Ability of Business Incubator (H1)

The need for assistances and enterprises advice grew important in the mid-1990s (InfoDev, 2009), which modify the model version from the first-generation incubators to become second-generation incubators. In addition to the typical office room, the second-generation incubators are characterized by shared facilities, proactive assistances, and mentoring, couching, and counselling assistances (Lalkaka, 2001). Scillitoe and Chakrabarti (2010) found that prior study contributes the opinion that counselling activities are an important model of enterprise guidance. They continue advice that more rapid guidance activities will increase the incubator manager to study better about the needs of the entrepreneur, and thus give more proactive commercialize assistance (Hackett & Dilts, 2004b), and the shifting of business know how and skill, either straightly or by assisting to the company to employ the incubator network effectively (Rice, 2002).

The variety of enterprise guidance arranged by a business incubator is fundamental in business incubation performance (Hackett & Dilts, 2008). Martin (1997) recommended that networking within the incubator can take place in a natural way if a physical zone is set aside for the intention of mail box, photocopying, drinking shop, and canteen facilities. Rice (2002) mentioned that high-technology entrepreneurial companies are often not familiar about business skills such as marketing, accounting, financial and human resources required to manage an enterprise. Abduh et al. (2007) mention that incubation management services provide a wide variety of professional commercial development assistance services including strategic planning, writing a business plan, and giving assistance through accounting, financial management, sales or marketing advice, legal advice, and enrich them on government policy, product development, and human resource assistance. Abetti (2004) emphasized that incubators require to focus their activities in coherence with incubates' requirement and market orientation by mapping out demand and market challenge; product development and manufacturing process; sales and marketing; budget and profit and loss reports; and personnel assistance, in order to speed up the expansion and increaser of the incubatees' enterprises.

Furthermore, Rice (2002) advised that the connection between the incubator manager and the tenant company is of several significance to the development of the business proposal. A study by Kuang et al. (2003) found in the context of academic incubators that incubator management must form closer binds with tenant companies to ensure incubator eminence. Rice (2002) mentioned incubator manager-tenant company dyads co-produce the incubation process, suggesting that the time intensity of business advice interruption have to be strategically applied by the incubator manager to the tenant companies, and that tenant companies have to be properly ready to employ the assistance and insights resulting from such interruption. On the basis of the above considerations, the formulation of hypothesis as follows:

**H1:** The greater the focus is on the abilities of business incubator to be moderated by the age of the facilities, the more likely the business incubator is to succeed due to good facilities (Gozali, 2018).

## 2.1.2 Incubator Governance (H2)

Vedel, Stephany, and Gabarret (2011) found that assistance and frequency of counselling between incubator managers and tenant companies do not have a positive result on economic achievement, particularly on job maker. Strategic management afforded at incubators could provide collaboration with tenant companies to elaborate through such things as their business framework, marketing strategy, funding method, patent and invention strategy, and product development (O'Neal, 2005). The degree of strategic management assistance in incubators explored by incubation researchers point to the significance of developing strategic collaboration in incubators, as emphasized by Agarwal (2002).

The level to which the incubator controls and assists tenant companies with the establishment of their enterprises, including assisting them to learn about risks concerning the resources infused in a commercialization, and about including the expense of potential (terminal) failure (Hackett & Dilts, 2004a). Hackett and Dilts (2004a) mention that the deep interaction of support afforded, comprehensiveness of support afforded, and the quality of the support afforded all characterized this component of business incubation activities. Deep interaction of support afforded' refers to ——the percentage of working hours related to controls and helps tenant companies (Hackett & Dilts, 2004a), while comprehensiveness of support provided is a method Hackett and Dilts (2004a) accommodated from Chrisman (1989), and it refers to the level to which strategic, operational, and administrative-related support are afforded by the incubator to the tenant companies. Finally, quality of support afforded contributes the relative meaning of the support afforded by the incubator to the tenant companies (McGrath, 1999; Rice, 2002; Hackett & Dilts, 2004a).

Service innovation, processes and management assistance have different advantages for business of the services area. service innovation is referred as a strategy that makes several advantages for companies, especially SMEs, then it is possible to enclose that enterprises that adopt and employ innovation process as part of their routine will have more opportunity of significant growth their level of company achievement. (Maldonado-Guzman, 2017). On the consideration of the above matters, then formulate the following hypothesis:

**H2:** The better the incubator's governance is moderated by credit and reward, the more likely the business incubator is to succeed(Gozali, 2018b).

## 2.1.3 Entry Criteria (H3)

Study on incubation selection factor remain to reproduce in search for the ideal set of criteria that would guide to elevated incubation achievement. Bergek and Norrman (2008) suggested a model for selection factor combining chosen primarily based on idea and chose primarily based on entrepreneur or the team. The authors proposed two basic method of incubate selection: (i) picking-the-winner; and (ii) survival of the fittest. In the first method, incubator managers try to examine a few successful enterprises before selection. In the second definition, incubator managers employ a more adaptive style of incubate selection and often take on a larger number of firms. This approach relies on the market and time to eventually determine success companies from failed companies. Therefore:

**H3:** The stronger the enforcement of tenant entry criteria, the higher the probability of business incubator success (Gozali, 2018b).

### 2.1.4 Exit Criteria (H4)

Peters, Rice and Sundararajan (2004) emphasized the importance of controlling, or guiding, which is related to as workshop and educational training offered, conferences, courses, either for a fee or free of charge to the incubates as factors associated with increasing incubate graduation level. Several studies invented that the degree of business advice provided at the incubators has a positive impact on the incubation process performance (Rice, 2002; Hackett & Dilts, 2004b; Bergek & Norrman, 2008).

**H4:** The stronger the enforcement of tenant exit criteria, the higher the probability of business incubator success (Gozali, 2018b).

### 2.1.5 Mentoring and Networking (H5)

Researches exhibit that frequent meeting with incubator management makes a better relationship and mainly donates to the tenant companies' and incubators' success (McAdam & Marlow, 2008; Scillitoe & Chakrabarti, 2010). From a social-capital, more frequent counselling interactions allow the creation of stronger binds that facilitate transfer of know how, skill and education between the incubator management and the tenant company. This includes entrepreneurship education from the incubator management, and for incubator management to study about the trend of the enterprises, thus making incubator manager to offer regard support (Scillitoe & Chakrabarti, 2010). They emphasized more frequent counselling interactions can lead to both better venture and technical advice.

The incubation reference invented several incubator studies that explored incubators' resources such as incubators' networks (Bøllingtoft & Ulhoi, 2005; Studdard, 2006); the quality of the management teamwork (Mian, 1997; Costa-David, Malan & Lalkaka, 2002); the style and the quality of the incubator's connection to a university (Tamasy, 2007); a professional service network and initial government funding (Lalkaka 1996); institutional support (Mian, 1997); and the incubators' image or dignity (Mian, 1997). It has been cited in the reference that incubates' reputation is enhanced because of their association with the incubator (Chrisman, Bauerschmidt & Hofer, 1998).

Many Researches show that mentoring and networking are the most significant factors determine a tenant's success (McAdam & McAdam, 2008). Regarding the improving efforts to develop the management commitment, dedication and engagement, educational learning and organizational culture and also advocates the reduction of those barriers such as: missing of management commitment, dedication and engagement, unrecorded or not properly defined processes and deficient of employee training, that determine the successful application of a framework. (Andreadis, 2017). The reference originated in the rational positioning thought focuses ultimately on community factors such as stakeholder activities, business framework, design goals and external constraints, but ignore the function role of internal resource obstacles on enterprise framework design (Wei, 2017).

In a similar vein, Todorovic and Moenter (2010) revealed that the networking process upgrade the efficiency and effectiveness of start-up companies, causing them to achieve aims and sustain improvement (McAdam & McAdam, 2006). However, improving an efficient network requires considerable resources, which requires a dedication individual to assist the networking process. Thus:

**H5:** The better the mentoring and networking of the business incubator moderated by good system of infrastructure, the more likely the business incubator is to succeed (Gozali, 2018b).

### 2.1.6 Funding and Support (H6)

Hamdani (2006) presented that highly utilized resources include assistance with commercial skill, marketing guidance, support about accounting or financial control, access to angels investor or business venture-capital, office room, and a library. Tenant company do not contingent solely on break event point and payback, they also contingent on the current stage of productivity of the firm (Perroni, 2017). Program cannot be generalized since assignment and community conditions, among other factors diverse for every organization and processes (Rocha-Lona, 2015). Financial management of new enterprise is an ultimate knowledge that is mostly cited as insufficient among new ventures (Lalkaka & Abetti, 1999; Beng Hui, Fernandez & Sio, 2011). The importance of this lacking among start-ups is that it is a ultimate problem of incubate fallout level rates (Beng Hui, Fernandez & Sio, 2011). Scope of financial management skill that start-ups have been examined to required help with are in the writing of grant business plan, preparing annual meeting and capital budgets, assessing and presenting on financial achievement report, and budgets controlling (Read & Rowe, 2003).

Mian (1997) and Vedovello and Godinho (2003) emphasized that incubators require to manage a function in supporting financial advice for the start-ups by giving them with data which they may need to linkage a network of partner venture. The consideration of covering the right mix of advocate services, which emphasizes financial skill, was provided by Böhringer (2006) who claimed that the right services would help the expansion of area and the sustainability of startup.

**H6:** The better the funding and support of the business incubator for its tenants is moderated by good system of infrastructure, the more likely the business incubator is to succeed (Gozali, 2018b).

### 2.1.7 Government Support and Protection (H7)

The incubator industry has developed through a partnership between incubators and local, nation and Commonwealth authorities, according to the fact of the Organization for Economic Co-operation and Development (Chandra and Chao, 2011 and OECD, 1999). A unique chance for creating business framework to fulfill the future difficulties of the development of commercialization, freight range and controlling economic performance, community and urban sustainability. The invention of explorative framework through the giving of information is allowed by this technology advancement. This activity has to be provoked as a government–industry collaboration where the enterprise design is managed by news supportive legislative and regulation model (Toh, 2009). From this perspective, a new hypothesis on government support and protection has been developed (H7)

**H7:** The better the support and protection from the government, the more likely the business incubator is to succeed (Gozali, 2018b).

### 2.1.8 University Regulation (H8)

In addition, regulators can improve the reward systems for faculty and researchers by acquainting new stimulant to cooperate with the businesses. Normally, faculty researchers are not granted in their dedication for cooperating with businesses, and in several nations it is even recorded as unethical to do so. Class training experience and publications improve to be dominant standard in profession record systems and salary level, and in several universities collaboration with industry is badly calculated and not measured in professional records. To solve this, a record by the Australian Advisory Council on Intellectual Property (Chandra & Chao, 2011) supports a reconsidering of the key performance indicators of public universities and research centers, creating scheme to improve the motivation of faculties and their researchers to cooperate with the industry. From this reference, a new hypothesis on the credit and reward system in University Regulation has been developed (H8).

**H8:** The better the university regulation is moderated by credit and rewards, the better the initiative programs and projects for business incubator success (university regulation) (Gozali, 2018b).

### 2.1.9 System Infrastructure (H9)

New business incubators do look to support and cover an infrastructure conducive to the establishment of incubates and arising nations (Hackett & Dilts, 2004b; Smilor, 1987). Universities either use government pattern and standard, private sector subsidizing or a mix of both to nurture and extend their 'third mission'. For the future, the goal should be a high scale of self-fluffiness of the university internal entrepreneurship assistance mechanism (OECD, 2010). Based on these literature studies, a new hypothesis on infrastructure and support system for a successful business incubator was formed (H9). All of the hypotheses have to be measured and analyzed, whereby the results will show either positive or negative relationships among all of the factors.

**H9:** The better the system and infrastructure are moderated by good system of infrastructure, the more likely the success of the business incubator (Gozali, 2018b).

## 2.2 Moderating Factors

### 2.2.1 Age of Facilities

A multitenant facility with onsite management that directs the business incubation program. An incubator facility (age and size) supports tenant companies proper leased room and adjustable rents. Collocating entrepreneurial companies in a business incubator facility makes more possibilities for clients to network, share experience, execute in a conducive circumstances for developing successful tenant performance (Lewis, 2011).

### 2.2.2 Good System of Infrastructure

ICT and incubators provide a solution to SME problems. ICT expands the goal and quickness of network communication: telephone, fax, voice mail, e-mail, and teleconferencing are technology empowers that facilitate organizational exchange so that illustrations, concepts,, information, skill, know how, and energy can be shared and transformed with bigger productivity, efficiency, and synergy and all at a cheap expenditure. ICT also exceeds geographic and political limits. It expands ready access to business intelligence and enterprise information such as sophisticated invention, technology advances, (Lewis, 2011), building physical infrastructure (Carayannis, 2006), The right foundational regulation and infrastructure allow the Triple Helix participants in the sophisticated system to work cooperatively in establishing graduated tenants (Wonglimpiyarat, 2016)

### 2.2.3 Credit and Rewards

Disregarding investment in the elements of innovation such as: faculties, academics and research centers (Carayannis, 2006). It is challenging for faculties to absorb that the technical establishment and promoting of their invention is better done to the experts (with their guidance and help). This is connected to the challenging part for a technology discover to admit that they will not accept all of the tribute, incentives, or huge rewards. (McAdam *et al.*, 2006). However, creating an efficient network needs considerable resources (effort, time, and money). Hence, the implementation of a process method to the management of these resources will make much needed assistance to stakeholders, especially business liaison managers, faculty entrepreneurs and academic senior management (McAdam *et al.*, 2006).

One approach for providing start ups-entrepreneurship is the process of establishing something different with value, by dedicating the necessary time and energy, assuming the necessary funding, psychic, and social consequences, and accepting the resulting rewards of monetary and self- satisfaction. The intrapreneur requires to be properly rewarded for the energy and effort extended in promoting the spin-off company. Extensive performance goals should be developed with rewards based on achieving these targets (Hisrich, 1988).

## 3. METHODOLOGY

This research used the qualitative and quantitative method, which involved sequential timing of the using of several methods. One approach is employed first, and the conclusion are used to select the sample, establish the instrument, or write the analysis for the other approach. Further application of establish designs were for dissimilar approaches of equal weight and sequential. Second method in data collection and procedure strategy, first, qualitative study and continued with quantitative study. The weight between the qualitative and quantitative study is ideally equal, although in practice one approach may yield priority to another.

The judgement to choose a proper approach in study hinges upon the goal of the research. It should be considerate by the research question (Marhall, 1996). Consequently, the use of the qualitative and quantitative approaches is deemed proper for this research. First, it seeks to examine the indicators and successful factors for business incubators for Indonesian public universities, then investigation of business incubator successful factors, and research framework performance through statistical analysis.

The survey questionnaire was developed based on literature review. A consolidated survey questionnaire consisted of different measurement scales and questions. Each related success factor was measured on a 1 to 5 Likert scale. The Likert scale was incorporated in the questionnaire as the respondents were asked to rate the importance of the factors relative to other factors. Further, the objective of the study was to establish those factors that have a relatively higher score. The quantitative study continued with reliability and validity tests which all the successful factors are valid and reliable (Gozali, 2018a), research hypotheses tests, and a structural model test. The research used the case study as a part of the qualitative method to study the differences among public university business incubators in Indonesia.

The qualitative study addressed from the literature review that was carried out by identification of the business incubator successful factors, and then the survey questionnaire was developed based on the face-to-face interview with Indonesian public university business incubator experts. The survey questionnaire has been validated by ten professors from six countries (i.e. United States of America, Scotland, Finland, Australia, Malaysia, and Indonesia) (Gozali, 2018). After the validation of survey questionnaire and correction process have been done, the final survey questionnaire was distributed to the respondents by face-to-face interview and through e-mail.

The quantitative study supplemented by data from in-depth, one-on-one interviews. The status of the quantitative aspects of the research is determined higher than the qualitative since the interviews with the expert were based on empirical data which was gathered first (Graff, 2016). Using a questionnaire for a large sample as the ultimately approach of quantitative data collection created this the survey of the research (Denscombe, 2007).

This research examined the result of identified business incubator successful factors, developed the survey questionnaire and the business incubator successful framework.

## 4. RESEARCH LOCATION AND SAMPLE

### 4.1 Research Location

For the actual research conduct, 66 respondents who are business incubator managers in Indonesian public universities were chosen from the following institutions, namely Institut Teknologi Bandung, Institute Teknologi Sepuluh November, Andalas University, Institut Pertanian Bogor, Diponegoro University, University of

Indonesia, Samratulangi University, Brawijaya University, Airlangga University, Riau University, Udayana University, Gorontalo University, Sebelas Maret University, Jambi University, North Sumatera University, Bandung Technopark, Padjajaran University, and Yogyakarta State University.

## 4.2 Research Sample

The sample used for this study consisted of business incubator managers in Indonesia public universities involved in the day–to-day operations of the incubator and graduated tenant companies. The business incubator managers as the sample or respondents would have the necessary insights and experiences of managing incubators and the relations within the incubator with tenant firms. the sample for this research covers 66 respondents, all of whom are business incubator managers from Indonesian Public Universities.

*Figure 1. An initial structural model for successful business incubators for Indonesian public universities (Gozali, 2018b)*

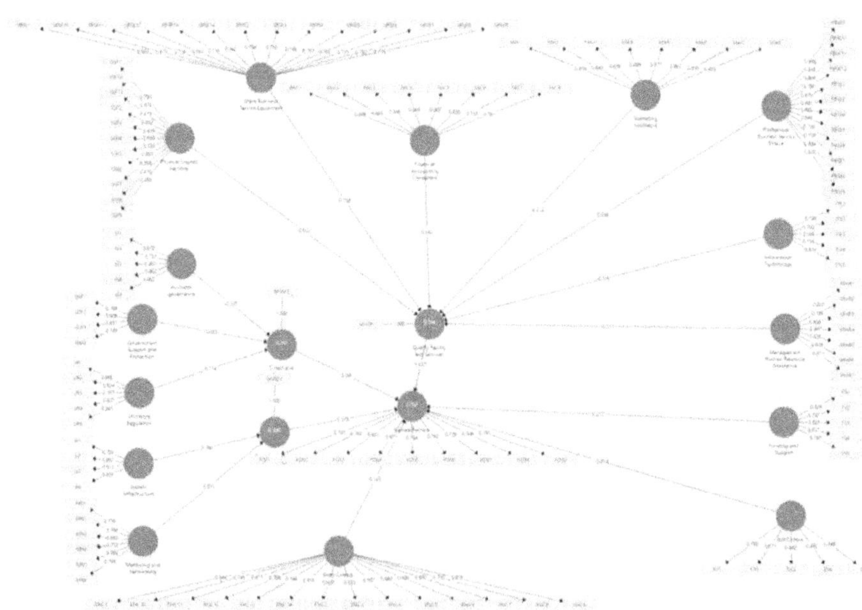

*Table 1. Structural model measurement for this research (Gozali, 2018b)*

| Hypotheses | Construct Relationship | Path Coefficient | t stat | p value |
|---|---|---|---|---|
| H1 | Information Technology -> Quality of Facility | 0.513 | **3.271** | **0.001** |
| H2 | Incubator Governance -> Credit and Rewards | -0.225 | 0.994 | 0.321 |
| H3 | Entry Criteria_ -> Successful factors | 0.148 | 1.332 | 0.183 |
| H4 | Exit Criteria -> Successful factors | -0.033 | 0.166 | 0.868 |
| H5 | Mentoring and Networking -> Good System of Infrastructure | 0.571 | **4.705** | **0.000** |
| H6 | Funding and Support_ -> Successful factors | 0.284 | **3.535** | **0.000** |
| H7 | Government Support and Protection -> Credit and Rewards | 0.023 | 0.121 | 0.904 |
| H8 | University Regulation | 0.714 | **3.607** | **0.000** |
| H9 | System Infrastructure -> Good System of Infrastructure | 0.180 | 1.288 | 0.198 |

## 5. FRAMEWORK AND ANALYSIS

Figure 1 shows the initial structural model of successful business incubator framework for Indonesian public universities of which the content has all of the initial indicators. All of the loading factors for each indicator shows various values.

After obtaining the strong relationship between constructs, the hypotheses have the results. The structural measurements are shown in Table 1. It is observed that hypotheses H2, H3, H4, and H7 are not supported (insignificant t-values) while the hypotheses H1 (just for Information Technology), H5, H6, H7 and H9 are supported (significant t-values).

Table 2 shows the result of hypotheses analysis indicates that several hypotheses was supported directly (i.e. Funding and Support), was supported partially (i.e. Information Technology and E-Commerce Assistance), were fully supported (i.e. Mentoring and Networking, University Regulation). The hypotheses were not supported such as: Ability of Business Incubator, Incubator Governance, Entry Criteria, Exit Criteria, Mentoring Networking, Government Support and Protection, System and Infrastructure. The supported successful factors for business incubator for Indonesian public universities obtain the strong model with moderating factors.

*Table 2. The result of hypotheses testing (Gozali, 2018b)*

| Hypotheses | Description | Result |
|---|---|---|
| H1 | The greater the focus is on the abilities of business incubator to be moderated by the age of the facilities, the more likely the business incubator is to succeed due to good facilities. | Supported Partially (Information Technology and E-com Assistance |
| H2 | The better the incubator's governance is moderated by credit and reward, the more likely the business incubator is to succeed | Not Supported |
| H3 | The stronger the enforcement of tenant entry criteria, the higher the probability of business incubator success. | Not Supported |
| H4 | The stronger the enforcement of tenant exit criteria, the higher the probability of business incubator success. | Not Supported |
| H5 | The better the mentoring and networking of the business incubator moderated by good system of infrastructure, the more likely the business incubator is to succeed. | Supported |
| H6 | The better the funding and support of the business incubator for its tenants is moderated by good system of infrastructure, the more likely the business incubator is to succeed. | Supported Directly |
| H7 | The better the support and protection from the government, the more likely the business incubator is to succeed. | Not Supported |
| H8 | The better the university regulation is moderated by credit and rewards, the better the initiative programs and projects for business incubator success (university regulation). | Supported |
| H9 | The better the system and infrastructure are moderated by good system of infrastructure, the more likely the success of the business incubator. | Not Supported |

*Table 3. Square of endogenous variables (Gozali, 2018b)*

| | **BI Model without Moderating Factors (see Figure 4.2)** | **BI Model with moderating Factors (see Figure 4.3)** |
|---|---|---|
| | R Square | R Square |
| Credit and Rewards | | 0.341 |
| Good System of Infrastructure | | 0.490 |
| Quality of Facility | | 0.294 |
| Successful factors | 0.744 | 0.782 |

*Figure 2. A structural model of successful factors for business incubators for Indonesian public universities without moderating factors (Gozali, 2018b)*

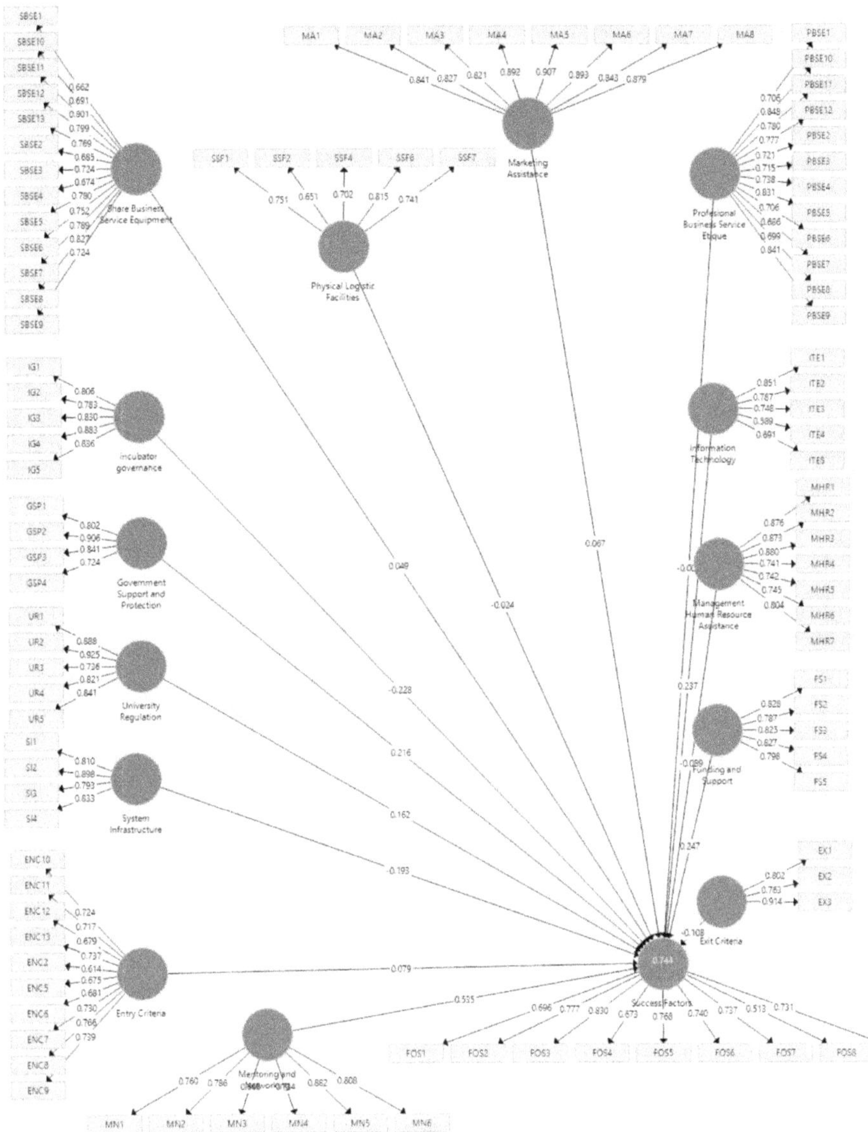

*Figure 3. A structural model of successful factors for business incubators for Indonesian public universities with moderating factors (Gozali, 2018b)*

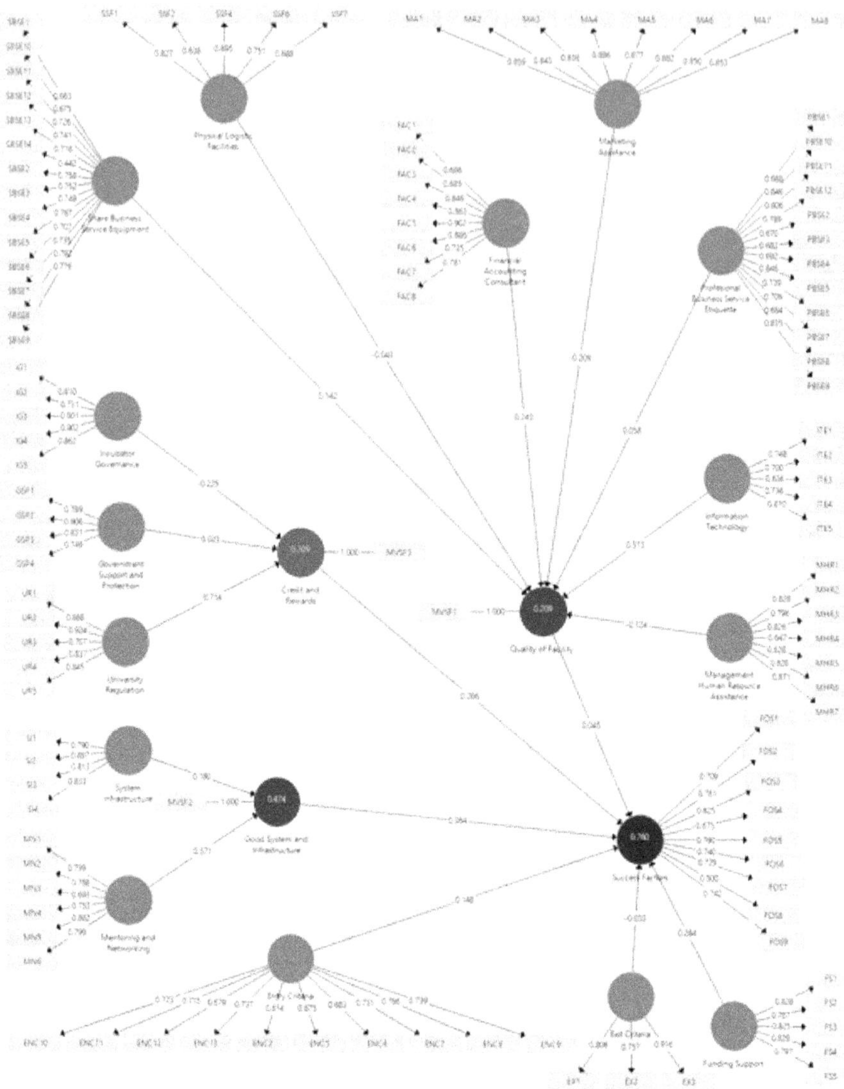

*Figure 4. A final framework of successful factors for business incubator for Indonesian public universities (Gozali, 2018b)*

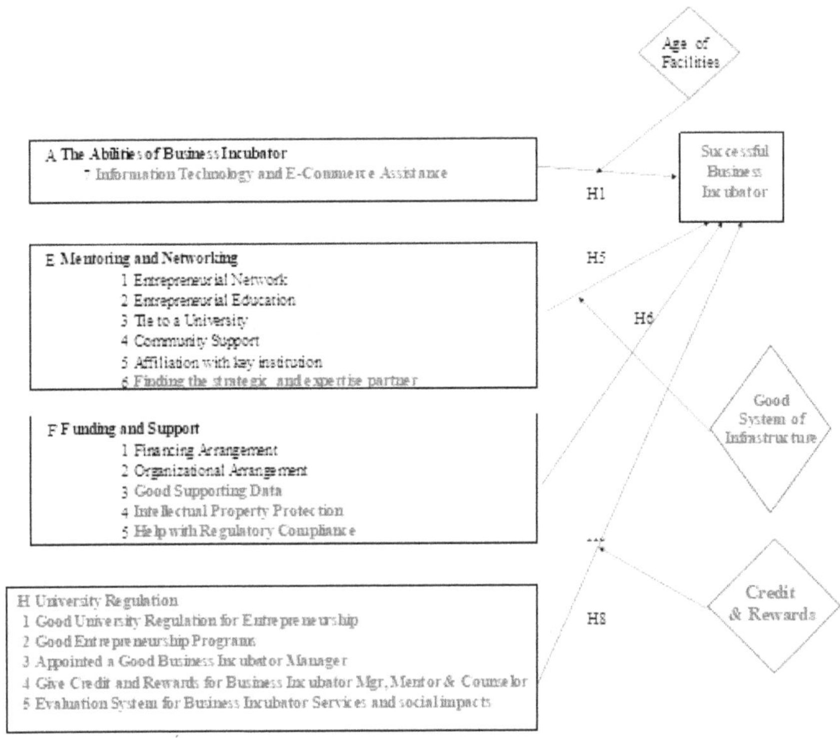


This study has four endogen variables, which are, credit and reward (R = 0.341), good system of infrastructure (R2=0,490), age of facility (R = 0.294) and business incubator successful factors (R2=0.782). According to Chin (1998), the R2 value of credit and reward and quality of facility are weak models. The good system of infrastructure are mostly in the moderate model category. The Successful Factors for business incubator is a strong model. Comparing the Successful Business Incubator Model with moderating factors and without moderating factors, the results show that the Successful Business Incubator model with moderating factors gives the best value of R Square because 0.782 is larger than 0.744 as shown in Table 3.

Figure 2 shows the structural model of successful factors for business incubators for Indonesian public universities without moderating factors and figure 3 shows the structural model of successful factors for business incubator for Indonesian public universities with moderating factors.

Figure 4 shows significantly a final framework with supported successful factors for business incubator for Indonesian public universities. The Framework with moderating factors results more strong value than the framework without moderating factors. And the final framework of successful factors for business incubator for Indonesian Public Universities finally achieved as a theoretical contribution.

## 6. RESULT

The first objective of the current research was to investigate the indicators and the factors for successful business incubators in Indonesian Public Universities using qualitative study by interviewing experts and validation from Business Incubator experts overseas to obtain several results. The result of qualitative study support the finding in literature review such as: government support and protection (funding problems, lack of government protection, short term government mindset), university regulation (lack of appreciation, no excellent entrepreneurship curriculum or programs, lack of support from professor and faculty), mentoring networking, incubator governance (lack of capable incubator manager), and system infrastructure. From the result of this research, 113 indicators have been discovered toward the successful business incubator factors for Indonesian public universities. From these 113 indicators, 9 successful factors with 3 moderating variables and 1 dependent variable are included.

The second objective of the current research was to examine critical successful factors that influence successful business incubators for Indonesian public universities by using statistical analysis by PLS and SPSS to obtain the results about critical successful factors of successful business incubators for Indonesian Public Universities. In testing the hypotheses, the result gave 5 critical successful factors that support the strong model or framework of successful business incubators for Indonesian public universities. The conclusion that was drawn from the research question indicated that government support has a role to play in an array of areas. University regulation is important in providing knowledge, resources and support to start-ups in Indonesia.

The third objective of the current research was to propose and develop successful business incubators for Indonesian public universities by using SEM analysis by PLS and SPSS to obtain results about the framework performance. The Framework with factors and moderating factors showed a strong model indication.

# 7. DISCUSSION AND CONCLUSION

## 7.1 Theoretical Contribution

This research makes a positive contribution to fill the gap and contributes to the development of theory in powerful ways. In particular, this research presents a structure model (Figure 3) and a final framework (Figure 4) for successful factors for business incubator for Indonesian public universities which is valuable to researchers, policymakers, and BI practitioners.

The new contribution is a business incubator framework which provides framework for the investigation of the under-researched phenomenon of business incubators in Indonesian public university. The conceptualization of the research design was guided and adapted from a previously developed framework by Verma (2004). It is worth investigating whether the successful factors for business incubators apply universally to all models/frameworks or if several factors need adaptation of their unique type of services offered.

The results of the quantitative study revealed the extent of the underlying factors impact on the business incubator framework. The significant factors from this business incubator framework are mentoring, networking, university regulation, funding and support and IT systems. In addition, one of the positive outcomes of frequent interactions in mentoring and networking (both formal and informal) could increase the understanding of incubate needs and in return, incubators are able to provide improved and tailored business assistance to the start-ups. The startup needs to produce performance reports to the incubator managers for financial support if the startup shows significant progress. The factors identified and scales that were developed have helped to display successful factors for the facilitation of the entrepreneurial process in Indonesia.

## 7.2 Methodological Contribution

Analysis of the results indicates that the multidimensional development of the research design in this research has produced the successful factors of the Indonesian public universities business incubator framework. The qualitative and quantitative study results provides insights into the underlying components of the business incubation process.

The mixed-methods research design involved the use of qualitative and quantitative study in business incubator study. The mixed-methods approach enhances the interpretation of significant research findings as exemplified in the previous chapter where qualitative and quantitative analysis resulted in improved understanding of the framework of Indonesian Public Universities Business Incubators.

The use of the mixed-method has become an increasingly recognized research approach that is effective in addressing complex research issues. The mixed-methods design enabled the researcher to overcome many of the limitations that constrain mono-method studies.

The qualitative component of study enabled conceptual development, which guided the crafting of questionnaire seeking to further understanding of business incubation factors. The quantitative study with statistical procedures and SEM proved to be a powerful tool in predicting the successful factors for business incubator in Indonesian public universities.

In previous business incubator research or studies applied the qualitative method for data analysis, in this research for methodological contribution applied both qualitative and quantitative methods for data analysis.

## 7.3 Practical Contribution

For practical contribution, the successful factors and framework could provide guidance for researchers, venture capitalists, incubator managers, and entrepreneurs in facilitating the business start-up process. The study is the first of its kind to empirically test successful factors within the Indonesian public universities. Incubators can use this study as a tool to examine their current strategies and accordingly invest resources in developing the business incubator successful factors.

The government also gets benefit from business taxes from the start-ups, and returns those benefits to develop many young start-ups to continue their business through many good business programs. The number of unemployment and criminal activities can also be reduced, and as a result, young people can elaborate their ability to build businesses in their expanded capabilities. Government also get benefit for set up new grand regulation for a new sustainable of national economic development strategy.

## 7.4 Limitation of the Research

The results of this research may be affected by a few factors that could limit its generalization. This includes the focus on incubator types (such as: agricultural incubators, information technology incubators, etc.), the location of the incubators and the size of the sample. The analysis relating to the indicators and factors is based on the data collected from the questionnaires and therefore may inherit self-reporting biases. This study may have aspired in trying to cover far too many business incubator indicators and factors and link them to a set of growth measures.

In conducting this research exercise, a series of difficulties occurred which need to be identified for future researchers in this field. A central concern involves the need to gain access to a large sample of incubator stakeholders, especially board of management participants, associations, government officials, etc.

Verma's business incubator framework adapted in this study delved deeply into government support protection, university regulation, system infrastructure factors and credit and reward system moderating factors.

A chief limitation of this study was the relatively small sample size of the quantitative study. As a result, the bootstrapping calculations cannot be in a complex category due to the limited number of respondents and the high complexity of the model (too many arrows). A further limitation involves the researcher's lack of experience in conducting and interpreting qualitative study. The selection of independent variables and definition of successful factors may be biased, as they are based on literature and practitioners' viewpoints.

The recommendations listed below are proposed as possible ways to improve this study. E-mail may not be the most effective medium for contacting top level respondents for research of this nature. It is recommended to conduct face-to-face interviews to obtain the respondent's answer. Business incubator seminars are the very best time to gather all the business incubator managers to be the best respondents.

The difficulty in gaining fully representative board members' involvement in this study suggests that cautious judgment should be exercised in any generalization of the results.

## 7.5 Recommendation for Future Works

Conducting a mixed-methods research design to explore an unclear area of research such as the business incubator practice in Indonesia produced a considerable amount of data which required intensive effort and time for analysis. A more detailed study on this aspect would be an interesting topic for future research. A study with a large sample size required in order to test the relationship among the various successful factors and how they influence an incubator's success. More comprehensive investigation into the themes and successful factors looked at here would prove to be valuable both to the discipline and to stakeholders working in the sector.

As an initial step, the survey instrument can be improved by adding other indicators of successful business incubator practice. In addition, this categorization may be something that requires verification by a governing body, as interviewees indicated that stratification and regulation may prove to be of value. It is recommended that regulation makers, whether within government bodies or organizational structures, use this information as a basis for evaluating and updating their outlook and the scope and emphasis of existing programs and as a planning tool for future investment. It is important that policies governing this work remain in line with the key objectives outlined in the National Development Plan.

Business incubators could implement the monitor performance of start-ups (incubates), improve business assistance, sharpen positive attributes, and improve reporting practices of business incubators.

Future research should focus on the added strategic value of business incubators to start-ups, and the impact on local and regional economics, particularly to examine this in combination with developing better business incubator services. The incubator industry is evolving like any other industry, and research is essential to observe any common themes and patterns that eluded our investigation based on cross-sectional perspective research.

These Indonesian Public University successful factors in the business incubator framework would be very different to Indonesian private universities business incubators. The scope of further research could examine in detail the successful factors in Private Universities, Vocational Education Programs, Science and Technology Parks, and ICT Business Incubators, and Technology Business Incubators.

Internationally, incubators have changed over time and so has the context. For business incubation to have a long-term future, the need for well-supported research models, supported by incubator stakeholders and potential sponsoring organizations, and the Government's full support is an urgent priority.

## ACKNOWLEDGMENT

Lesa Mitchell, Roesdiman Soegiarso, Eduard Tjahjadi, Eko Harry Susanto, Asril Fitri Syamas, Andi Gani, Jann Hidayat, Meika Syahbana, Zainai Bin Mohamed, Samira Albati Kamaruddin, Docki Saraswati, Eko Harry Susanto, Agustinus Purna Irawan, Lamto Widodo, Adianto, Harto Tanujaya, Reiny Iriana, Reynaldo Hutomo, Novia, Erickson, Vanny Minanda.

## REFERENCES

Abduh, D' Souza, Quazi, & Burley. (2007). Investigating and Classifying Clients' Satisfaction with Business Incubator Services. Managing Service Quality: An International Journal, 17(1).

Abetti, P. A. (2004). Government-Supported Incubators in the Helsinki Region, Finland: Infrastructure, Results, and Best Practices. *The Journal of Technology Transfer*, *29*(1), 19–40. doi:10.1023/B:JOTT.0000011179.47666.55

Agarwal, S. P. (2002). *Strengthening technology incubation system for creating high technology-based enterprises in Asia and the Pacific*. Economic and Social Commission for Asia and the Pacific.

Andreadis, E., Garza-Reyes, J. A., & Kumar, V. (2017). Towards a conceptual framework for value stream mapping (VSM) implementation: An investigation of managerial factors. *International Journal of Production Research*, *55*(23), 7073–7095. doi:10.1080/00207543.2017.1347302

Babbar, S., Behara, R. S., Koufteros, X. A., & Huo, B. (2017). Emergence of Asia and Australasia in operations management research and leadership. *International Journal of Production Economics*, *184*, 80–94. doi:10.1016/j.ijpe.2016.11.018

Bergek, A., & Norrman, C. (2008). Incubator Best Practice: A Framework. *Technovation*, *28*(1–2), 20–28. doi:10.1016/j.technovation.2007.07.008

Böhringer, A. (2006, June). *Knowledge specialization and transfer: a knowledge based framework for business incubation*. Copenhagen: DRUID Summer Conference.

Bøllingtoft, A., & Ulhøi, J. P. (2005). The Networked Business Incubator—Leveraging Entrepreneurial Agency? *Journal of Business Venturing*, *20*(2), 265–290. doi:10.1016/j.jbusvent.2003.12.005

Campbell, C. (1989). Change agents in the new economy: Business Indubators and Economic Development. *Economic Development Review (Schiller Park, Ill.)*, *7*(3), 56–57.

Campbell, C., Kendrick, R., & Samuelson, D. (1985). Stalking the Latent Entrepreneur. *Economic Development Review (Schiller Park, Ill.)*, *3*(2), 43–48.

Carayannis, E. G., & von Zedtwitz, M. (2005). Architecting GloCal (global–local), Real-Virtual Incubator Networks (G-RVINs) as Catalysts and Accelerators of Entrepreneurship in Transitioning and Developing Economies: Lessons Learned and Best Practices from Current Development and Business Incubation Practices. *Technovation*, *25*(2), 95–110. doi:10.1016/S0166-4972(03)00072-5

Chandra, A., & Chao, C. A. (2011). Growth and Evolution of High-Technology Business Incubation in China. *Human Systems Management*, *30*(1-2), 55–69. doi:10.3233/HSM-2011-0739

Cheng, S., & Schaeffer, P. V. (2011). Evaluation Without Bias: A Methodological Perspective on Performance Measures for Business Incubators. *Région et Développement*, *33*, 211–225.

Chrisman, J., Bauerschmidt, A., & Hofer, C. W. (1998). The Determinants of New Venture Performance - An extended model. *Entrepreneurship Theory and Practice*, *23*(1), 5–29. doi:10.1177/104225879802300101

Chrisman, J. J. (1989). Strategic, administrative, and operating assistance: The value of outside consulting to pre-venture entrepreneurs. *Journal of Business Venturing*, *4*(6), 401–418. doi:10.1016/0883-9026(89)90010-4

Cornelius, B., & Bhabra-Remedios, R. (2003). *Cracks in the Egg: Improving Performance Measures in Business Incubator Research*. Academic Press.

Costa-David, J., Malan, J., & Lalkaka, R. (2002). Improving Business Incubator Performance Through Benchmarking and Evaluation: Lessons Learned from Europe. *Materiały*, *16*, 28–04.

Denscombe, M. (2007). *The good research guide*. McGraw-Hill Education.

Dietrich, F., Harley, B., & Langbein, J. (2010). *Development Guidelines for Technology Business Incubators*. Academic Press.

Gozali, L. (2015). A Framework of Successful E-Business Incubator for Indonesian Public Universities. *The Asian Journal of Technology Management*, *8*(2), 118–132.

Gozali, L. (2018b). *Framework Towards a Successful Business Incubator for Indonesian Public Universities* (Ph.D thesis). Universiti Teknologi Malaysia.

Gozali, L., Masrom, M., Zagloel, T. Y. M., & Haron, H. N. (2016). A Framework of Successful Business Incubators for Indonesian Public Universities. *International Journal of Technology*, *7*(6), 1086–1096. doi:10.14716/ijtech.v7i6.3351

Gozali, L., Masrom, M., Zagloel, T. Y. M., Haron, H. N., & Tjahjadi, E. (2018a). A Framework toward Successful Business Incubator for Indonesian Public Universities: A Pilot Review. *International Conference of Industrial Engineering Operation Management 2018.*

Graff, J. C. (2016). Mixed methods research. *Evidence-Based Practice*, 47.

Hackett, A. M., & Dilts, D. M. (2008). Inside the Black Box of Business Incubation. *The Journal of Technology Transfer*, *33*(5), 439–471. doi:10.100710961-007-9056-9

Hacketts, S. M., & Dilts, D. M. (2004a). A Real Options - Driven Theory of Business Incubation. *The Journal of Technology Transfer*, *29*(1), 41–54. doi:10.1023/B:JOTT.0000011180.19370.36

Hacketts, S. M., & Dilts, D. M. (2004b). A Systematic Review of Business Incubation Research. *The Journal of Technology Transfer*, *29*(1), 55–82. doi:10.1023/B:JOTT.0000011181.11952.0f

Hamdani, D. (2006). *Conceptualizing and measuring business incubation*. Statistics Canada, Science, Innovation and Electronic Information Division.

Hisrich, R. D. (1988). New business formation through the enterprise development center: A model for new venture creation. *IEEE Transactions on Engineering Management*, *35*(4), 221–231. doi:10.1109/17.7444

Hui, Beng, D., Fernandez, E. A. V., & Sio, D. B. K. (2011). Privatization of Business Incubation: Initiatives to Achieve sustainability and success. *Policy Brief*, *3*(1), 1–4.

Infodev, I. T. U. (2009). *ICT regulation toolkit*. Academic Press.

Kuang, C. F., Hsiao, H., Luong, L., Lin, G. C. I., & Wu, N. (2003). Development of a New Self-sufficient Model for University Incubator. *International Journal of Innovation and Incubation*, *1*(1), 33–50.

Lalkaka, R. (2001). Best practices in business incubation: Lessons (yet to be) learned. In *International Conference on Business Centers: Actors for Economic and Social Development*. Academic Press.

Lalkaka, R., & Abetti, P. A. (1999). Business incubation and enterprise support systems in restructuring countries. *Creativity and Innovation Management*, *8*(3), 197–209. doi:10.1111/1467-8691.00137

Lalkaka, R., & Bishop, J. (1996). *Business Incubators in Economic Development – An Initial Assessment in Industrialising Countries*. New York, NY: United Nations Development Programme.

Lewis, D. A. (2001). Does Technology Incubation Work? A Critical Review. Washington, DC: Economic Development Administration. US Department of Commerce.

Lewis, D. A., Harper-Anderson, E., & Molnar, L. A. (2011). *Incubating success. Incubation best practices that lead to successful new ventures*. Ann Arbor, MI: Institute for Research on Labor, Employment, and Development.

Maldonado-Guzman, G., Garza-Reyes, J. A., Rocha-Lona, L., & Kumar, V. (2017, September). Service innovation and performance in Mexican service SMEs. In *IFIP International Conference on Advances in Production Management Systems* (pp. 230-239). Springer. 10.1007/978-3-319-66926-7_27

Marshall, M. N. (1996). Sampling for qualitative research. *Family Practice, 13*(6), 522–526. doi:10.1093/fampra/13.6.522 PMID:9023528

Martin, F. (1997). Business incubators and enterprise development: Neither tried or tested? *Journal of Small Business and Enterprise Development, 4*(1), 3–11. doi:10.1108/eb020975

McAdam, M., Galbraith, B., McAdam, R., & Humphreys, P. (2006). Business processes and networks in university incubators: A review and research agendas. *Technology Analysis and Strategic Management, 18*(5), 451–472. doi:10.1080/09537320601019578

McAdam, M., & Marlow, S. (2008). A Preliminary Investigation into Networking Activities within the University Incubator. *International Journal of Entrepreneurial Behaviour & Research, 14*(4), 219–241. doi:10.1108/13552550810887390

McAdam, M., & McAdam, R. (2008). High Tech Start-ups in University Science Park Incubators: The Relationship between the Start-up's Lifecycle Progression and Use of the Incubator's Resources. *Technovation, 28*(5), 277–290. doi:10.1016/j.technovation.2007.07.012

McGrath, R. G. (1999). Falling forward: Real options reasoning and entrepreneurial failure. *Academy of Management Review, 24*(1), 13–30. doi:10.5465/amr.1999.1580438

Mian, S. A. (1997). Assessing and Managing the University Technology Business Incubator: An Integrative Framework. *Journal of Business Venturing, 12*(4), 251–285. doi:10.1016/S0883-9026(96)00063-8

Mirzanti, I. R., Simatupang, T. M., & Larso, D. (2015). Mapping on Entrepreneurship Policy in Indonesia. *Procedia: Social and Behavioral Sciences, 169*(0), 346–353. doi:10.1016/j.sbspro.2015.01.319

Molnar, L., Adkins, D., Yolanda, B., Grimes, D., Sherman, H., & Tornatzky, L. (1997). *Business Incubation Works*. Athens, Ohio: NBIA Publications.

Motoyama, Y., & Wiens, J. (2015). *Guidelines for local and state governments to promote entrepreneurship*. Ewing Marion Kauffman Foundation. doi:10.2139srn.2580321

O'Neal, T., & Kulonda, D. (2005). The Role of Technology Incubators in Engineering Education. In *ASEE Southeast Section Conference 2005*. University of Tennessee at Chattanooga.

OECD. (1999). *Business incubation: international case studies*. OECD Publishing.

OECD. (2010). *High-Growth Enterprises –What Governments Can Do to Make a Difference*. OECD.

Perroni, M. G., da Costa, S. E. G., de Lima, E. P., & da Silva, W. V. (2017). The relationship between enterprise efficiency in resource use and energy efficiency practices adoption. *International Journal of Production Economics*, *190*, 108–119. doi:10.1016/j.ijpe.2016.08.023

Peters, L., Rice, M., & Sundararajan, M. (2004). The role of incubators in the entrepreneurial process. *The Journal of Technology Transfer*, *29*(1), 83–91. doi:10.1023/B:JOTT.0000011182.82350.df

Read, L., & Rowe, D. (2003). *The key building blocks: A guide to business incubation and incubator management part II*. UKBI.

Rice, M. P., & Matthews, J. (1995). *Growing New Ventures, Creating New Jobs: Principles and Practices of Successful Business Incubation*. Westport, CT: Quorum Books.

Rocha-Lona, L., Alvarez-Reyes, S. E., Garza-Reyes, J. A., & Kumar, V. (2015). An integrated approach of Six Sigma and QSAM methodologies for a pharmaceutical company: A shipment improvement process. *International Journal of Lean Enterprise Research*, *1*(3), 266–283. doi:10.1504/IJLER.2015.071743

Scillitoe, J. L., & Chakrabarti, A. K. (2010). The Role of Incubator Interactions in Assisting New Ventures. *Technovation*, *30*(3), 155–167. doi:10.1016/j.technovation.2009.12.002

Smilor, R. W., Gibson, D. V., & Dietrich, G. B. (1990). University Spin-Out Companies: Technology Startups from UT–Austin. *Journal of Business Venturing*, *5*(1), 63–76. doi:10.1016/0883-9026(90)90027-Q

Studdard, N. L. (2006). The effectiveness of entrepreneurial firm's knowledge acquisition from a business incubator. *The International Entrepreneurship and Management Journal*, *2*(2), 211–225. doi:10.100711365-006-8685-z

Tamasy, C. (2007). Rethinking Technology-Oriented Business Incubators: Developing a Robust Policy Instrument for Entrepreneurship, Innovation, and Regional Development? *Growth and Change*, *38*(3), 460–473. doi:10.1111/j.1468-2257.2007.00379.x

Toh, K. T. K., Nagel, P., & Oakden, R. (2009). A business and ICT architecture for a logistics city. *International Journal of Production Economics*, *122*(1), 216–228. doi:10.1016/j.ijpe.2009.05.021

Utami, R. M., & Lantu, D. C. (2014). Development competitiveness model for small-medium enterprises among the creative industry in bandung. *Procedia: Social and Behavioral Sciences*, *115*, 305–323. doi:10.1016/j.sbspro.2014.02.438

Vedel, B., Stephany, E. & Gabarret, I. (2011). *Selection Strategy, Assistance and Performance of Incubation Structures: A French Study*. Paper presented to ICSB.

Vedovello, C., & Godinho, M. (2003). Business incubators as a technological infrastructure for supporting small innovative firms' activities. *International Journal of Entrepreneurship and Innovation Management*, *3*(1/2), 4–21. doi:10.1504/IJEIM.2003.002215

Verma, S. (2004). *Successful factors for Business Incubators: An Empirical Study of Canadian Business Incubators*. Ottawa, Canada: Eric Sprott School of Business, Carleton University.

Wei, Z., Song, X., & Wang, D. (2017). Manufacturing flexibility, business model design, and firm performance. *International Journal of Production Economics*, *193*, 87–97. doi:10.1016/j.ijpe.2017.07.004

Wonglimpiyarat, J. (2016). Government policies towards Israel's high-tech powerhouse. *Technovation*, *52*, 18–27. doi:10.1016/j.technovation.2016.02.001

# Section 2

# Technology and Applications for Improving User Experience

# Chapter 5
# They Know What You Will Do Next Click

**Serra Çelik**
https://orcid.org/0000-0003-4916-3989
*Istanbul University, Turkey*

## ABSTRACT

*This chapter focuses on predicting web user behaviors. When web users enter a website, every move they make on that website is stored as web log files. Unlike the focus group or questionnaire, the log files reflect real user behavior. It can easily be said that having actual user behavior is a gold value for the organizations. In this chapter, the ways of extracting user patterns (user behavior) from the log files are sought. In this context, the web usage mining process is explained. Some web usage mining techniques are mentioned.*

## INTRODUCTION

Today, the Internet is an essential part of daily life. Thanks to the Internet, people easily connect, buy anything they want, share their thoughts. Selling services and goods through online positively affect businesses. At this point, the web sites which are well designed have a massive advantage to other ones. Most companies use their commercial web sites as a connector tool with their customers. They can easily see what customers need and then, can improve their selling and marketing strategies.

The Internet users leave their traces while are navigating on a web site. These traces are called user path, which is the essential user-generated data source. Discovering user path, businesses gain valuable information for their marketing strategies, detect potential risks, take the strategic decisions. Marketing managers aim to attract potential

DOI: 10.4018/978-1-7998-1879-3.ch005

customers (visitors, users) to their websites, understand their purchasing behavior, and ensure the continuity of their customers. A potential customer has been attracted to the site and commences a visit, the next set of issues centers on what transpires while the visitor is there (Bucklin, 2008). Therefore, database marketing may help a firm to better market to their customers (McCarty & Hastak, 2007). The aim here is to analyze the customer's transaction and behavioral data to predict the future behavior of the customer, such as the purchase (Hughes, 2005). Database marketing is efficient because of the use of data-mining techniques based on machine learning that allows marketers to segment their customer databases better.

While users are browsing the pages on the Internet, purchasing products/services, sharing on social media, the tracks they leave are stored on the web servers. Data of customers' browsing behaviors include among others customers' IP address, operating system, keywords used for searching, URL of Web pages visited, time of page visit, and length of time spent on a Web page. The data, which is produced and stored on the web, can be analyzed by Web mining. Web mining is the process of valuable information discovery from Web data.

Web usage mining can be defined as the application of data mining techniques to discover usage patterns from Web data (Srivastava, Cooley, Deshpande, & Tan, 2000). Web usage mining refers to the discovery of user access patterns from Web usage logs, which record every click made by each user, and applies many data mining algorithms. One of the critical issues in Web usage mining is the pre-processing of clickstream data in usage logs to produce the correct data for mining. Web usage mining aims to study user clicks and their applications to e-commerce and business intelligence. The objective is to capture and model behavioral patterns and profiles of users who interact with a Web site. Such patterns can be used to understand the behaviors of different user segments better, to improve the organization and structure of the site, and to create personalized experiences for users by providing dynamic recommendations of products and services (Liu, 2007). The generated data which are called web log data is stored in the web servers. Web Usage Mining uncovers the hidden patterns underlying the Web Log Data. These patterns represent user browsing behaviors which can be employed in detecting deviations in user browsing behavior in web-based applications where data privacy and security is of utmost importance. Knowledge Discovery from Web Log Data has an essential role in serving the needs of web-based applications.

The aim of this chapter is seeking ways of determining user behavior. Predicting customers' interests are based on the data gathered from many customers. The current behavior of visitors (customers) will be maintained in the future. The similarity between users is used to predict the levels of product preference. The Web usage mining process proposed by Srivastava et al. (2000) for web usage mining will be mentioned, and the techniques frequently used to predict user behavior (in the estimation of the next click of the user) will be explained.

## WEB DATA

Today's primary data sources are electronic transactions, web site logs, and sensor information (Ohlhorst, 2013). These data sources can be divided into two as structured or unstructured. Structured data can be defined as traditional data, consists of classified information, and easy to analyze. Unstructured data such as text, sensor data, audio, video, clickstreams, log files have come forth the era of big data. Unstructured data sources are made up mainly of streaming data coming from social media platforms, mobile applications, location services, and Internet of Things technologies. At this point, it can be said Web is the most significant data source now. The data on the Web is distributed, heterogeneous, unstructured, and also massive and grows exponentially every day (Xu, Zhang, & Li, 2011).

Srivastava et al. (2000) define the types of data for Web Mining as content, structure, usage, and user profile. Content is the real data that is designed to convey to the users. Structure is the data which describes the organization of the content. Usage data represents the pattern of usage of Web pages, such as IP addresses, page references, and the date and time of accesses. User Profile provides demographic information about users, which include registration data and customer profile information, of the Web site (Srivastava, Cooley, Deshpande, & Tan, 2000). For discovering web user behavior, three types of web data, which are web logs, web pages, and the web hyperlinks structure, is used (Velasquez & Palade, 2008).

The primary data sources are the server log files, which include Web server access logs and application server logs in Web usage mining. Additional data may be available from client-side or proxy-server (Thakare & Gawali, 2010). Web log file analysis began with the purpose to offer to Web site administrators a way to ensure adequate bandwidth and server capacity to their organization. This field of study made significant advances with time, and now e-companies seek ways to use Web log files to obtain information about visitor profiles and buyers activities (Agosti & Di Nunzio, 2007; Facca & Lanzi, 2005). For each request from a user's browser to a web server, a response is generated automatically, called a web log file, log file, or web log. This response takes the form of a simple single-line transaction record that is appended to an ASCII text file on the web server (Markov & Larose, 2007, p.148). Web logs are based on the structure given by W3C activity. When a web page is accessed, the HTML code is interpreted by the browser. A register is then created in the web log file for the accessed page as well as for each object referred on that page (Velasquez & Palade, 2008).

Web logs usually contain basic information such as name and IP of the remote host, date and time of the request, the request line exactly as it came from the client. A line in a Web server log represents the request made to the server (Prasad, Reddy, & Acharya, 2010). Common Log Format structure can be described (Joshi, Joshi, & Yesha, 2003):

```
{ ⟨domain⟩ ⟨identity⟩ ⟨authorization⟩ ⟨date⟩ ⟨method⟩ ⟨URL⟩
 ⟨protocol⟩ ⟨status⟩ ⟨size returned⟩ }
```

Common Log Format contains Host, Client ID, User login, Date and time of Request, HTTP Request, Status, Page size, Referrer, and User agent. In Table 1, Common Log Format fields are defined in detailed..

## WEB USAGE MINING

Data mining is the process of discovering interesting patterns and knowledge from large amounts of data. The data sources can include databases, data warehouses, the Web, other information repositories, or data that are streamed into the system dynamically (Han, Kamber, & Pei, 2012, p.8). The term Web Mining, first used by Etzioni (1996), is defined as applying data mining techniques on web data to discover knowledge.

Kosala and Blockeel (2000) categorized Web mining into three areas of interest based on which part of the Web to mine: Web content mining, Web structure mining, Web usage mining. Web content mining describes the discovery of useful information from the Web contents/documents. Discovering consumer sentiments using customer reviews on Web is a good result of Web content mining (Liu, 2007). Web structure mining deals with the linking structure of Web sites, based on graph mining algorithms and natural language processing techniques (Xu et al., 2011). In Web usage mining, the goal is to analyze and model user profiles and behavioral patterns associated with a website (Liu, 2007). The term web use mining was first termed Cooley et al. (1997) as the automatic discovery of user access patterns from Web servers.

Web usage mining consists of three stages, as shown in Figure 1, namely preprocessing, pattern discovery, and pattern analysis (Srivastava et al., 2000). The most crucial stage is the preprocessing step as to transform raw log data into a specific format so that particular data mining algorithm can be applied efficiently (M. Srivastava, A. Srivastava, & Garg, 2019). The common preprocessing tasks are data cleaning and filtering, de-spidering, user identification, session identification, and path completion (Markov & Larose, 2007; Prasad et al., 2010).

*Table 1. Common log format fields*

| Fields | Definitions |
|---|---|
| IP Address / Remote Host | This field consists of the Internet IP address of the remote host making the request, such as 192.168.2.20. |
| Identity | Identity information supplied by the customer. |
| Authuser | used when the SSL (Security Socket Layer) is activated. The client may use this field in order to receive or send confidential information. |
| Time | date and time when a web object is requested by the web browser to the web web server. |
| Request | The HTTP request field consists of the information that the client's browser has requested from the web server. |
| Status Code | an integer code which shows the request status. The status code field provides a three-digit response from the web server to the client's browser, indicating the status of the request, whether or not the request was a success, or if there was an error, which type of error occurred.<br>• Successful transmission (200 series, indicates that the request from the client was received, understood, and completed)<br>o 200 success<br>o 201 created<br>o 202 accepted<br>o 204 no content<br>• Redirection (300 series, indicates that further action is required to complete the client's request)<br>o 301 moved permanently<br>o 302 moved temporarily<br>o 303 not modified<br>o 304 use cached document<br>• Client error (400 series, indicates that the client's request cannot be fulfilled, due to incorrect syntax or a missing file)<br>o 400 bad request<br>o 401 unauthorized<br>o 403 forbidden<br>o 404 not found<br>• Server error (500 series, indicates that the web server failed to fulfill what was apparently a valid request)<br>o 500 internal server error<br>o 501 not implemented<br>o 502 bad gateway<br>o 503 service unavailable |
| Bytes / Transfer Volume | The transfer volume field indicates the size of the file (web page, graphics file, etc.), in bytes, sent by the web server to the client's browser. |
| Referrer | a text string sent by the client that indicates the original source of a request or link. |
| User-Agent | the name and version of the client software making the request. |

The raw web logs are not ready to data mining in first place thus data preprocessing is one of the time-consuming stages of Web usage mining process and generally takes more than 60% of the whole Web usage mining process (Srivastava et al.,

*Figure 1. Web usage mining process*

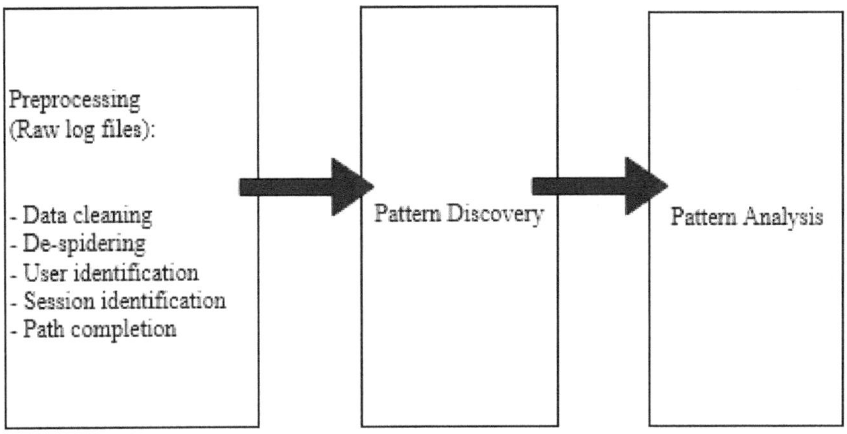


2019). This process is critical to the successful extraction of useful patterns from the data and can be referred to as data preparation (Liu, 2007, p.450).

The data recorded in server logs reflect the access of a Web site by multiple users, and only the IP address, agent, and server side click-stream are available to identify users and server sessions. However, it is vital to notice that the data collected by server logs may not be entirely reliable because some page views may be cached by the user's browser or by a proxy server. In a Web server log, all requests from a proxy server have the same identifier, even though the requests potentially represent more than one user. Also, due to proxy server level caching, multiple users throughout an extended period could view a single request from the server. The Web server can also store other kinds of usage information such as cookies, which are markers generated by the Web server for individual client browsers to automatically track the site visitors (Batista & Silva, 2001).

The first step of the pre-processing stage is data cleaning and filtering. Data cleaning refers to the removal of irrelevant entries and attributes that are not required for the Web usage mining process. The requests for image files and other irrelevant files associated with requests for a particular page are filtered out from the web log file by exploring the page extensions (e.g., .gif, .html, .xbm, .htm, .pdf, .exe, .zip, .txt, .wp, .jpg, .ps) in the web log file (Prasad et al., 2010; Markov & Larose, 2007).

Web log files are text files and have fields like time, as mentioned before. With data cleaning process, variables, which are necessary for web usage mining, extract. Time field has two variables, the date and the time. Request field contains the request method, the uniform resource identifier (i.e., page or document requested), the protocol version, and optionally, some search information. Next step of data cleaning is time stamp creating. The time stamp is useful for estimating the duration of the user's visit to the Web site and for maintaining the sequence of web requests across days (Markov & Larose, 2007, p.159).

De-spidering the web log file is the second step of the pre-processing stage. Web search engines need the most current information is presented to their users. Therefore, the search engines dispatch automatic web bots such as spiders, and crawlers, to web sites. The web bots appear as user visitors on log files. In web usage mining, the goal is to reveal how individuals (human) use the Website; thus, those nonhuman activities should be eliminated (Markov & Larose, 2007, p.163). De-spidering can be based on heuristic (Berendt & Spiliopoulou, 2000) or classification methods (Tan & Kumar, 2000).

The third step of the pre-processing stage is user identification. The goal of this step is to identify each distinct user (Prasad et al., 2010). In the user identification phase, a user is recognized, and its activities are grouped and written into a user activity file (Liu, 2007). The remote host field, or IP address field, may in principle be used to identify users. However, the widespread use of proxy servers, corporate firewalls, and local caches renders problematic the use of the IP address as a substitute for user identification (Markov & Larose, 2007). User identification is the process of as so ciating page references, even those with the same IP address, with different users. The site topology is required in addition to the server logs to perform user identification (Cooley, Mobasher, & Srivastava, 1999).

Session identification is the fourth step of the pre-processing stage. Session identification is the process of segmenting the access log of each user into individual access sessions (Prasad et al., 2010). The concept of a user session is crucial because it corresponds to what is often considered to be a visit to a site (Markov & Larose, 2007, p.168). The user session is a sequence of Web pages clicked by a single user during a specific period. A user session is usually dominated by one specific navigational task, which is exhibited through a set of visited relevant pages that contribute significantly to the mission conceptually. The navigational interest on one particular page is represented by its significance weight value, which is dependent on user visiting duration or click number (Xu et al., 2011, p.15).

In sessionization, users' activities are divided into multiple sessions because a user can visit a website more than once that results in numerous sessions for the same user (Liu, 2007; Prasad et al., 2010). Session identification takes all of the page references for a given user in a log and breaks them up into user sessions (Cooley et al., 1999). For session identification, discovered from empirical findings, a 30-min threshold for total session duration has been recommended and used in many applications (Spiliopoulou, Mobasher, Berendt, & Nakagawa, 2003).

The last step of the pre-processing stage is path completion. Path completion refers to the inclusion of relevant page access records that are missing in the access log due to the browser and proxy server caching. Methods that try to overcome this problem include the use of cookies, cache busting, and explicit user registration (Prasad et al., 2010). Heuristic methods based on referrer log and site topology are employed in (Cooley et al., 1999).

After the pre-processing stage, the pattern discovery stage is performed. Pattern discovery focuses on predictions about the user will visit next and the information they are looking for. Statistical and data mining methods (clustering algorithms, association rules, classification algorithms, sequential patterns) are applied to discover patterns.

In the pattern discovery stage, the goal is to capture, model, and analyze the behavioral patterns and profiles of users interacting with a Web site. The discovered patterns are usually represented as collections of pages, objects, or resources that are frequently accessed or used by groups of users with common needs or interests (Liu, 2011). This stage will be detailed in the next section.

In the final stage of the process is pattern analysis. Not all of the patterns uncovered in the pattern discovery stage would be considered unusual or useful. In this stage, the goal is to filter out uninteresting rules or patterns found in the previous stage. Visualization techniques are helpful in application domains expert to analyze the discovered patterns (Batista & Silva, 2001).

Additionally, visualization techniques are used for a more straightforward interpretation of the results. Using these results in association with content and structure information concerning the Web site there can be extracted useful knowledge for modifying the Web site according to the correlation between the user and content groups (Eirinaki & Vazirgannis, 2003). At the end of the process, user models that can be used as input to applications such as recommendation engines, visualization tools, and Web analytics and report generation tools.

## EXTRACTION OF USER BEHAVIOR

In this section, the pattern discovery and analysis techniques employed in Web usage mining are discussed. Primary using purpose of Web usage mining is to extract user interest and characteristics as Web log data has valuable information about the user's interest. From a simple web page use of statistics to advanced web mining algorithms, user behavior analysis is complicated and time-consuming (Joshi & Krishnapuram, 2000). The user behavior in a web site is represented by page sequence (the sequence of pages visited and registered in the web log files), page content, and spent time (time spent by the user on each page) (Velásquez & Palade, 2008).

In Web usage mining, several data mining techniques can be used. Data mining techniques need to take into consideration the unique characteristics of Web data and perform pattern discovery tasks accordingly. Web data that related to web user behavior has to be processed to create feature vectors, whose components will depend on the particular implementation of the mining algorithm to be used and the preference patterns to be extracted (Velásquez & Palade, 2008).

By using Web usage mining, the user behavior can be modeled, and, thus, with modeling web user behavior, consumer (customer, visitor) purchase behavior can be understood, market sharing can be increased, visitors can be transformed to customers, and customer loyalty can be raised. In marketing, understanding consumer purchase behavior is a fundamental principle for achieving active market participation (Kotler & Armstrong, 2000).

In Web usage mining, after the pre-processing process, exploratory data analysis is used to gain knowledge about visitor behavior. Information about most frequently accessed pages, number of visits (session) actions, average time per page, the average length of a path through a site, common entry and exit points are calculated easily by statistical analysis. The statistical analysis of pre-processed session data constitutes the most common form of analysis. In this context, data is aggregated by predetermined variables such as days, sessions, visitors, or domains (Liu, 2011). Also, basic graphs can be used to inspect the interrelationships among the variables.

When the exploratory analysis is finished, the modeling stage begins. Modeling techniques can be divided into two as supervised or unsupervised. In unsupervised techniques (unsupervised learning), there is no target variables, algorithm searches for patterns among the variables. Clustering is the most commonly used unsupervised technique. On the other hand, in supervised techniques (supervised learning), there is a particular target variable, and the algorithm is given many examples where the value of the target variable is provided, so that the algorithm may learn which values of the target variable are associated with which values of the predictor variables. All classification methods are supervised methods (Markov & Larose, 2007).

In this section, some main modeling techniques, such as association rules, clustering, and classification, will be mentioned.

## Association Rules

Association rule mining is a technique for finding frequent patterns, associations, and correlations among sets of items. Once the user sessions are extracted from server logs, association rules can be generated from this set of user sessions which contain remote host, user id, and a set of URL's (Gündüz-Ögüdücü, 2011; Jagan & Rajagopalan, 2015). Once the user sessions are extracted from server logs, association rules can be generated from this set of user sessions (Gündüz-Ögüdücü, 2011). Association rules are used to reveal correlations between pages accessed together during a server session. Such rules indicate the possible relationship between pages that are often viewed (visited) together even if they are not directly connected and can reveal associations between groups of users with specific interests and purchase patterns (Eirinaki & Vazirgiannis, 2003; Iváncsy & Vajk, 2006; Liu, 2011). For example, one can find correlations such as the following (Gündüz-Ögüdücü, 2011, p.36):

(40% of users visit the Web page with URL /home/page1, and the Web Page with URL

/home/page2 in the same user session, 30% of users who accessed the Web page with URL /home/products, also accessed /home/products/computers).

The basic idea in association rules is to find significant correlations among a large data set. A typical example of this technique is the "purchasing analysis," which uses customer buying habits to discover associations among purchased items. The discovery rules are formalized as if < X > then < Y > expressions (Velásquez & Palade, 2008). Association rule analysis enables Web sites to organize the site content more efficiently or to provide effective cross-sale product recommendations (Liu, 2011). In other words, in web usage mining, association rules are used to discover pages that are visited together quite often (Jagan & Rajagopalan, 2015). The significant advantage of using Association rules in web usage mining, it is required to satisfy user-specified minimum support and confidence (Zhou & Huang, 2014).

The advantages of algorithms are to enable guiding for web site reformation and improve the system performance (Dogne, Jain, & Jain, 2015). Association rules generate enormous rules that cause researchers to spend ample time and expertise to discover fascinating ones (Geetharamani, Revathy, & Jacob, 2015). It can be said association rules algorithms suffer from storage and run time issues because of these algorithms mine all of the various page sets based on minimum support threshold and all possible association rules based on minimum confidence threshold (Malarvizhi & Sathiyabhama, 2016).

The most commonly used Association rules algorithms are Apriori and FP-Growth (Frequent Pattern Growth) (Jagan & Rajagopalan, 2015; Anitha & Isakki, 2016). Based on these basic algorithms, many algorithms have evolved in the research literature (Han & Kamer, 2004). Apriori algorithm is the best-known association rule mining algorithm to mine association rules (Geetharamani et al., 2015). The advantages of using apriori algorithm are used large item set property, easily parallelized, and easy to implement. The limitation of the algorithm is costly to handle a massive number of candidate sets (Kumar & Rukmani, 2010).

FP-Growth algorithm calculates frequent items sets from large data sets. The advantage of FP-Growth is compared to Apriori is that it uses only two times data scans is, therefore, often valid even on large data sets (Anitha & Isakki, 2016). FP-Growth algorithm uses a compact data structure and eliminates repeated database scans (Kumar & Rukmani, 2010). As web logs are often extensive in size but sparse in density, the efficiency of the frequent pattern mining algorithm is essential (Dharmarajan & Dorairangaswamy, 2016).

## Clustering

Clustering is widely used in Web mining. With clustering, items can be grouped according to similar characteristics. In the context of Web mining, there are two kinds of clusters to be discovered: page clusters and user clusters. Page clustering identifies groups of pages that seem to be conceptually related according to the users' perception (Eirinaki & Vazirgiannis, 2003). Also, Web pages can be clustered into groups that may represent a particular topic (Liu, 2011). Clustering of user records (sessions or transactions) is one of the most commonly used analysis tasks in Web usage mining (Liu, 2011). User clustering results in groups of users that seem to behave similarly when navigating through a Web site. Such knowledge is used in e-commerce to perform market segmentation but is also helpful when the objective is to personalize a Web site (Eirinaki & Vazirgiannis, 2003; Jagan & Rajagopalan, 2015).

Similarity measure plays an essential role in clustering. To perform Web page clustering, similarity measures often determine the distance between a pair of Web pages based on the feature vectors describing them (Gündüz-Ögüdücü, 2011). There are several approaches to data clustering. These approaches can be categorized as partitioning methods, hierarchical methods, density-based methods, grid-based methods, model-based methods, high-dimensional clustering, and constraint-based clustering (Han & Kamber, 2012).

Clustering is usually used with other data mining techniques in Web page prediction. For accurately predicting the next page of an active user session, the clustering methods that work on sequence data are more appropriate (Gündüz-Ögüdücü, 2011). Further analysis of user groups based on their demographic attributes (such as age, gender, income level) may lead to the discovery of valuable business intelligence (Liu, 2011).

In the web usage mining domain, k-means and Fuzzy C-means algorithms are the most used clustering algorithms. The advantage of the K-Means Algorithm is that it is simple to use. The disadvantages of the algorithm are that outliers profoundly influence them, that they work only with numerical data, and that the inappropriate number of clusters can produce incorrect results (Agarwal & Saxena, 2019).

Fuzzy C-Means Algorithm called a soft version of k-means is where clusters are formed based on fuzzy belongingness and closeness of each data set. The significant advantage of the algorithm is that it does not require predefined centroids, and the disadvantage is that it requires more time for computation (Agarwal & Saxena, 2019).

## Sequential Patterns

Sequential patterns try to discover which items are followed by another set of items (Jagan & Rajagopalan, 2015). Sequential pattern discovery is an extension of association rules mining in that it reveals patterns of co-occurrence incorporating the notion of time sequence (Eirinaki & Vazirgiannis, 2003). In Web server logs, a visit of a user is recorded over some time. A time stamp can be attached either to the user session or to the individual page requests of user sessions. Sequential pattern discovery techniques enable the Web mining system can determine temporal relationships among data items such as the following (Gündüz-Ögüdücü, 2011, p.44): (30% of users who visited /home/products/dvd/movies, had visited /home/products/games within the past week, 40% of users request the page with URL / home/products/monitors after visiting the page /home/products/computers).

Sequential patterns discover the user's navigation behavior (Joshila Grace, Maheswari, & Nagamalai, 2011; Liu, 2011). By using this approach, the trends of the activity of the users can be determined, and predictions to the next visited pages can be calculated (Iváncsy & Vajk, 2006). Thus, Web marketers can predict future visit patterns, which will be helpful in placing advertisements aimed at certain user groups (Liu, 2011).

The advantages of using Sequential patterns are useful user trend can be discovered, predictions concerning stay pattern can be made, improving website steering, personalize advertisements, dynamically reshuffle link structure and adopt web site inside to individual client necessities or to provide clients by automatic recommendations that best costume customer profiles (Dogne et al., 2015).

## Markov Models

Markov model is compact, simple, expressive, and based on a well-established theory (Nigam, Tokekar, & Jain, 2015). Markov model can be used as an approach to model navigational activities in the Web site. Each page view can be represented as a state, and the transition probability between two states can represent the likelihood that a user will navigate from one state to the other. In the first-order Markov Model, each state corresponds to a web page, and each pair of viewed web page corresponds to a state transition. Two artificial states i.e., start and final, are incorporated in the model. In the second-order Markov Model, each state corresponds to the sequence of two viewed web pages and so on. (Nigam et al., 2015). This representation allows for the computation of several useful user or site metrics. The goal of the approach is to predict the next user action based on a user's previous surfing behavior (Liu, 2011). In Web page recommendation, Markov models can be used with clustering techniques (Cadez, Heckerman, Meek, Smyth, & White, 2003).

## Classification

Classification is a process that maps a data item into one of several predefined classes (Eirinaki & Vazirgiannis, 2003). Classification aims to learn a classification function (classifier) from data that are labeled with predefined classes or categories. The resulting classifier is then applied to classify future data instances into these classes. Because the data instances used for learning (training data) are labeled with predefined classes, the method is called supervised learning (Liu, 2011). In the Web domain, one is interested in developing a profile of users belonging to a particular class or category. This requires the extraction and selection of features that best describe the properties of given the class or group. Classification can be done by using supervised learning algorithms such as decision trees, naive Bayesian classifiers, k-nearest neighbor classifiers, Neural Networks Classifier (Artificial Neural Networks), and Support Vector Machines (Dogne et al., 2015). It is also possible to use previously discovered clusters and association rules for the classification of new users (Liu, 2011). Classification techniques play an important role in Web usage mining for modeling the users according to various predefined metrics. For example, given a set of user transactions, the sum of purchases made by each user within a specified period of time can be computed, or classifiers are used to classify Web pages according to users navigation and then classified pages are used to reorganize the Web site structure (Fu, Creado, & Ju, 2001). A classification model can then be built based on this enriched data to classify users into those who have a high propensity to buy and those who do not, taking into account features such as users' demographic attributes, as well their navigational activities (Liu, 2011). Classification on web logs may lead to the discovery of exciting patterns such as the following (Jafari, Sabzchi, & Irani, 2014): (Users from state or government agencies who visit the site tend to be interested in the page /company/lic.html., 60% of users, who placed an online order in /company/products /Music, were in the 18-25 age groups and lived in Chandigarh)

In the context of e-commerce, classification algorithms help to model purchasing behavior. The web users do not always visit the website with a purchase intention; they can also visit for benchmarking the products or price. Moe (2003) investigates whether different patterns of web site shopping or usage behavior can be found in clickstream data and, if so, whether they might be related to purchasing. An e-commerce site visitors can be clustered into one of four shopping strategies: directed buying (visitors exhibit goal-directed search behavior, have a specific product in mind when entering the site and are likely to exit the store has made a purchase), search/deliberation (these visitors also exhibit goal-directed search behavior, but have only a general product category in mind. They may need multiple visits to gather more information before making a purchase), hedonic browsers (users in this category tend to enter the site without a product or category in mind. Purchases that may occur are therefore likely to be impulse buys), knowledge-building (this group has no intention of buying and is simply visiting the online store to gather information). This situation causes the number of purchases to be considerably smaller than the number of visits; in other words, the customer response rate remains low. Therefore, the purchase model will not offer the right predictions. Using resampling methods is a way for over that problem in data mining.

## Recommender Systems

A recommender system is a system which accepts recommendations from people as inputs, aggregates them, and directs them to appropriate recipients as recommendations (Resnick & Varian, 1997). Recommender systems suggest interesting items in addition to filtering undesired objects.

Data mining methods are complicated and time-consuming, so user modeling models are formed off-line. The user's preferences information comes from several sources as user queries, the content of Web pages that a user requires during a visit, the structure of the Web site, and Web log data. Web usage mining is often used to reveal a user's interest. Recommendation process has an additional stage, unlike the web use mining process. The stages of the recommendation process are preprocessing, pattern discovery, pattern analysis, and recommendation stage. While preprocessing, pattern discovery and analysis steps of web usage mining process are conducted off-line, recommendation step is carried on-line (Gündüz-Ögüdücü, 2011). In recommendation step, the Web server should keep track of the active user session. The recommendation model generates recommendations as a set of links after each request of the active user.

Recommender systems are widely used on the Web for recommending products and services to users. Most e-commerce Web sites have such systems. These systems help users deal with the information overload by giving them personalized recommendations, can help consumers find products they wish to purchase, improve cross-sell by suggesting additional products and increase customer loyalty (Schafer, Konstan, &Riedl, 2001). There are two basic approaches to recommendations. Content-based recommendations (the user will be recommended items similar to the ones the user preferred in the past) and Collaborative filtering (the user will be advised things that people with similar tastes and preferences liked in the past) (Liu, 2011).

## Web Query Mining

A visitor who looks for information is to use a search engine. With typing a query, the visitor communicates with a search engine. The query consists of a few keywords; thus, Web pages whose content matches these keywords can be found. Query logs are files that record queries submitted by users along with the results returned by the search engine and the results that have been clicked by the user. Query logs contain a historical record of the activities of the users of a search engine. The query's content is usually stored in the web log registers; thus, query log data is considered a kind of Web usage data. A sessionization process can use the keywords executed during the user's session. Such web mining algorithms can be applied to extract significant patterns about the users' information needs. This process is also known as Web query mining (Velásquez & Palade, 2008; Liu, 2011).

## CONCLUSION

Living in the data age brings some technological imperatives to the organizations. Individuals can navigate on the Web with their mobile devices (smartphones, tablets) at any time, and anywhere. Organizations must be on the Web and stay in touch with their customers. Today, most of the profitable organizations are digital-born. It can be said that only the companies, which are using data well, will be survived. At this point, it is worth mentioning about digital transformation. The term "transformation" expresses the comprehensiveness of the actions that need to be taken when organizations are faced with new technologies. Thus, a digital transformation typically involves a company-wide digital (transformation) strategy, which goes beyond functional thinking and holistically addresses the opportunities and risks that originate from digital technologies. A digital transformation strategy guides the organization in its journey toward being digitally transformed (Singh & Hess, 2017). The financial performance of the organizations, which are good at using their data, more than their competitors is at least three times better (EIU, 2013). Living in the data age is forced to change business structures and cultures (Bange, Handford, & Janoschek, 2014). The traditional organizations, which are aware of the value of their data, have started to convert their business processes and production methods into digital. Even when traditional goods and services are not being replaced with digital alternatives, information technology is firmly integrated as a tool in basic research, product design, sourcing, manufacturing, logistics, sales, marketing, and service (Downes & Nunes, 2013). In this context, it can be said Web log data is a treasure for companies. The Web users (visitors) can act like someone else when posting text or photos to websites or platforms; however, Web log data is real because the data is usual situations of the users. In this new environment, the fight to retain customers or catch new ones is a decisive factor for a company's success in the digital market (Velasquez & Palade, 2008). The amount of log data is typically massive, depending on the type of organizational infrastructure and applications running on it. The log data can be used for monitoring servers, improving business and customer intelligence, building recommendation systems, fraud detection, and much more.

Web usage mining can be grouped into two main categories: mining user navigation patterns, to improve the web site structure and content, and user modeling in adaptive interfaces, or personalization. The latter is key to the creation of web-based platforms, which need personalized systems for improving the relationship with their users (Velásquez & Palade, 2008). Web usage mining aims to study user clicks and their applications to e-commerce and business intelligence. The objective is to capture and model behavioral patterns and profiles of users who interact with a Web site. Such patterns can be used to understand the behaviors of different user segments better, to improve the organization and structure of the Web site (Liu, 2011).

With the massive growth of the Web, there is an enormous volume of dynamic, distributed, heterogeneous, structured, unstructured, and high dimensional data available on Web. Data pre-processing is considered as a time-consuming and complex stage in the Web usage mining process due to the vast and noisy nature of log data. Web usage mining is an essential tool for realizing more personalized, user-friendly, and business-optimal Web services. Web usage data will continue to grow. User transactions on the Web, not limited to their clicks during sessions on regular Web sites, but also their queries and entire interactions with search engines, and even online Ads that they encounter.

Web usage mining is a powerful tool for corporations that invest in the e-business sector. The application of Web usage mining techniques to data gathered from customers' online activity helps them to acquire business intelligence by providing high-level knowledge in the form of rules and patterns that describe consumer navigational and purchasing behavior (Buchner & Mulvenna, 1998). Thus consumer profiles and market segmentation can be achieved, giving these companies a competitive advantage. Even in the case of smaller organizations or individuals, the outcome of log analysis and Web usage mining can help them improve the performance of their systems, identify their Web site's visitors, and even customize their Web site making it more efficient and user-friendly. Therefore, there exists a large variety of Web analytics' products, ranging from free traffic analysis tools to integrated CRM (Customer Relationship Management) solutions (Eirinaki & Vazirgiannis, 2003).

In Web Usage Mining, several issues must be taken into consideration, including decisions to be made during data filtering and processing, user and session identification, and page view identification (Eirinaki & Vazirgiannis, 2003). The main issue is how to determine which technique is most suited for the current problem setting. As in most of the application domains, also in Web Usage Mining, there is not a unique answer for such a question. Facca and Lanzi (2005) define four applications area as Personalization of Web Content, Pre-fetching, and Caching, Support to Desing, E-commerce.

In summary, web log data helps organizations make decisions and personalize their websites (Dhandi & Chakrawarti, 2016). Web usage patterns can be used to gather business intelligence to improve Customer attraction, Customer retention, sales, marketing, and advertisement, cross-sales (Dharmarajan & Dorairangaswamy, 2016). With Web Usage mining, organizations can establish a good and healthy relationship with customer by giving them the exact information that they want, can understand the needs of the customer better and they can act in response to customer needs faster, can retain customers by presenting information relevant to them (Dhandi & Chakrawarti, 2016).

The most used pattern discovery methods are association rules mining, clustering, classification, sequential pattern analysis, and Markov models. Pattern discovery methods are dealing with the enormous volumes of data and the scalability issue. Scalability means the rate of memory and time needed for the task, according to the parameters that affect the performance of the algorithms, such as the number of users or pages (Aldekhail, 2016). The storage space and runtime requirements of these methods increase due to a large number of patterns for sequential pattern and association rules, and the large number of states for Markov models (Jafari et al., 2014). Predicting the next web page is not sufficient; evaluation of prediction models is vital because every model has its pros and cons. Prediction results will be helpful if high prediction accuracy is achieved with minimum complexity, which is depended on the prediction model (Nigam et al., 2015).

## REFERENCES

Agarwal, A., & Saxena, A. (2019). An accountability of various clustering techniques for ımprovement of page searching process and analysis. *International Conference on Sustainable Computing in Science, Technology & Management (SUSCOM-2019)* 10.2139srn.3354444

Agosti, M., & Di Nunzio, G. (2007). Web log mining: A study of user sessions. *Proceedings of the 10th DELOS Thematic Workshop on Personalized Access, Profile Management, and Context Awareness in Digital Libraries (PersDL).*

Aldekhail, M. (2016). Application and significance of web usage mining in the 21st century: A literature review. *International Journal of Computer Theory and Engineering*, *8*(1), 41–47. doi:10.7763/IJCTE.2016.V8.1017

Anitha, V., & Isakki, P. (2016). A survey on predicting user behavior based on web server log files in a web usage mining. *2016 International Conference on Computing Technologies and Intelligent Data Engineering (ICCTIDE'16)*, 1-4. 10.1109/ICCTIDE.2016.7725340

Bange, C., Handford, M., & Janoschek, N. (2014). *Information culture leveraging the power of collective intelligence for better decision making. BARC Research Study*. BARC Institute.

Batista, P., & Silva, M. (2001). Mining web access logs of an on-line newspaper. *The Proceedings of 12th International Meeting of the Euro Working Group on Decision Support Systems.*

Berendt, B., & Spiliopoulou, M. (2000). Analyzing navigation behaviour in web sites integrating multiple information systems. *The International Journal on Very Large Data Bases*, *9*(1), 56–75. doi:10.1007007780050083

Buchner, A., & Mulvenna, M. D. (1998). Discovering internet marketing intelligence through on line analytical web usage mining. *SIGMOD Record*, *27*(4), 54–61. doi:10.1145/306101.306124

Bucklin, R. (2008). Marketing models for electronic commerce. In B. Wierenga (Ed.), *Handbook of Marketing Decision Models* (pp. 327–369). Springer. doi:10.1007/978-0-387-78213-3_10

Cadez, I., Heckerman, D., Meek, C., Smyth, P., & White, S. (2003). Model-based clustering and visualization of navigation patterns on a web site. *Data Mining and Knowledge Discovery*, *7*(4), 399–424. doi:10.1023/A:1024992613384

Chakrabarti, S., Dom, B. E., Gibsony, D., Kleinberg, J., Kumar, R., Prabhakar, R., ... Tomkins, A. (1999). Mining the link structure of the world wide web. *IEEE Computer*, *32*(8), 60–67. doi:10.1109/2.781636

Cooley, R., Mobasher, B., & Srivastava, J. (1997). Web mining: Information and pattern discovery on the world wide web. In *Proceedings of the 9th IEEE International Conference on Tools with Artificial Intelligence (ICTAI'97)* (pp. 558-567). Computer Society. 10.1109/TAI.1997.632303

Cooley, R., Mobasher, B., & Srivastava, J. (1999). Data preparation for mining world wide web browsing patterns. *Knowledge and Information Systems*, *1*(1), 5–32. doi:10.1007/BF03325089

Dhandi, M., & Chakrawarti, R. J. (2016). A comprehensive study of web usage mining. *2016 Symposium on Colossal Data Analysis and Networking (CDAN)*. 10.1109/CDAN.2016.7570889

Dharmarajan, K., & Dorairangaswamy, M. A. (2016). Web usage mining: Improve the user navigation pattern using FP-growth algorithm. *Elysium Journal of Engineering Research & Management*, *3*(4), 1–4.

Dogne, V., Jain, A., & Jain, S. (2015). Evolving trends and its application in web usage mining: A survey. *International Journal of Soft Computing and Engineering*, *4*(6), 98–101.

Downes, L., & Nunes, P. F. (2013). Big bang disruption. *Harvard Business Review*, *91*(03), 44–56.

Eirinaki, M., & Vazirgannis, M. (2003). Web mining for web personalization. *ACM Transactions on Internet Technology*, *3*(1), 1–27. doi:10.1145/643477.643478

EIU. (2013). *Fostering a data-driven culture*. A report from the Economist Intelligence Unit. Retrieved from https://eiuperspectives.economist.com/technology-innovation/fostering-data-driven-culture

Etzioni, E. (1996). The world wide web: Quagmire or gold mine. *Communications of the ACM*, *39*(11), 65–68. doi:10.1145/240455.240473

Facca, F. M., & Lanzi, P. L. (2005). Mining interesting knowledge from weblogs: A survey. *Data & Knowledge Engineering*, *53*(3), 225–241. doi:10.1016/j.datak.2004.08.001

Fu, Y., Creado, M., & Ju, C. (2001). Reorganizing web sites based on user access patterns. In *Proceedings of the Tenth International Conference on Information and Knowledge Management* (pp. 583–585). ACM Press. 10.1145/502585.502698

Geetharamani, R., Revathy, P., & Jacob, S. G. (2015). Prediction of users webpage access behaviour using association rule mining. *Sadhana*, *40*(8), 2353–2365. doi:10.100712046-015-0424-0

Gündüz-Ögüdücü, S. (2011). *Web page recommendation models, theory and algorithms. Synthesis Lectures on Data Management #10*. Morgan & Claypool Publishers.

Han, J., Kamber, M., & Pei, J. (2012). *Data mining: Concepts and techniques* (3rd ed.). Elsevier.

Hughes, A. (2005). *Strategic database marketing: The masterplan for starting and managing a profitable, customer-based marketing program.* McGraw-Hill.

Iváncsy, R., & Vajk, I. (2006). Frequent pattern mining in web log data. *Acta Polytechnica Hungarica, 3*(1), 77–90.

Jafari, M., Sabzchi, F. S., & Irani, A. J. (2014). Applying web usage mining techniques to design effective web recommendation systems: A case study. *ACSIJ Advances in Computer Science: an International Journal, 3*(02), 78–90.

Jagan, S., & Rajagopalan, S. P. (2015). A survey on web personalization of web usage mining. *International Research Journal of Engineering and Technology, 2*(01), 6–12.

Joshi, A., & Krishnapuram, R. (2000). On mining web access logs. *Proc. of the 2000 ACM SIGMOD Workshop on Research Issues in Data Mining and Knowledge Discovery*, 63–69.

Joshi, K., Joshi, A., & Yesha, Y. (2003). On using a warehouse to analyze web logs. *Distrubuted and Parallel Databases,* (13), 161-180.

Joshila Grace, L. K., Maheswari, V., & Nagamalai, D. (2011). Web log data analysis and mining. In *CCSIT 2011, Part III, CCIS 133* (pp. 459–469). Springer-Verlag Berlin Heidelberg.

Kosala, R., & Blockeel, H. (2000). Web mining research: A survey. *SIGKDD Explorations: Newsletter of the Special Interest Group on Knowledge Discovery & Data Mining, 2*(1), 1–15. doi:10.1145/360402.360406

Kotler, P., & Armstrong, G. (2000). *Marketing: An introduction.* Upper Saddle River, NJ: Pearson Prentice Hall.

Kumar, B. S., & Rukmani, K. V. (2010). Implementation of web usage mining using apriori and fp growth algorithms, *Int. J. of Advanced Networking and Applications, 1*(06), 400–404.

Liu, B. (2007). *Web data mining exploring hyperlinks, contents, and usage data* (1st ed.). Springer.

Liu, B. (2011). *Web data mining exploring hyperlinks, contents, and usage data* (2nd ed.). Springer.

Malarvizhi, S. P., & Sathiyabhama, B. (2016). Frequent pagesets from web log by enhanced weighted association rule mining. *Cluster Computing*, *19*(1), 269–277. doi:10.100710586-015-0507-z

Markov, Z., & Larose, D. (2007). *Data mining the web uncovering patterns in web content, structure, and usage*. Wiley-Interscience.

McCarty, J., & Hastak, M. (2007). Segmentation approaches in data-mining: A comparison of RFM, CHAID, and logistic regression. *Journal of Business Research*, *60*(6), 656–662. doi:10.1016/j.jbusres.2006.06.015

Moe, W. W. (2003). Buying, searching, or browsing: Differentiating between online shoppers using in-store navigational clickstream. *Journal of Consumer Psychology*, *13*(1&2), 29–39. doi:10.1207/153276603768344762

Nigam, B., Tokekar, S., & Jain, S. (2015). Evaluation of models for predicting user's next request in web usage mining. *International Journal on Cybernetics & Informatics*, *4*(1), 1–13. doi:10.5121/ijci.2015.4101

Ohlhorst, F. (2013). *Big data analytics turning big data into big money*. John Wiley & Sons, Inc.

Prasad, G. S., Reddy, N., & Acharya, U. (2010). Knowledge discovery from web usage data: A survey of web usage pre-processing techniques. In *Information Processing and Management BAIP 2010, CCIS 70* (pp. 505–507). Springer-Verlag Berlin Heidelberg. doi:10.1007/978-3-642-12214-9_88

Resnick, P., & Varian, H. R. (1997). Recommender systems. *Communications of the ACM*, *40*(3), 56–58. doi:10.1145/245108.245121

Schafer, J. B., Konstan, J. A., & Riedl, J. (2001). E-commerce recommendation applications. *Data Mining and Knowledge Discovery*, *5*(5), 115–153. doi:10.1023/A:1009804230409

Singh, A., & Hess, T. (2017). How chief digital officers promote the digital transformation of their companies. *MIS Quarterly Executive*, *16*(01), 1–17.

Spiliopoulou, M., Mobasher, B., Berendt, B., & Nakagawa, M. (2003). A framework for evaluation of session reconstruction heuristic in web usage analysis. *INFORMS Journal on Computing*, *15*(2), 171–190. doi:10.1287/ijoc.15.2.171.14445

Srivastava, J., Cooley, R., Deshpande, M., & Tan, P. (2000). Web usage mining: Discovery and applications of usage patterns from web data. *SIGKDD Explorations*, *1*(2), 12–23. doi:10.1145/846183.846188

Srivastava, M., Srivastava, A. K., & Garg, R. (2019, February). Data preprocessing techniques in web usage mining: A literature review. In *Proceedings of International Conference on Sustainable Computing in Science, Technology and Management (SUSCOM)*. Amity University Rajasthan. 10.2139srn.3352357

Tan, P., & Kumar, V. (2000). *Modeling of web robot navigational patterns*. Army High Performance Computing Research Center.

Thakare, S. B., & Gawali, S. (2010). A effective and complete preprocessing for web usage mining. *International Journal on Computer Science and Engineering*, *2*(3), 848–851.

Velasquez, J. D., & Palade, V. (2008). *Adaptive web sites a knowledge extraction from web data approach*. Amsterdam: IOS Press.

Xu, G., Zhang, Y., & Li, L. (2011). *Web mining and social networking techniques and applications*. Springer Science+Business Media.

Zhou, X., & Huang, Y. (2014). An improved parallel association rules algorithm based on mapreduce framework for big data. In *11th International Conference on Fuzzy Systems and Knowledge Discovery*, (pp. 284–288). 10.1109/FSKD.2014.6980847

## ADDITIONAL READING

Alpaydin, E. (2014). *Introduction to machine learning* (3rd ed.). MIT Press.

Guerbas, A., Addam, O., Zaarour, O., Nagi, M., Elhajj, A., Ridley, M., & Alhajj, R. (2013). Effective web log mining and online navigational pattern prediction. *Knowledge-Based Systems*, *49*, 50–62. doi:10.1016/j.knosys.2013.04.014

Kwon, K., & Kim, C. (2012). How to design personalization in a context of customer retention: Who personalizes what and to what extent? *Electronic Commerce Research and Applications*, *11*(2), 101–116. doi:10.1016/j.elerap.2011.05.002

Larose, D.T. (2005). *Discovering knowledge in data an introduction to data mining*. John Wiley & sons, Inc., Publication.

Suguna, R., & Sharmila, D. (2012). Association rule mining for web recommendation. *International Journal on Computer Science and Engineering*, *4*(10), 1686–1690.

## KEY TERMS AND DEFINITIONS

**ASCII Text File:** ASCII (American Standard Code for Information Interchange) text file is a text file which in which each byte represents one character according to the ASCII code. This text file may be comma-delimited, space-delimited, or tab-delimited, and enables electronic communication.

**Click-Stream:** The sequence of page visits executed by a particular user navigating through a website.

**Cookies:** The cookie is a text string sent from a website and stored on the user's computer by the user's web browser while the user is browsing.

**HTML:** HTML is the standard markup language for web pages.

**Hypertext Transfer Protocol (HTTP):** HTTP is designed to enable communications between clients and servers.

**W3C:** W3C (The World Wide Web Consortium) is an international community that develops open standards to ensure the long-term growth of the web.

# Chapter 6
# An Assessment of Federated Machine Learning for Translational Research

**Manoj A. Thomas**
*The University of Sydney Business School, Australia*

**Diya Suzanne Abraham**
*Virginia Commonwealth University, USA*

**Dapeng Liu**
https://orcid.org/0000-0002-8822-656X
*The University of New South Wales, Australia*

## ABSTRACT

*Translational research (TR) is the harnessing of knowledge from basic science and clinical research to advance healthcare. As a sister discipline, translational informatics (TI) concerns the application of informatics theories, methods, and frameworks to TR. This chapter builds upon TR concepts and aims to advance the use of machine learning (ML) and data analytics for improving clinical decision support. A federated machine learning (FML) architecture is described to aggregate multiple ML endpoints, and intermediate data analytic processes and products to output high quality knowledge discovery and decision making. The proposed architecture is evaluated for its operational performance based on three propositions, and a case for clinical decision support in the prediction of adult Sepsis is presented. The chapter illustrates contributions to the advancement of FML and TI.*

DOI: 10.4018/978-1-7998-1879-3.ch006

## INTRODUCTION

Business focus have been evolving over the centuries, from production and marketing to innovation and intelligence. Over the past decade, the transforming power of artificial intelligence (AI) has dramatically and profoundly impacted industries. AI applications and innovations offer business entities as many new opportunities as challenges. AI simulates human intelligence and reasoning process in machines (e.g., computers, online services and smart devices), learning (e.g., acquisition and transformation of data, recognition of rules, patterns, and building models), and reasoning (using identified patterns and models to predict, conclude or approximate the optimal).

Machine learning (ML), the backbone of AI, can learn from big data and generate models that business can take advantage to conduct business analytics and prediction. ML consists of algorithms for data-driven predictions or decisions. Well known applications of ML include speech recognitions, recommender systems, and more recently, autonomous vehicles (Kuderer, Gulati, & Burgard, 2015). Traditionally, ML involves the use of powerful computational platforms for applying data mining techniques and learning algorithms on large datasets.

In healthcare, many realistic scenarios involve smaller datasets and rare events, and therefore traditional ML is not as effective in such contexts. A preferred approach is to have algorithms that can interact with agents and optimize learning through these interactions, where the agents can themselves be human (Holzinger, 2016). This implies that in scenarios where datasets are relatively small, establishing a symbiotic relationship between the computational and human agents is more relevant than the traditional approach of relying on large datasets.

Human agents using edge devices such as mobile phones and wearable technologies are examples of user interactions that can be used in ML. Furthermore, user interactions have the advantage of natural labelling of data thus reducing a search space of exponential possibilities drastically by heuristic selection of samples (Holzinger, 2016).

Traditional systems that support ML are not optimized for leveraging human agents. The common approach is to consolidate datasets on a server or cloud on which training in conducted and then the algorithms are productionalized. However, this approach has setbacks. It assumes ubiquitous access to the cloud, continuous streaming of data, high throughput, and network dependence. Additionally, the models stored in the cloud tend to be highly generalized as it does not take into consideration the imbalance in user interactions. In this research, we propose a federated ML approach to address these challenges. We define Federated machine learning (FML) as the pipelining of machine learning and data analytic tools by incorporating multiple data sources, and intermediate data analytic processes and products to output high quality results that support knowledge discovery and decision making. It aims to move ML closer to the end user. This paper will demonstrate how FML may be applied in the context of translational research (TR).

TR is the harnessing of knowledge from basic science and clinical research to develop new drugs, devices, and treatment options (Sussman, Valente, Rohrbach, Skara, & Ann Pentz, 2006; Woolf, 2008). National Institute of Health identifies two related cycles of TR (Payne, Embi, & Sen, 2009; Sung et al., 2003). First, it is the translation of discoveries from the laboratory to clinical research involving human trials and studies. Second, it is the translation of findings from clinical research to enhance best practices. The cycles thus involve the translation of knowledge and evidence from the "bench" (i.e., the laboratory) to the "bedside" (i.e., clinical and public health practice), and reciprocally from the "bedside" to the "bench" (Payne et al., 2009; Woolf, 2008). Translational informatics (TI) is a sister domain to TR, and concerned with the application of informatics theories and methods to TR.

Rapid advances in ML and data analytics provide novel opportunities to apply FML in TI. Doing so may improve the quality of interaction of models with human agency, identify and gather data from rare and unique events, and limit dependency on cloud-based services. FML could bring computational and analytical solutions closer to the decision maker.

This study aims to achieve three objectives. First, we build upon TR concepts to propose an architecture for FML. We do so by highlighting why current cloud-client approaches are not suited for FML. Second, we evaluate the proposed architecture for its operational performance based on three propositions. Third, we envisage how FML may facilitate knowledge discovery and clinical decision support in TI. This study makes two contributions. First, our research demonstrates that IS scholarship has much to contribute to the advancement of disciplines such as TR and bioinformatics. Based on the knowledge discovery and clinical decision support needs of TR (Payne et al., 2009; Sung et al., 2003), we used design science methodology to develop a platform to federate ML. The proposed platform leverages low power servers to mediate the execution of the ML algorithms. Second, the study represents a natural progression in the adaptation and use of IS methods, models, and frameworks to address many data and knowledge challenges in TI. By presenting a case for clinical decision support in the prediction of adult Sepsis, we show how socio-technical systems that incorporate machine learning and data analytic tools are applicable for knowledge discovery and decision making.

The rest of this study is organized as follows. We first provide a background on TR and TI, including a review of related literature. We then highlight the limitations of the current cloud-client approach for ML, and present an architecture for FML. Based on three propositions, we then conduct an evaluation of the operational performance of the proposed architecture. The practical and theoretical contributions are discussed, and we conclude by paper by presenting directions for future research.

## BACKGROUND AND CONCEPTS

### Translational Research (TR)

TR is crucial to the healthcare industry. Literature provides many definitions of TR across various disciplines (Westfall, Mold, & Fagnan, 2007). TR is commonly represented in two stages. The first stage is translating knowledge, techniques or mechanisms from basic sciences to the development of new treatments. The second stage is translating findings from clinical trials to clinical and public health practice (Woolf, 2008). Often, a reciprocal stage is also included in TR, where observations from practice is intended to inform basic science (Figure 1). In a nutshell, TR is the interface between basic science and clinical practice, the goal of which is to produce a promising treatment that can be "brought to market" and ultimately to improve public health (Woolf, 2008). It is the effective translation of advances in basic science into new approaches for the prevention, diagnosis, and treatment of diseases. TR is of critical importance and essential to improving health (Fontanarosa & DeAngelis, 2002). TR is an established priority at many leading scientific institutions such the National Institutes of Health (NIH), research universities, and health systems engaged in biomedical research programs.

### Translational Informatics (TI)

TI focuses on the application of informatics theories, emergent technologies, and socio-technical methods for TR (Chen, Qian, Yan, & Shen, 2013). TI enables the 1) acquisition of knowledge and information, 2) representation of knowledge and information in an actionable format, 3) semantic integration and aggregation of data to support the discovery and validation of knowledge models, 4) dissemination of knowledge through public engagement (Harris et al., 2009; Payne et al., 2009). The importance of TI in drug development, preclinical trials, and clinical decision support is well recognized (Rubio et al., 2010). Pharmaceutical industries heavily invest in TI to advance new tools and devices applicable to healthcare (Beaulah, Correll, Munro, & Sheldon, 2008; Lee et al., 2000). Nevertheless, research is sparse on effective approaches and strategies to support its multiple functions.

Surprisingly, not much research has been published in IS related to TI. To better understand the role of IS in addressing challenges related to TI, we conducted a detailed review of literature. We started with a full-text search of the PROQUEST database

*Figure 1. Translational research cycle (Payne et al., 2009; Sung et al., 2003)*

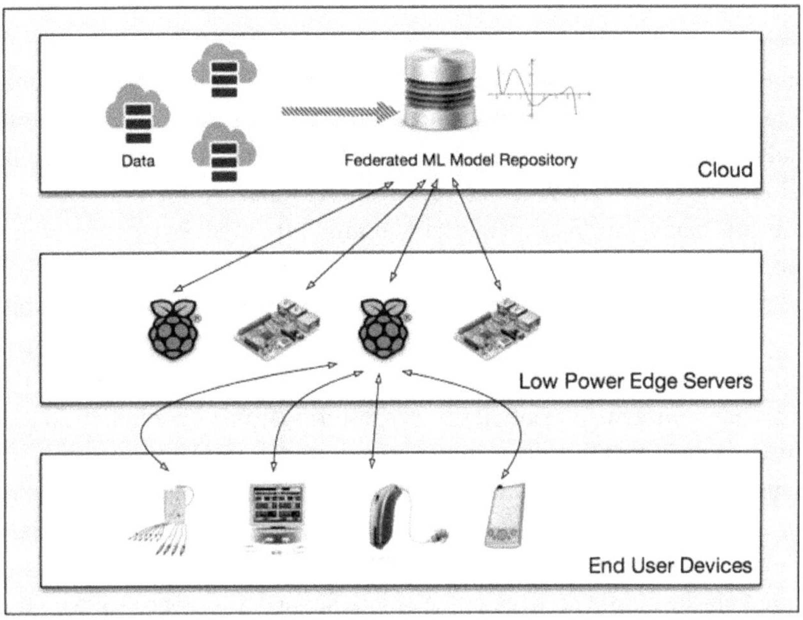


to identify studies published in journals from the IS journal ranking page of the Association of Information Systems (AIS). We searched for the terms "Translational Research" or "Translational Informatics." Within IS, we were interested in studies that have adopted the design science methodology. Therefore, we further filtered the results using the search term "Design Science." Our search yielded 27 studies published between 2002 and 2017, among which 19 were dissertations and thesis. Among the remaining eight articles, only one study (Gholami, Watson, Molla, Hasan, & Bjørn-Andersen, 2016) was published in an IS journal. This research, published in the Journal of the Association of Information Systems, merely mentions that IS research could follow the tradition of TR to test and publish discoveries beneficial to practice. It does not develop or evaluate DS artifacts related to TI. In the TI domain, we found many studies that have adopted and used IS methods, models, and frameworks. Examples include pattern recognition (Baldi & Long, 2001; Pan, 2002), data mining (Becker et al., 2003), data visualization (Porollo, Adamczak, & Meller, 2004), data representation and annotation (Koski, Gray, Lang, & Burger, 2005), and decision-making (Beck et al., 2008; Castaneda et al., 2015; Lindgren, 2008; O'neill, Dluhy, & Chin, 2005). Our review of literature indicates that there is much potential for IS scholarship related to the advancement of TI. In this research, we present one such case where machine learning, data analytics, and cloud computing are for clinical decision support.

## THE NEED FOR FML

Today, a majority of model training, testing, and inferencing takes place in the cloud. For example, when a user activates *Apple Siri* or *Ok Google* on the smartphone for natural language (NL) based search and action, the human voice is transmitted to the cloud for processing via ML models. Once the inferencing process is completed in the cloud, the results are returned to the smartphone.

There are many advantages of maintaining ML models using a centralized model repository in the cloud. It enables knowledge engineers to more easily access and update the models with better models. Scaling and execution of complex and process intense models are easier in the cloud under high demand conditions. There are also limitations to managing ML models in the cloud. They include time lag effects, privacy concerns, and latency issues resulting from unpredictable data spikes transmitted to and from the cloud.

Strong arguments can be made to position previously trained and tested models closer to the end user (say, on a locally available edge server or even the user's smartphone) than in the cloud. First, if inferencing is done locally, less data is transmitted via the public network to the cloud. Second, this approach would reduce latency, as inferencing takes place at the local edge server and results delivered more quickly to the end user. Third, it offers improved potential to integrate visual and audio intelligence in the modelling process. Many spatial dimensions of audio data (e.g., timbre, texture, emotion, phonetics) and video data (e.g., stimulus numerosity, primary coding, contextual discrimination, structure scanning, cue integration, and cognitive processing) (Siegel & Fischl, 1971) and their relationships are minimally captured or lost outside the immediate confines of where the data is generated. The inherent limitations of faithfully capturing and transferring these data dimensions limit knowledge engineers from including these features in ML models. Audible and visible intelligence can be enriched by proximally locating edge devices and sensors closer to the source of data. Fourth, the approach improves data privacy and security as the data remains local.

Nevertheless, if inferencing capabilities are to be managed locally on low power edge devices, sufficient consideration of the underlying architecture is desired to ensure respectable model performance, reliability, and response times. It is to be noted that, edge servers do not eliminate the need for cloud-based capabilities. In fact, training and testing of models would still require the power and flexibility of the cloud. Yet, the advantage of distributed intelligence is gained when the local edge devices complement cloud capabilities in the form of federated machine learning.

## FML FOR TRANSLATIONAL RESEARCH

In the healthcare industry, the prevalent system architecture is the client/server model (Coulouris, Dollimore, & Kindberg, 2005; Fielding & Taylor, 2000). With the growing popularity of the cloud-based solutions, the emergent model of system architecture aligns more towards variations of cloud-related services such as SaaS, PaaS, and DaaS (Newman, 2015; Singleton, 2016). In this new paradigm of cloud-based computing, ML models are stored and managed in the cloud (Li, Thomas, & Osei-Bryson, 2017) which are then accessed and used by the application on the end user's device. By contrast, the objective of our proposed architecture is to move machine learning closer to edge devices and end users, and thereby relying less on constant access to massive data and intelligence stored in the cloud.

Payne et al. (2009) proposed a conceptual framework to incorporate major categories of information collected, managed and disseminated through TR. Specifically, it targets the integration of information and data spanning distinct users, data sources, and evidence silos. Modeling and machine learning approaches may then be used for the analysis of data and extraction of knowledge for clinical decision support. The framework emphasizes the use of socio-technical approaches in TI research. We use this framework as the foundation to propose an architecture for FML.

Our review of literature did not find any research that has addressed FML. The closest reference is a Google Research Blog on collaborative machine learning without centralized training data (McMahan & Ramage, 2017). The blog entry refers to the use of Tensorflow technology to enable mobile phones collaboratively learn a shared prediction model. The approach suggests decoupling ML from data stored in the cloud. Although details presented in the blog entry are vague, it is ascertained that the proposed federated learning approach is optimized for modern high-end Android devices, and enables smarter ML models, lower latency, and less power consumption.

As indicated earlier, TI is at a nascent stage when it comes to applying newer technologies such as ML and data analytics. Current industry practice is the use of cloud-client architecture model, which is heavily network dependent and requires integrated data models and real-time communication with the cloud. Table 1 compares the prevalent industry practice with the preferred approach for FML across the dimensions of architecture, service model, access, machine learning, data model, and security.

A common architecture prevalent today is the cloud-client model, where model training is performed on the cloud-based server, and accessible by the end user devices. However, as the architecture is completely cloud dependent it requires consistent Internet connectivity and involve higher associated costs. A localized low-cost low power edge server that serves as a buffer between the end user and the cloud server is preferred as it reduces dependency on the cloud. For the service model dimension, the industry norm is one of an integrated approach. This involves integrated data and models stored in the cloud. A federated system would reduce this dependence, as well as decouple data from the models. For the access dimension, the industry practice is one of a fixed-point access resulting in limited portability, and for network dimension, the industry standard is network dependent due to high throughput requirements. ML is typically based on real-time communication with the cloud, but a buffering layer could provide model interaction without the need for continuous connectivity with the cloud. Security dimension in the integrated data model is typically a centralized management approach. The drawback is that it is more vulnerable. The proposed way forward is a data blind approach, where the data is decentralized and federated.

A supporting architecture for FML should provision the following capabilities. First, although most users may have ubiquitous access to Internet and data, the architecture should be respectful of this fact, particularly in scenarios where Internet throughput is limited, and data access is expensive. Second, the architecture should enable seamless collection and management of privacy-sensitive multi-dimensional dataset from the end users. The data movement to and from end user devices should be in a compressed and granular format. Third, model training and update in the cloud should not require dependence on a continuous stream of data from the end user devices. Forth, the architecture should not assume independently and identically distributed (IID) data. The idea arises from realistic scenarios whereby different type of users interact with the algorithms in different ways, and thus the action of one user may not be interchangeable with another user. Furthermore, there may be more data from one type of user (e.g., a clinical trial participant), and less from another type of user (e.g., a bioinformatics researcher). This imbalance should not skew model training or overfit models for common use cases. On the contrary, the differential between individual use cases and the master model should be anonymized and blurred through an averaging process, while respecting the privacy of the individual user. This reconciliation also ensures that the models are not thrown off by outlier user cases. Fifth, the data analytics pipeline should be extensible and replicable, such that it moves machine learning closer to edge devices and end users. Sixth, federate machine learning should utilize natural labeling of end user data in the model training and knowledge discovery process (here, natural labeling implies the identification of use cases based on user's interaction with applications, thus eliminating the need for human intervention for labeling unique use cases).

## FML ARCHITECTURE

We propose an architecture for FML as shown in Figure 2. The architecture aims to address the preferred directions as stated in the previous sections. It consists of three tiers, namely cloud, low power edge server, and end user devices.

At the top is the cloud tier, which utilizes the cloud for data storage, model generation, and model management. Anonymized data is stored in the cloud and reconciled models are maintained in the FML model repository. The repository to store the models is built using MongoDB. The cloud layer maintains a clear separation between the FML model repository and the user data. It thus offers enhanced data security by controlling and limiting how the models use the data for training and testing. Furthermore, the trained and tested models are pushed to the low power edge servers in the middle tier. The users interact with the models on the low power edge servers in the middle tier and have no direct access to the cloud layer. This adds an additional layer of security between the end user and the data stored in the cloud.

**The Low Power Edge Servers**: In the middle tier provide affordable computing capabilities. The models are stored at this layer for consumption by the end user

*Table 1. Prevalent industry practice and preferred approach for FML*

| Dimension | Industry Practice (Current Approach) | Shortcomings of Current Approach | Preferred Approach |
|---|---|---|---|
| Architecture | Cloud/client model | Constant cloud dependency; Higher cost of cloud services | Minimize cloud dependency; Low-cost low power servers closer to end user |
| Service Model | Integrated service | Reliance on data and intelligence stored in the cloud | Federated |
| Access | Fixed point access | Limited portability | Portability |
| | Network dependency | Dependence on high throughput | Network agnostic layers |
| Machine Learning | ML requires real-time communication with cloud | ML fully dependent on cloud connectivity | Smarter ML models independent of cloud access |
| Data Model | Integrated data models | Centralized data and model management | Decentralized data decoupled from the model repository |
| Security | Centralized | Higher vulnerability | Data blind |

*Figure 2. Federated machine learning architecture*

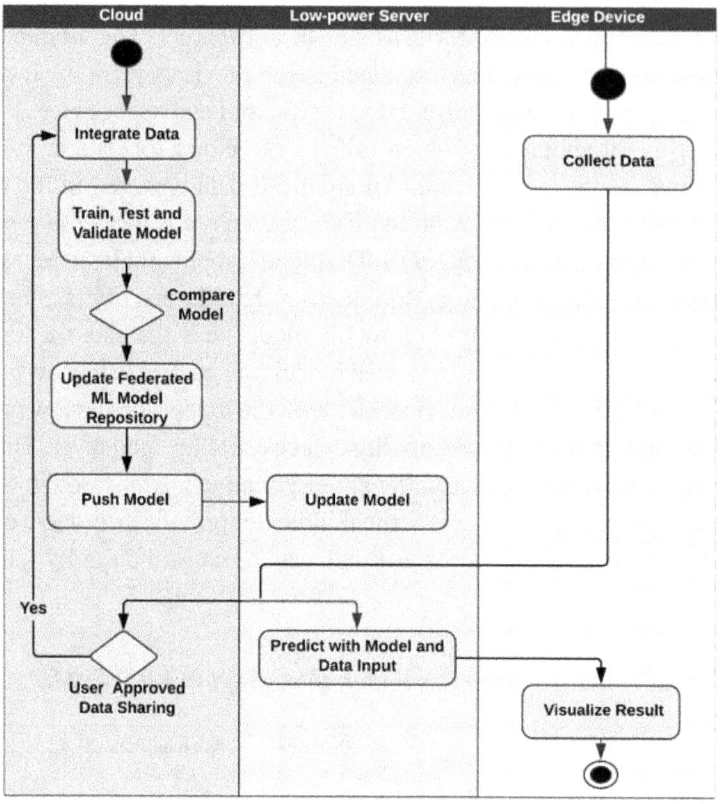


applications. Low cost servers make them more attractive for deployment in resource constrained environments such as a healthcare clinic at a rural location. They have the advantage of portability as would be required in an ambulatory setting. Additionally, they reduce dependency on the cloud as only the relevant and application specific models are stored at this layer. End user devices access the models directly from this layer over a localized network, and without the need for Internet connectivity to the cloud. The low power edge servers periodically access the cloud and receive the newly updated model. This approach largely reduces cloud access and data transfer frequency, thereby resulting in moderated cloud dependency.

Google has pioneered the use of advanced deep learning approaches such as Tensorflow, but its execution on the end user devices is possible only on new and powerful mobile devices. This may not be taken for granted. In many instances, TI requires data gathering and ML algorithms to run closer to the end user devices and sensors. The low power edge servers serve as a middle layer to mediate the execution of the computationally intense ML algorithms. It eliminates the need for reliance on ML algorithms stored in the cloud.

**End User Devices** are the primary means by which the subjects interact with the models. They may vary in terms of size, purpose, and utility ranging from mobile devices to smart sensors and wearable technology devices. These devices are registered to the low power edge servers in the middle tier. Data exchange between the end user devices and low power edge servers are encrypted and anonymized. Privacy-sensitive data is moved from the end user device to the middle tier in a compressed and granular format. The middle tier temporarily stores the newly collected user data before it is passed to the cloud tier.

The architecture implicitly provides the ability to federate ML in two distinct ways. First, it accommodates for the difference in the way different users interact with the models. For example, the architecture can determine if more data from one use case may unfavorably skew the model outcome for another use case scenario. Second, the architecture federates the model training and update required over time. Data from the individual user interaction and model feedback are collected, combined, and transferred from the middle tier to the cloud tier. Reconciliation of master models is managed in the cloud tier, averaging out individual variations, while preserving the privacy of the user. Master models are then saved to the FML model repository. The workflow for our proposed FML architecture is depicted in Figure 3.

## EVALUATION

Table 1 identified five dimensions for the evaluation of FML, namely architecture, service model, access, machine learning, and security. In this research, we present an evaluation of the architecture for its operational performance.

A key difference between our proposed architecture and traditional approaches is the introduction of the low power edge servers in the middle tier. To assess the feasibility of our proposed architecture, we tested three propositions related to operational performances.

*Figure 3. Federated ML workflow*

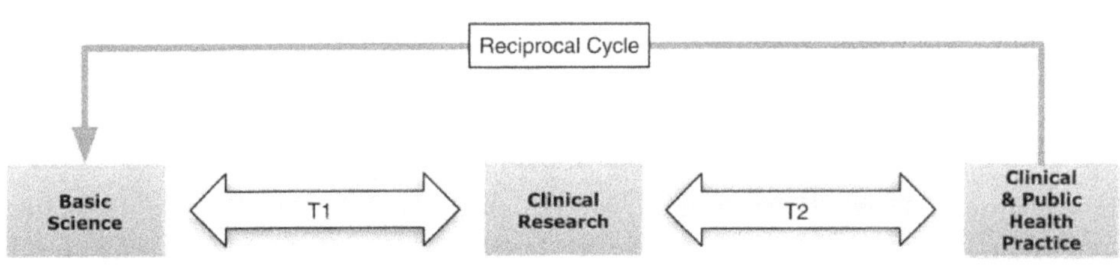

**P1:** *For small dataset, performance of the middle tier (i.e., the low power edge servers) is comparable to high end servers.*
**P2:** *For large dataset, performance of the middle tier (i.e., the low power edge servers) is comparable to high end servers.*
**P3:** *For large dataset, performance the middle tier (i.e., the low power edge servers) will be within tolerable and acceptable range.*

To conduct a comparative assessment of the operational performance, we used a diabetics dataset consisting of 10 years of clinical care data from 130 US hospitals. The data consists of over 50 features representing patient and hospital outcomes. Model engineers may choose a specific subset of features from the dataset to generate machine learning models based on predetermined clinical decision support requirement. For this evaluation, we selected eight features and one dependent variable, 'readmit.' Since we were interested in model prediction capabilities of the low power edge servers in the middle tier across different size dataset for different ML algorithms, we created two datasets of sizes 1000 and 100,000. We then reviewed the operational performance (i.e., time to prediction) for four ML model prediction methods, namely Logistic regression, Naive Bayes, and two types of Support Vector Machines. We tested two types of SVM (gamma=2, C=1; and kernel=linear, C=0.025;) so that we could determine how the architecture performed with variations in the model parameters. It is to be noted that, in our proposed approach, federated model training is conducted only in the cloud tier. This is similar to current approaches that utilize cloud base ML. What distinguishes our proposed architecture is that model prediction is conducted at the middle tier, which is in contrast to current approaches where the prediction is normally undertaken at the cloud layer. As discussed earlier, the latter requires dependence on data and intelligence stored in the cloud, which may not be taken for granted in TR studies.

1. SVM-Support Vector Classification ($SVM^{1}$: gamma=2, C=1; and $SVM^{.025}$: kernel=linear, C=0.025)
2. Cloud server: Virtual Machine with 2.2 GHz Intel Core i7 processor, 16GB RAM, 1600 MHz DDR3 HD
3. Low power edge server: Raspberry Pi 3 with 1.2 GHz ARM Cortex processor, 1GB RAM, 900 MHz 64Gb SD card

For the small sample dataset (n = 1000), our experiments indicated that the model prediction response for the low power edge servers ranged from 0.571ms for LR to 1.143ms for NB. The prediction time for cloud servers ranged from 0.204ms for SVM.025 (kernel=linear, C=0.025) to 0.548ms for LR. For small datasets, the prediction time for LR is comparable for both low power edge server and cloud

*Table 2. Performance of low power edge server*

| Sample Size (n) | Architecture | Training/Prediction Time to Prediction | Execution Time (ms) | | | |
|---|---|---|---|---|---|---|
| | | | LR | NB | SVM[1] | SVM.025[1] |
| 1000 | Federated ML architecture | Training (cloud server[2]) | 7.982 | 1.166 | 207.353 | 120.115 |
| | | Prediction (low power edge server[3]) | 0.571 | 1.143 | 0.947 | 0.789 |
| | Cloud Server | Prediction (cloud server) | 0.548 | 0.273 | 0.345 | 0.204 |
| 100000 | Federated ML architecture | Training (cloud server[2]) | 534.727 | 28.657 | 269016 | 108763 |
| | | Prediction (low power edge server[3]) | 0.586 | 1.139 | 35.672 | 10.271 |
| | Cloud Server | Prediction (cloud server) | 1.292 | 0.371 | 8.281 | 1.387 |

server. The response times for the remaining ML algorithms are slower for low power edge server. For the larger sample dataset (n = 100000), the low power edge server yielded a wider range of model prediction time. For example, the prediction time for LR was 0.586ms whereas the prediction time for SVM (gamma=2, C=1) was 35.672ms. Therefore, the results do not confirm P1 and P2. To assess whether the operational performance is tolerable and acceptable, we used guidelines from Nah (2004), and set a benchmark response time of 2000ms as a tolerable response time for information retrieval. The experiment results for both small and large sample dataset were well within the acceptable and tolerable response time benchmark of 2000ms, thus confirming P3.

# DISCUSSION OF FINDINGS AND IMPLICATIONS

## Findings

The results of our experiment showed that low power edge servers perform slower in comparison to the operational performance of cloud server. Considering the significantly powerful configuration of the cloud server this may not come as a surprise. Nevertheless, it is interesting to note that the low power edge servers performed well within the acceptable and tolerable levels for information retrieval resulting from model predictions. The ML algorithms at the middle tier were implemented using Miniconda[1]. As a scaled down version of Anaconda, Miniconda provides support for only a subset of ML algorithms. In addition to the ML algorithms that we presented in Table 2, other popular ML algorithms supported by Miniconda include

Linear Discriminant Analysis (LDA), K-Nearest Neighbors (KNN), Decision Trees (DT), and Neutral network. We do not include the evaluation of these models in our research due to space limitations.

Our experiments also indicated that all ML algorithms we tested out-performed our operational benchmark (Nah, 2004). Thus, it is fair to conclude that end users will not be affected by the marginally lower performance of the middle tier. Our study showed that low power edge servers in our proposed design equivalent prediction capabilities as that of cloud-based service at a much lower cost (the cost of a Raspberry Pi server is $30 compared to the $370 monthly data transfer cost of a cloud-based solution used in our experiment). Thus, we argue that the benefits of FML far outweighs the limitations of the low power edge servers.

## Practical and Theoretical Contributions

Sepsis is defined as a life threatening organ dysfunction caused by an uncontrolled host response to infection (Hotchkiss et al., 2016). Efforts to use machine learning to predict sepsis in the Intensive Care Unit have achieved positive results (Desautels et al., 2016). However, they rely mostly on vital data reported in Electronic Health Records (EHR). Variables typically used for developing ML models for the assessment of adult sepsis include age, sex, acuity, blood pressure, pulse rate, pain scale, and temperature. The vitals currently used in ML models are not collected in real time constant but gathered through periodic human interventions and recorded in EHR. The discrete data is thus limiting for real time monitoring and prediction of sepsis. The FML architecture is well positioned to address these challenges by collecting data from different sources (e.g., wearable and proximity sensors), and providing clinical decision support in a network agnostic manner.

The solution is portable to various settings such as a hospital triage, ambulatory clinics, and even the patient's home. The FML approach for FML leverages low power edge servers to mediate the communication between cloud and the human agency. A low power server is housed at the sepsis assessment location. A patient wearing sensors generates streaming data which is run against the sepsis prediction model stored on the low power server. The prediction outcome is presented to the human agent for clinical decision support. If the predictions for a particular patient is determined to be inaccurate beyond a threshold, it is an indication that there is an imbalance in the data. Data from that specific low power server is then sent to the cloud to retrain the model. Data from different individuals are collected, combined, and averaged across all similar use-cases. The difference between the prediction for individuals and the master model are aggregated to train and update the models in the cloud. Changes associated with each individual is anonymized and blurred through the averaging process, while respecting the privacy of individual user. The reconciliation ensures that the sepsis prediction model is not thrown off by outlier use-cases.

Although socio-technical perspectives are emphasized in TR (Payne et al., 2009; Sung et al., 2003), a review of related literature indicates that it has been mostly left unaddressed. A plausible explanation is the lack of expertise essential to integrate emergent technologies and socio-technical perspective in TR. Our study illustrates how IS scholarship may inform and fill this crucial gap. Specifically, we demonstrated how IS scholarship related to machine learning and data analytics were brought together to benefit TI through FML. In this study, we demonstrated how FML architecture may be used in TI. The proposed architecture can be adapted to other research areas targeting knowledge discovery and decision making.

### Limitations and Future Directions

It could be argued that the middle tier introduces an added level of system complexity. It may seem that employing low power edge servers may negatively impact performance and scalability of the data analytics pipeline. However, integrating new technologies such as micro-services (Thönes, 2015) address these concerns. We limit our discussion on this topic due to space considerations and set aside related discussion for future studies. Another limitation of our study is that we used existing Diabetics dataset for our evaluation. We did not use end user data in our experiment. Future research will utilize end user data for a more comprehensive assessment of the federate ML architecture.

## CONCLUSION

Advancements in technologies now enable ML models to reside in smart devices (e.g., mobile phone) that we carry with us and use daily. For instance, technology firms have started to develop AI chips (Lu, 2019) that would support FML on various edge devices in the future. An example is the Alibaba AI chips can be used in business operations (e.g., product search and automatic translation on e-commerce sites), personalized recommendations, advertising, and intelligent customer services (Kharpal, 2019).

Although big data analytics and machine learning are driving digital transformation across all industries, TI has lagged in its adoption and use. Knowledge management challenges in TI include imbalance of data, privacy and confidentiality concerns, proprietary data distributed across multiple end user devices, and extensive reliance on cloud. In this research, we propose a federated machine learning platform that leverages low power servers to mediate the execution of the ML algorithms. The proposed approach aims to move machine learning closer to edge devices and end users. Positioning machine learning closer to the end-user enables the human agency to interact with ML models without constant access to massive data and intelligence stored in the cloud. Applications of TI in areas such as clinical trials and clinical practice can greatly benefit from the scalability, cost effectiveness, and improved quality of decision support from the FML architecture.

The learning performance in AI innovations heavily relies on data, the privacy and security of which is always a serious concern and challenge. The FML approach presented in this chapter demonstrates how data is can be isolated from ML models to ensure higher levels of privacy and security (Yang, Liu, Chen, & Tong, 2019). In other words, with FML approach, business entities (e.g., enterprises, individuals) contribute to the centralized ML models and take advantage of the centralized models in the federated environment, whereas entity level is decentralized. This approach reduces the extend of data compromised in the unlikely event of a security breach. China's Cyber Security Law that was enacted in 2017 forbids companies from sharing individual user information when conducting data transactions with partnering organizations. FML methods provide a viable solution to address this regulation through the federation of ML models among business partners and keeping original data private. Take Google Chrome as a more specific example – Chrome collects user profiles and browsing history to rank and recommend web search results. With the help of FML, users need not share hot data about their browsing behaviors but only a personal ranking output weighted by the local data needs to be exposed.

Digital transformation through FML has great potential. Future applications of FML would be to support crowd-sourcing processes, where citizens are able to contribute ML outcomes from their devices in a federated manner rather than the current practice of sharing data to attain a centrally determined outcome.

## SUMMARY POINTS

- Sparse research in Translational Research (TR) on effective approaches and strategies to support its multiple functions.
- Unique Machine Learning (ML) scenarios in Healthcare (e.g., rare events and small datasets) are difficult to model.
- Preference for ML models that interact with agents (humans, edge devices, wearables, sensors) and optimize learning.
- Emerging trend towards moving ML models closer to end-users.
- Federated Machine Learning (FML) for pipelining ML and data analytic tools by incorporating multiple source datasets, intermediate data analytic

processes, and products to output high quality results that support knowledge discovery and decision making.
- FML evaluated using Diabetes dataset shows benefits outweighs limitation of low power edge servers.

## REFERENCES

Baldi, P., & Long, A. D. (2001). A Bayesian framework for the analysis of microarray expression data: Regularized t-test and statistical inferences of gene changes. *Bioinformatics (Oxford, England)*, *17*(6), 509–519. doi:10.1093/bioinformatics/17.6.509 PMID:11395427

Beaulah, S. A., Correll, M. A., Munro, R. E., & Sheldon, J. G. (2008). Addressing informatics challenges in Translational Research with workflow technology. *Drug Discovery Today*, *13*(17-18), 771–777. doi:10.1016/j.drudis.2008.06.005 PMID:18617017

Beck, J. M., Ma, W. J., Kiani, R., Hanks, T., Churchland, A. K., Roitman, J., ... Pouget, A. (2008). Probabilistic population codes for Bayesian decision making. *Neuron*, *60*(6), 1142–1152. doi:10.1016/j.neuron.2008.09.021 PMID:19109917

Becker, K. G., Hosack, D. A., Dennis, G., Lempicki, R. A., Bright, T. J., Cheadle, C., & Engel, J. (2003). PubMatrix: A tool for multiplex literature mining. *BMC Bioinformatics*, *4*(1), 61. doi:10.1186/1471-2105-4-61 PMID:14667255

Castaneda, C., Nalley, K., Mannion, C., Bhattacharyya, P., Blake, P., Pecora, A., ... Suh, K. S. (2015). Clinical decision support systems for improving diagnostic accuracy and achieving precision medicine. *Journal of Clinical Bioinformatics*, *5*(1), 4. doi:10.118613336-015-0019-3 PMID:25834725

Chen, J., Qian, F., Yan, W., & Shen, B. (2013). Translational biomedical informatics in the cloud: Present and future. *BioMed Research International*. PMID:23586054

Coulouris, G. F., Dollimore, J., & Kindberg, T. (2005). *Distributed systems: Concepts and design*. Pearson Education.

Desautels, T., Calvert, J., Hoffman, J., Jay, M., Kerem, Y., Shieh, L., ... Barton, C. (2016). Prediction of sepsis in the intensive care unit with minimal electronic health record data: A machine learning approach. *JMIR Medical Informatics*, *4*(3), e28. doi:10.2196/medinform.5909 PMID:27694098

Fielding, R. T., & Taylor, R. N. (2000). *Architectural styles and the design of network-based software architectures* (PhD dissertation). University of California.

Fontanarosa, P. B., & DeAngelis, C. D. (2002). Basic science and translational research in JAMA. *Journal of the American Medical Association, 287*(13), 1728–1728. doi:10.1001/jama.287.13.1728 PMID:11926900

Gholami, R., Watson, R. T., Molla, A., Hasan, H., & Bjørn-Andersen, N. (2016). Information systems solutions for environmental sustainability: How can we do more? *Journal of the Association for Information Systems, 17*(8), 521–536. doi:10.17705/1jais.00435

Harris, P. A., Taylor, R., Thielke, R., Payne, J., Gonzalez, N., & Conde, J. G. (2009). Research electronic data capture (REDCap) - A metadata-driven methodology and workflow process for providing translational research informatics support. *Journal of Biomedical Informatics, 42*(2), 377–381. doi:10.1016/j.jbi.2008.08.010 PMID:18929686

Holzinger, A. (2016). Interactive machine learning for health informatics: When do we need the human-in-the-loop? *Brain Informatics, 3*(2), 119–131. doi:10.100740708-016-0042-6 PMID:27747607

Hotchkiss, R. S., Moldawer, L. L., Opal, S. M., Reinhart, K., Turnbull, I. R., & Vincent, J.-L. (2016). Sepsis and septic shock. *Nature Reviews. Disease Primers, 2*(1), 16045. doi:10.1038/nrdp.2016.45 PMID:28117397

Kharpal, A. (2019). *Alibaba unveils its first A.I. chip as China pushes for its own semiconductor technology.* Retrieved from https://www.cnbc.com/2019/09/25/alibaba-unveils-its-first-ai-chip-called-the-hanguang-800.html

Koski, L. B., Gray, M. W., Lang, B. F., & Burger, G. (2005). AutoFACT: An auto matic functional annotation and classification tool. *BMC Bioinformatics, 6*(1), 151. doi:10.1186/1471-2105-6-151 PMID:15960857

Kuderer, M., Gulati, S., & Burgard, W. (2015). *Learning driving styles for autonomous vehicles from demonstration.* Paper presented at the Robotics and Automation (ICRA), 2015 IEEE International Conference on. 10.1109/ICRA.2015.7139555

Lee, J. J., Hong, W. K., Hittelman, W. N., Mao, L., Lotan, R., Shin, D. M., ... Papadimitrakopoulou, V. M. (2000). Predicting cancer development in oral leukoplakia: Ten years of translational research. *Clinical Cancer Research, 6*(5), 1702–1710. PMID:10815888

Li, Y., Thomas, M. A., & Osei-Bryson, K.-M. (2017). Ontology-based data mining model management for self-service knowledge discovery. *Information Systems Frontiers, 19*(4), 925–943. doi:10.100710796-016-9637-y

Lindgren, H. (2008). Decision support system supporting clinical reasoning process-an evaluation study in dementia care. *Studies in Health Technology and Informatics, 136*, 315. PMID:18487750

Lu, H. (2019). *Xinhua Headlines: New trends driving China's AI industry* Retrieved from http://www.xinhuanet.com/english/2019-08/31/c_138353742.htm

McMahan, B., & Ramage, D. (2017). *Federated Learning: Collaborative Machine Learning without Centralized Training Data*. Retrieved from https://research.googleblog.com/2017/04/federated-learning-collaborative.html?m=1

Nah, F. F.-H. (2004). A study on tolerable waiting time: How long are web users willing to wait? *Behaviour & Information Technology, 23*(3), 153–163. doi:10.1080/01449290410001669914

Newman, S. (2015). *Building microservices: designing fine-grained systems*. O'Reilly Media, Inc.

O'neill, E. S., Dluhy, N. M., & Chin, E. (2005). Modelling novice clinical reasoning for a computerized decision support system. *Journal of Advanced Nursing, 49*(1), 68–77. doi:10.1111/j.1365-2648.2004.03265.x PMID:15610383

Pan, W. (2002). A comparative review of statistical methods for discovering differentially expressed genes in replicated microarray experiments. *Bioinformatics (Oxford, England), 18*(4), 546–554. doi:10.1093/bioinformatics/18.4.546 PMID:12016052

Payne, P. R., Embi, P. J., & Sen, C. K. (2009). Translational informatics: Enabling high-throughput research paradigms. *Physiological Genomics, 39*(3), 131–140. doi:10.1152/physiolgenomics.00050.2009 PMID:19737991

Porollo, A. A., Adamczak, R., & Meller, J. (2004). POLYVIEW: A flexible visualization tool for structural and functional annotations of proteins. *Bioinformatics (Oxford, England), 20*(15), 2460–2462. doi:10.1093/bioinformatics/bth248 PMID:15073023

Rubio, D. M., Schoenbaum, E. E., Lee, L. S., Schteingart, D. E., Marantz, P. R., Anderson, K. E., . . . Esposito, K. (2010). Defining translational research: implications for training. *Academic Medicine: Journal of the Association of American Medical Colleges, 85*(3), 470.

Siegel, A. I., & Fischl, M. (1971). Dimensions of visual information displays. *The Journal of Applied Psychology*, *55*(5), 470–476. doi:10.1037/h0031777 PMID:5113200

Singleton, A. (2016). The Economics of Microservices. *IEEE Cloud Computing*, *3*(5), 16–20. doi:10.1109/MCC.2016.109

Sung, N. S., Crowley, W. F. Jr, Genel, M., Salber, P., Sandy, L., Sherwood, L. M., ... Getz, K. (2003). Central challenges facing the national clinical research enterprise. *Journal of the American Medical Association*, *289*(10), 1278–1287. doi:10.1001/jama.289.10.1278 PMID:12633190

Sussman, S., Valente, T. W., Rohrbach, L. A., Skara, S., & Ann Pentz, M. (2006). Translation in the health professions: Converting science into action. *Evaluation & the Health Professions*, *29*(1), 7–32. doi:10.1177/0163278705284441 PMID:16510878

Thönes, J. (2015). Microservices. *IEEE Software*, *32*(1), 116–116. doi:10.1109/MS.2015.11

Westfall, J. M., Mold, J., & Fagnan, L. (2007). Practice-based research: "Blue Highways" on the NIH roadmap. *Journal of the American Medical Association*, *297*(4), 403–406. doi:10.1001/jama.297.4.403 PMID:17244837

Woolf, S. H. (2008). The meaning of translational research and why it matters. *Journal of the American Medical Association*, *299*(2), 211–213. doi:10.1001/jama.2007.26 PMID:18182604

Yang, Q., Liu, Y., Chen, T., & Tong, Y. (2019). Federated machine learning: Concept and applications. *ACM Transactions on Intelligent Systems and Technology*, *10*(2), 12. doi:10.1145/3298981

## ENDNOTE

1 http://repo.continuum.io/miniconda/Miniconda3-latest-Linux-armv7l.sh

# Chapter 7
# Factors That Determine Users' Switching Intention From a Debit/Credit Card to Adopting E-Wallet as a Payment Method

**Michael Lo**
https://orcid.org/0000-0003-2448-0240
*President University, Indonesia*

**Adhi Setyo Santoso**
https://orcid.org/0000-0001-6857-2708
*President University, Indonesia*

## ABSTRACT

*This study investigates the switching intention and actual behavior of e-wallet users in Greater Jakarta by using the push-pull and mooring model. There were 33 constructs distributed in Greater Jakarta. Using a quantitative technique, this research used non-probability sampling. The samples collected consisted of 357 valid responses. PLS-SEM and bootstrapping were conducted to check the validity, reliability, and hypothesis testing. Based on the results, low efficiency and personal experience had no significant influence towards switching intention. There was a significant influence from switching cost, social image, security, and alternative attractiveness towards switching intention. Lastly, there was a significant influence of switching intention towards actual behavior. The results gained from the study generated managerial and theoretical implications, which later can be used for the e-wallet provider to improve its services.*

DOI: 10.4018/978-1-7998-1879-3.ch007

## INTRODUCTION

Since the beginning of time, the payment system has been an inseparable tool in our daily lives. The payment system can be defined as a system that allows for fund transfers and circulation (Lamberte). In order to have an effective system, the payment system must include three preconditions. One is to have a standardized procedure to transmit payments between users. The second is to have a secure and reliable system that is always available. Last but not least, the system affordability (price of the system) should also be taken into account (National Bank of Serbia).

The aim of this study is to investigate the factors that influence the switching intention and actual behavior of debit/ credit card users to adopt E-wallet as their payment system. The independent variables of this research are low efficiency, switching cost, personal experience, social image, security, and alternative attractiveness. Switching intention is a mediating variable from all of the independent variables towards the actual behavior of E-wallet usage. This study is created with the purpose of examining the factors that affect users' willingness to switch from a debit/ credit card to E-wallet. More specifically, this study examines which factors actually affect user willingness to switch from a debit/ credit card to E-wallet, whether E-wallet is being used as a substitute for a debit/ credit card in actual usage, whether there is a relation between the push, pull, and mooring effect towards switching intention, and whether there is a relation between switching intention and actual usage.

## BACKGROUND

Back in 1885, Ernst George Ravenstein analyzed that human migration is influenced by push and pull effects. Push factors are those that persuade a person to shift of one's own accord, and in most of the cases, individuals are forced to migrate because there is a risk that the individual has to bear if a person does not migrate. On the other hand, pull effects are the positive outcomes gained if people choose to move over to another option. For example, another option may have better promotion, services, or development (Lee, 1966). But using a push-pull model does not accurately show the intention of individual migration. Numerous studies have shown that psychological and normative elements also impact the individual decision to switch. In 1995, Moon proposed a PPM method and put mooring elements in the formula to explain migrations. The PPM method is used as the most advanced method today since it also reminds researchers to recognize that cultural, moral, and personal issues are also important elements to be noticed in people's migrations (Bansal, Taylor, & James, 2005).

## E-WALLET POTENTIAL

From time to time, humans develop and try to make life easier. Many forms have been taken into practice such as trading by bartering and then shifting towards a cash payment known as money. Money was created to make the trades more efficient and convenient. Money also takes various forms in terms of currency. The necessity to seek a more efficient method of payment is gradually being emphasized by many countries as one of the impacts of this digital era (Daniel, Swartz, & Fermar, 2004). Money itself is used for many economic activities such as functioning as a unit of measurement and as a payment instrument. The development of money has occurred in the past decades in order to minimize transaction fees that are created from doing transactions. For example, back when money was still in the form of coins, a transaction with a huge amount and a far location would cost a lot of effort and time to complete (Odior & Banuso, 2012).

The payment system will continue to evolve throughout time. A payment system itself is a foundation that supports all economic activities, and the communities will require more practical systems with better safety and efficiency (Nakajima, 2012). When it comes to completing transactions, people will clearly choose a more convenient way (Legters, 2013). A study by Humphrey (2004) showed that in a country with an advanced economy such as China and the United States, the usage of cash to do transactions in the retail sphere has been dropping since 1980 (Humphrey, 2004). Over the past years, offline payment systems have been modified by technological advancements (E-wallet) which generates several big E-wallet provider such as Google pay, Alipay, Apple pay, Microsoft Wallet, and etc.

Until recently, the majority of transactions done by customers on the Internet used either credit cards or debit cards. But sometimes there is a minimum purchase requirement when a user wants to use a debit/ credit card (Rivest, 1997). There are a lot of instances of fraud with debit/ credit cards (Pal, Herath, De, & Rao, 2018). Then, there is also a minimum balance, especially for a debit card. Most importantly, the debit and credit card system and procedures in completing a transaction led the financial institutions to be unable to reach smaller businesses such as stands, food trucks, and others (Dinakaran, 2016). With credit and debit cards having a downside, in which they have a transaction cost involved in every transaction, it makes E-wallet as an interesting and more convenient way to complete transactions. E-wallet is consistently being used more often to complete payments (Francisco, Juan, & Francisco, 2014). Therefore, the focus of this study that will be discussed is: Will debit/ credit card users be willing to switch to E-wallet to perform their transactions?

# LITERATURE REVIEW

## Actual Behavior

According to the theory of planned behavior, actual behavior has a direct connection to behavioral intention. Actual behavior is considered one of the ultimate dependent variables in the new technology acceptance and adoption model (Kim, 2012; Kim & Kang, 2016). Past research conducted by Ajzen (1991) and then used by Kim *et al.* (2016) defined actual behavior as different individual responses as the result of a situation given to each individual.

## Switching Intention

Switching intention is defined as the end line or an exit decision made by the customer to end a relationship towards a specific service provider (Stewart, 1998; Calvo-Porral & Lévy-Mangin, 2015). Intention to switch also refers to a negative attitude given by the consumer which leads to leaving a certain service for a substitute (Oliver, 1997; Calvo-Porral & Lévy-Mangin, 2015). Switching intention is also defined as an intention to betray an existing relationship that connects the current provider with the consumer. The decision solely depends on the consumer's decision on whether to stop buying or using the service given by the provider (Shen & Li, 2010; Liu & Technology, 2015).

## Low Efficiency

According to Shen (2012), low efficiency is one unique element that drives users away from a system-based service such as e-commerce. He also stated that the goal of such a system-based service is to complete the objectives of the service in the most efficient way available. Then, low efficiency can be considered as one of the push factors that will affect users' perceptions of whether they will continue to use the services of the provider or not (Li, 2018).

Past research (Olsina Santos, 1999) stated that efficiency is defined as the performance and the accessibility of a program. This definition also includes the ability of a certain program to achieve a good rating for the performance and responsive of the program (Ellahi & Bokhari, 2013).

## Switching Cost

Usually, switching cost constructs are used as an indicator to measure loyalty. Lam (2004) and Wu *et al.* (2014) believe that with an increase in switching costs, there will be lower switching behavior or intention from the users. These facts lead to an explanation of switching costs in other service industries as having different purposes and constructs.

Switching cost is defined as the cost linked to changing a service or product provider (Dick & Basu, 1994; I-Cheng, Chuang-Chun, & Kuanchin, 2014). The variables behind a switching cost may vary from economic, physical, psychological, and emotional sacrifices that might occur before, after, and during the shifting period (Kim, Shin, & Lee, 2006; I-Cheng, Chuang-Chun, & Kuanchin, 2014). These costs may be real costs, perceived monetary costs, or non-monetary costs (Kotler & Andreasen, 1996; I-Cheng, Chuang-Chun, & Kuanchin, 2014).

Switching costs can be differentiated into three major categories which are sunk costs, setup costs, and continuity costs (Patterson & Smith, 2003). Sunk costs refer to costs that have already occurred and cannot be recovered such as investments. Users are constricted to a certain situation where they will lose an investment that they have already made if a certain service is stopped (Rusbult, 1980; I-Cheng, Chuang-Chun, & Kuanchin, 2014). Setup costs include the initial or adoption costs needed to switch to a new alternative option/service. Lastly, continuity costs are costs incurred by the future benefits that need to be sacrificed if one chooses to switch to another option (Patterson & Smith, 2003).

## Personal Experience

Personal experience is perceived as an experience a person gets after completing transactions or shopping experiences that lead one to gain information and knowledge (Yoon, 2012). Personal experience is also described as the degree where users' attitudes and behaviors are determined by their information processing capability (Clark, Abela, & Ambler, 2005). The personal experience element is defined as a user's ability to perceive their past experience since the outcome of one's personal experience may differ even though the process is the same. At the same time, users' personal experiences also reduce the uncertainty when it comes to their actual behavior because with experiences, they can adjust their expectations and their decisions on whether to purchase or not (Campo & Breugelmans, 2015).

Consumer behavior is influenced by users' personal experiences in terms of switching in e-commerce and social commerce. Past research (Li, 2018) revealed that in analyzing the probability of E-commerce switching to social commerce. The study explored how the PPM model affects switching intention. In the process, it showed that there's a connection from personal experience toward social benefit, self-presentation, and efficiency on switching intention.

As cited in Hsieh *et al. (2012),* personal experience affects future users' actual behavior since users will recall their past experiences when it comes to making future decisions. With the different experiences in switching of each individual, personal experience is put into mooring factors.

## Social Image

Social image is defined as the desired image in a social perspective which is the result of communicating and interacting with others (Goffman, 1967; Francisco, Juan, & Francisco, 2014). It is also described as a desire that drives people to develop, sustain, or avoid misfortune towards their status/image that is related to their social activities (Chung, Stoel, & Ren, 2012). Social image can also be understood as the level where an individual believes that there is an importance for oneself to use the innovation because of others' perspectives (Chong, Darmawan, & Lee, 2011; Francisco, Juan, & Francisco, 2014).

Fenollar and Cuestas (2010) stated that social image itself is an important element to determine the social image outcome, and it is also a crucial variable that affects new products since a social image serves as a guideline for people's perspectives and acts as a driving force for those new products.

A study conducted by Chung, Stoel, and Ren (2012) analyzed non-users' perceptions of an online community that were affected by age differences. The study measured the variables that influenced non-user perceptions of the online community. It was mentioned that the variable that had the highest impact of intention to use was the external influence which was based on the social image and subjective norms.

Past research by Francisco, Juan, and Francisco (2014) discussed the antecedents of the adoption of a new mobile payment system. With the moderating effect of age, the study aimed to test a framework to determine whether there were relations and influences, and the importance of certain variables on the framework towards the adoption of the new mobile payment system.

Social image was found to be an important determinant for the intention to use and adoption of mobile payment services (Francisco et al., 2014). Social image is an important element to determine the adoption of a new product or service (Fenollar & Cuestas, 2010).

## Security

Researchers from the past have shown that security plays a huge role in e-services such as m-banking, E-wallet, and Internet banking (Casalo, Flavian, & Guinaliu, 2007; Susanto, Chang, & Ha, 2016). More specifically, the effects of security usually affect the adoption of a technology in its early stages. It is clear to say that the user perception of security is one of the major concerns for the acceptance of a new model/technology (Centeno, 2004; Shih & Fang, 2006; Tan & Teo, 2000; Susanto,

Chang, & Ha, 2016). It was stated by Chen (2012) in his research that security and privacy have a positive impact towards users' satisfaction and trust when it is done correctly. The research was conducted with having smartphone banking services as the context. Users that have experienced good security and privacy when using smartphone banking services will tend to have a better satisfaction in relation to the system that they believe has a good security and privacy system (Cheng, 2014). The process of accepting and having a good perception of the security system and privacy leads to an improvement toward the possibility of the system being used in the future (Lin, Wu, & Tsai, 2005; Susanto, Chang, & Ha, 2016).

Because of the advancements of technology, there is significant growth in the amount of data the providers will possess. It leads to uncertainty and fear due to the lack of confidentiality in the users' information. It leads to personal information security being an important factor in determining whether the users will adopt a new technology (Lai & Wang, 2015). Security is one of the major reasons for switching since according to a recent study conducted by TD Bank, risk of payment fraud itself is the number one concern for 44% of financial industry in 2019 which is 14% increase in just 12 months (ThreatMetrix, 2019)

## Alternative Attractiveness

A past study conducted by Ping Jr. (1993) and mentioned by Ghasrodashti (2017) defined alternative attractiveness as the availability of a possible similar or higher satisfaction of another substitute relationship. Alternative attractiveness is also connected to the customer perception and awareness of alternative options in comparison to the current situation/relation (Ghasrodashti, 2017). Alternative attractiveness is influenced by customers' awareness of other options/alternatives. Therefore, the customers end up switching their relations/options once they become aware of an alternative. The attractiveness instead is due to the availability of a superior product or service (Ghasrodashti, 2017). Otherwise, if customers are not aware of alternative options, they will tend to stay in the relationship even when it is less beneficial than the other choices.

## Research Gap

Research regarding the switching of debit/ credit cards to E-wallet in Indonesia has never been conducted before, according to Emerald Insight and Google Scholar for the past 10 years. Most of the studies using the PPM method discuss switching behavior and customer satisfaction in mobile services (Calvo-Porral & Lévy-Mangin, 2015); E-commerce switching to social commerce (Li, 2018); virtual migration for social networking sites (I-Cheng, Chuang-Chun, & Kuanchin, 2014); and switching behavior in the airline industry (Jung & Han, 2015).

A study conducted by Li (2018) used the Push-Pull Mooring Method as its framework. The PPM method was also used in this study to the extent of user switching behavior from debit/ credit cards to E-wallet. This study mostly used the same model and variables of research carried out by Li (2018). Since there were different objects and contexts of the study, there were some alterations and additions of variables.

## METHODOLOGY

Information regarding the theoretical framework, hypotheses, operational definitions of variables, instrument, sampling, and data collection are included in this chapter. In addition, a complete explanation regarding the analysis method of descriptive and inferential analysis, reliability, and validity check are available below.

### Theoretical Framework

This research was conducted in order to know the willingness to switch from a debit/ credit card to E-wallet. In order to help readers to understand this research, a theoretical framework is provided as shown in the figure below.

### Hypothesis Development

The researcher developed seven hypotheses based on the theoretical framework of this research as stated below:

#### Push Effect on Switching Intention

With the advancement of technology, users demand more effective and efficient ways to live. Users seek more convenient and effortless methods to complement their shopping behavior (Shen, 2012). Low efficiency will affect switching intention since when customers cannot gain information easily, there will be an increase in their cost. Low efficiency Shas a significant impact on switching intention, based on a study by Li (2018) which was conducted by using SmartPLS Version 2.0 with a total number of 382 people participating in this study; 25 responses were found to be invalid, resulting in 357 usable questionnaires. The female respondents comprised 68% of the sample. Nearly 71% of the respondents had a university education. Approximately 72% of the respondents were between 20 and 40 years of age, and

*Figure 1. Framework*

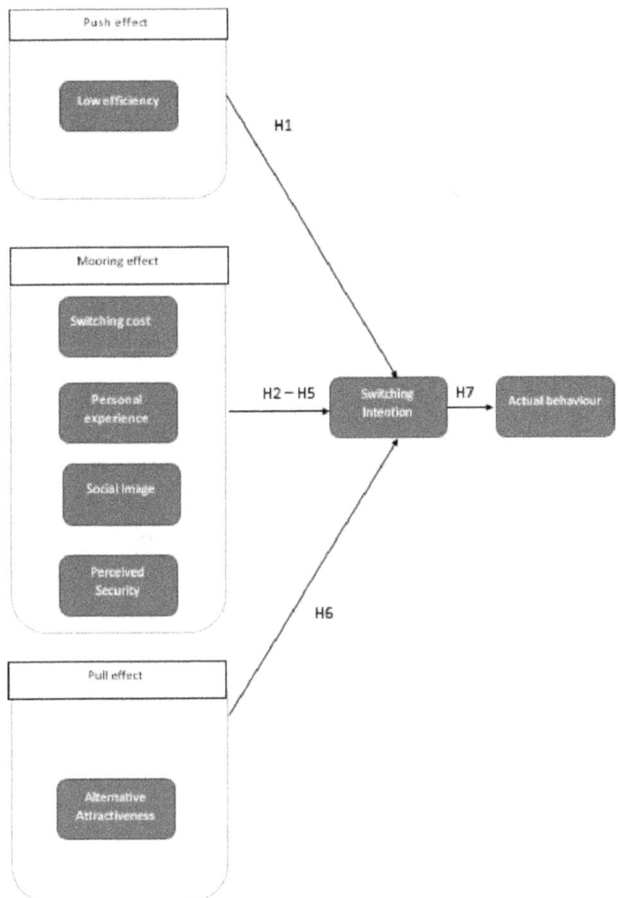


over 63% had more than 3 years of online shopping experience. Approximately 58% reported spending an average of US$36–$70 each time they purchased an item from an online store. Thus, this research proposes the following hypothesis:

**H1:** Low efficiency influences switching intention.

## Mooring Effect on Switching Intention

In the Push-Pull Mooring Model (PPM model), it shows that switching cost is an important factor in mooring factors. The higher the switching cost, it leads to a lower switching intention, and the users may not switch at all (Jung & Han, 2015).

Switching cost has a significant impact on switching intention, which was discovered by I-Cheng *et al.* (2014). The study was conducted using standardized PLS path coefficients in testing the structural model using the SEM technique. A two-stage analysis method was implemented for the data analysis: a measurement model and a structural model. A total of 237 questionnaires were returned, with 218 questionnaires being regarded as valid after eliminating duplicate copies, copies with missing values, or other invalid copies.

Another study conducted by Ghasrodashti (2017) also showed a noticeable influence of switching cost on switching intention. The study was conducted by using Smart PLS and SPSS 16 to analyze descriptive statistics. Finally, 402 questionnaires were obtained. According to the literature, a sample size of 402 respondents is suitable. Thus, this study puts forward:

**H2:** Switching cost influences switching intention.

Personal experience has a different impact on each user's willingness to switch in the future (Kim, 2012). The users' future decisions can be determined by their past experiences in using the services (Farah, 2017). Jung *et al.* (2017) demonstrated that switching cost affects switching intention.

Personal experience has a significant impact on switching intention, based on a study conducted by Li (2018) which was conducted by using SmartPLS Version 2.0 with a total number of 382 people participating in this study; 25 responses were found to be invalid, resulting in 357 usable questionnaires. The female respondents comprised 68% of the sample. Nearly 71% of the respondents had a university education. Approximately 72% of the respondents were between 20 and 40 years of age, and over 63% had more than 3 years of online shopping experience. Approximately 58% of them reported spending an average of US$36–$70 each time they purchased an item from an online store.

Another study conducted by Hsieh *et al.* (2012) also showed that there is a positive relation between personal experience as a mooring factor and switching intention. The study was conducted by using SmartPLS 2.0 to assess the psychometric properties of the scale and the structural model. This latent structural equation modeling technique uses a component-based approach and supports the integration of both measurement and structural models. The online survey lasted for a four-week period during the spring of 2011; the convenience sample yielded 329 respondents, all of whom were registered on Facebook. We excluded 10 respondents who lacked experience with blog usage, for a total of 319 valid participants for the data analysis. Among these respondents, 157 (49.2%) were men and 162 (50.8%) were women. The vast majority (91%) were younger than 30 years of age, and 97.5% had at least a college degree. Then this study proposes the following:

**H3:** Personal experience influences switching intention.

Social image has been used many times in TAM and its later modification. Social image has proven to be an important factor when it comes to new products or services (Fenollar & Cuestas, 2010). There is a positive relation between social image and switching intention based on a study conducted by Francisco *et al.* (2014). The size of the final sample was 2012 valid questionnaires, using a quota sampling method based on users' characteristics. The researcher tested the research hypotheses in the literature review using a structural equation model (SEM). The data analysis was carried out following a two-stage procedure. It analyzed the validity of the scales in the first place and performed a multi-group analysis in the second place, in order to check the associations that were put forward theoretically.

On the other hand, in a study by Han & Hyun (2012), it shows that there are no significant influences of social image towards switching intention. The study was conducted by using the Statistical Package for Social Sciences (SPSS) 16 and AMOS 5 to analyze the data. A total of 620 questionnaires were distributed to restaurant patrons, and 389 questionnaires were returned. After 22 unusable surveys were discarded, 367 remained (valid response rate = 59.19%). Three extreme outliers were excluded, resulting in 364 samples for data analyses. Then, this study hypothesizes:

**H4:** Social image influences switching intention.

Strong security needs to be established to ensure users that there will be no leaking of users' personal information. If there is a strong mooring factor such as security, it will lead to a higher switching intention (Lai & Wang, 2015).

A study conducted by Lai (2015) shows that there is a positive relation between security and switching intention with a total of 206 valid responses retained: 50.5% were females, 51.5% had chronic diseases, and 55.8% lived in an urban area. Their ages ranged from 45 to 75 years old. In this study, the Visual PLS software program was used for the PLS analysis, while the analysis of the model itself used a confirmatory factor analysis (CFA). Thus, this study postulates:

**H5:** Perceived security influences switching intention.

## Pull Effect on Switching Intention

The importance of alternative attractiveness is to give insights into an alternative for the users to compare with the current relationship. Having insufficient information of an alternative will definitely lead to non-switching behavior (Sharma & Patterson, 2000). It was stated that when information about an alternative is given to users, only then will the users end up switching if they find there is a better relation possible from the alternative (Ghasrodashti, 2017).

Past research conducted by Ghasrodashti (2017) showed that there is a significant influence of alternative attractiveness towards switching intention. A total of 402 respondents was suitable for the study, and they were descriptively analyzed using SPSS 16. The study was conducted by using Partial Least Squares (PLS) to evaluate the validity and reliability of the scales of the study. Then, to validate the adequacy of the PLS path model globally, a global fit measure (GoF) was used.

A study by I-Cheng *et al.* (2014) also revealed that there is a significant influence of alternative attractiveness towards switching intention. This study used the PLS method of SEM (SmartPLS) to conduct the data analysis, because the PLS method can handle formative constructs and highly complex predictive models. A total of 237 questionnaires were returned, with 218 questionnaires being regarded as valid after eliminating duplicate copies, copies with missing values, or other invalid copies. The sample comprised 46.8% male and 53.2% female respondents. In total, 24.8% of the respondents were below the age of 20 years, 66.9% of the respondents were between the ages of 20 and 40 years, and 8.3% of the respondents were above the age of 40 years. Thus, this study puts forward:

**H6:** Alternative attractiveness influences switching intention.

## Switching Intention on Actual Behavior

A relationship between switching intention and actual behavior has been demonstrated in some theories such as TAM (Davis, 1989), UTAUT (Venkatesh, Morris, & Davis, 2003), and the Expectation Confirmation Model (Bhattacherjee, 2001). The model used in this research (PPM model) shows that there is a relation between switching intention and actual behavior.

Switching intention and actual behavior have a positive relationship which can be seen in previous research conducted by Li (2018) who conducted research about switching to E-commerce by using SmartPLS Version 2.0 with a total number of 382 people participating in this study; 25 responses were found to be invalid, resulting in 357 usable questionnaires. The female respondents comprised 68% of the sample. Nearly 71% of the respondents had a university education. Approximately 72% of the respondents were between 20 and 40 years of age, and over 63% had more than 3 years of online shopping experience. Approximately 58% reported spending an average of US$36–$70 each time they purchased an item from an online store.

Bansal *et al.* (2005) also proved that there is a positive relation between switching intention and actual behavior which was conducted by using a Confirmatory Factor Analysis (CFA), using LISREL 8 with a maximum-likelihood (ML) estimation. A final sample of 680 completed surveys (356 from auto-repair service consumers, and 324 from hairstyling services) was obtained. More than 70 percent of the replacements involved missing data of less than 1 percent, with a maximum replacement of 2.9 percent for one variable. Hence, this study suggests:

**H7:** Switching intention influences actual behavior.

## OPERATIONAL DEFINITION OF VARIABLES

### Instrument

The utilization of Google Form was applied for the questionnaire, and it was distributed through personal social media. An online questionnaire can bring up the possibility of targeting the wrong respondents; hence, screening questions are needed in order to screen the wrong respondents. Screening questions were placed at the beginning of the questionnaire, so that only the right respondents filled in the rest of the questionnaire to get the valid data. There were two screening questions applied in this research, listed as follows:

1. Have you ever used E-wallet (Gopay, OVO, T-cash, etc.)?
2. Have you ever used E-wallet to pay for something other than transportation (food, product, etc.)?

After a respondent passed the screening questions, it meant that he or she was the right respondent and could proceed to complete the rest of the questionnaire.

### Sampling

This research determined the sampling criteria in order to meet the research requirements. The target respondents for this research were those who had experience in using E-wallet to pay for something besides transportation. This research used non-probability sampling and random sampling as the chosen methods. Random sampling means that each person has an equal chance of being selected to fill out this questionnaire. The questionnaire utilized Google Form as the tool, and the link was shared to 400 contacts through personal social media (Line, WhatsApp, Instagram, etc.). Sugiyono (2015) conveyed that when there is no exact number of the total population or if it is categorized as an infinite population, the number of the sample for the 5% error is equal to 300 samples.

*Table 1. Operational definition of variables*

| Measurement | Adjusted from | Sources | Construct |
|---|---|---|---|
| 1. Processing a transaction using a debit/ credit card isn't efficient (payment processing, time consuming)<br>2. A debit/ credit card payment isn't convenient (has a physical form)<br>3. Using a debit/ credit doesn't provide a genuine service<br>4. I don't think the fees from a debit/ credit card are worth paying for<br>5. Overall, a debit/ credit card doesn't help me more than E-wallet | 1. Is it worth to pay for<br>2. Helps to improve my driving skills<br>3. Processing transactions on PChome is efficient<br>4. The PChome search function is quick<br>5. PChome provides customized applications | (Li, 2018) (Höltl & Trommer, 2013) | Low Efficiency |
| 1. It would cost me a lot of money to switch from a debit/ credit card to E-wallet<br>2. It would take me a lot of effort to switch from a debit/ credit card to E-wallet<br>3. It would take me a lot of time to switch from a debit/ credit card to E-wallet<br>4. There'll be no problem if I change from a debit/ credit card to E-wallet<br>5. I feel it's necessary to change from a debit/ credit card to E-wallet | 1. In general, switching airlines would require a lot of time<br>2. In general, switching airlines would require a lot of effort<br>3. Switching to another airline would involve a monetary loss due to a higher price for a seat<br>4. In general, switching to another airline would be a hassle<br>5. I prefer not to switch because I think I will lose the benefits received from the brand" | (Bansal, Taylor, & James, 2005) (Han, Kim, & Hyun, 2011) (Hou, Chern, Chen, & Chen, 2011) (Jung & Han, 2015) | Switching Cost |
| 1. In the past, I usually shopped using a debit/ credit card<br>2. In the past, I seldom changed my payment method<br>3. Nowadays, I still use a debit/ credit card<br>4. I know how to use a debit/ credit card<br>5. I understand how a debit/ Credit card works when completing a transaction | 1. In the past, I usually shopped online<br>2. I know everything about online shopping stores<br>3. I can understand almost all the aspects of the services I purchase from my adviser<br>4. I possess good knowledge of financial planning services and products<br>5. I can understand my adviser's techniques and strategies very well | (Sharma & Patterson, 2000) (Hsieh, Hsieh, Chiu,, & Feng, 2012) (Li, 2018) | Personal Experience |
| 1. The people in my environment who use E-wallet are more educated than those who do not use it<br>2. The people in my environment who use E-wallet have a superior profile<br>3. Using E-wallet is a status symbol in my environment<br>4. I prefer to use E-wallet when I am with a group of friends<br>5. I use E-wallet because it makes a good impression on others | 1. The people in my environment who use this type of tool are more prestigious than those who do not use it<br>2. The people in my environment who use this type of tool have a superior profile<br>3. Using this type of tool is a status symbol in my environment<br>4. Shopping for fashion products makes a good impression on others<br>5. I prefer to go shopping with a group of friends | (Venkatesh V. &., 2008) (Francisco, Juan, & Francisco, 2014) | Social Image |
| 1. I think E-wallet has a better mechanism to ensure the safe transmission of its users' information<br>2. I don't feel any discomfort when performing a transaction using E-wallet<br>3. I know that my financial information is secure when I use E-wallet<br>4. I don't worry that someone can access my financial information when I use E-wallet<br>5. Overall, E-wallet is a safe payment method | 1. I think this smartphone banking service has mechanisms to ensure the safe transmission of its users' information<br>2. I worry about the abuse of my financial information when I use Fintech<br>3. My financial information is not secure when I use Fintech<br>4. I worry that someone can access my financial information<br>5. Overall, this smartphone banking service is a safe place to transmit sensitive information | (Casalo, Flavian, & Guinaliu, 2007; Lee M., 2009; Susanto, Chang, & Ha, 2016; Ryu, 2018) | Security |
| 1. All considered, E-wallet would be less costly than a debit/ credit card<br>2. E-wallet provides better services<br>3. E-wallet is more convenient and easy to use<br>4. E-wallet has more benefits that a debit/ credit card<br>5. I would feel more satisfied with the services of E-wallet than I am with a debit/ credit card | 1. All in all, another adviser would be less costly than the present adviser is<br>2. A new adviser would provide a full range of services<br>3. A new adviser is located closer to me as compared to the current adviser<br>4. A new adviser would benefit me more than my current adviser in achieving my goals<br>5. I would feel more satisfied with the services of a new adviser than I am with my current adviser | (Sharma & Patterson, 2000) | Alternative Attractiveness |

*continued on following page*

*Table 1. Continued*

| Measurement | Adjusted from | Sources | Construct |
|---|---|---|---|
| 1. I am considering switching from a debit/ credit card to E-wallet<br>2. The likelihood of me switching from a debit/ credit card to E-wallet is high<br>3. I intend to use E-wallet as my priority payment method<br>4. I am determined to switch from a debit/ credit card to E-wallet<br>5. I would not continue using a debit/ credit card if it's possible | 1. Improbable.....Probable<br>2. Unlikely.....Likely<br>3. No chance.....Certain<br>4. I have an intention to switch my smartphone service's company<br>5. I have no intention to use my earlier smartphone company's services | (Calvo-Porral & Lévy-Mangin, 2015) (Li, 2018) | Switching Intention |
| 1. I gradually use E-wallet from time to time<br>2. I use E-wallet more than a debit/ credit card<br>3. I rarely use a debit/ credit card since I started using E-wallet<br>4. I do most of my transactions on E-wallet rather than a debit/ credit card<br>5. I use E-wallet to help me on a daily basis | 1. Regarding usage frequency, I usually use Kidshome every day<br>2. I spend more time on Kidshome than on PChome | (Li, 2018) | Actual Behavior |

# Data Collection

## Procedures

This research focused on quantitative research. Collecting data through Google Form and sharing it to social media contacts is the method that is used in order to get data from the targeted respondents directly or also known as primary data. Then, the gathered data was measured using SmartPLS 3.0. Online questionnaires can be used as a tool to gather data regarding respondents' opinions (Paten, 2016). It has many advantages: easy to be distributed widely, cost-efficient, and quick in collecting the results (Debois, 2016). A six-point Likert Scale is used as the measurement for each construct where 1 means "strongly disagree" and 6 means "strongly agree". By using an even number, it can avoid having respondents be neutral.

## Validity and Reliability

### *Validity*

There are 4 types of validity based on a study by Taherdoost (2016), which are face validity, criterion validity, content validity, and construct validity. Face validity refers to the degree which analyzes operational construct subjectively. Content validity itself is the degree to which items reflect the content universe to which the instrument will be generalized. Criterion validity consists of predictive validity, concurrent validity, and postdictive validity. Taherdoost (2016) defined construct validity as how well you perceive or transform a concept, idea, or behavior which is a construct into a functioning and operating reality, the operationalization. Construct validity consists of convergent validity and discriminant validity.

The standards for what makes a "good" α coefficient are entirely discretionary and depend on one's theoretical knowledge of the scale in the questions. α coefficients that are less than 0.5 are more likely to be unacceptable (Goforth, 2015). Outer loadings of 0.5 are considered as acceptable while outer loadings of 0.7 are considered as highly satisfactory. The Average Variance Extracted (AVE) needs to meet a minimum of 0.5 or higher to be considered as satisfactory. Discriminant validity can be assessed by Fornell and Larcker's criterion. It is done by examining the cross-loading factors of each variable. The requirement itself is the cross loading on its own construct should be higher than the other cross loading in the same construct.

### *Reliability Test*

The reliability of PLS-SEM is examined with the use of Cronbach's α and composite reliability (CR). Cronbach's α is defined as a measure that is used to assess and calculate the reliability and consistency of a set of test items. Cronbach's α needs to meet 0.5 to be considered as reliable (Sari, 2015). Then composite reliability refers to the true value of the reliability of a construct. It is known as a preferred alternative to Cronbach's α and it needs to meet a minimum of 0.7 to be considered as reliable (Oei, 2016).

## Respondents' Profile

After passing the screening questions, the respondents proceeded to fill in their profiles, starting from their name, gender, age, occupation, educational level, monthly income, and E-wallet usage frequency. The typed their name, and then chose the gender whether male or female. Age was given 5 options, which were: <15, 15-20, 21-25, 26-30, and >30 years old. Four options were provided for occupation: student, employee, unemployed, and others. Four options were also given for educational level: high school or below, Undergraduate Degree, Bachelor's Degree, and Master's Degree or above. Four options were listed for monthly income: less than IDR 1.5 million, 1.5 million – 3 million, 3.1 million – 5 million, and more than 5 million. Last but not least, 4 options were provided for E-wallet usage frequency: once, 2-4, 5-8, and more than 8 times a month.

## Descriptive Analysis

A six-point Likert scale was used as a descriptive analysis in this research. There were 6 points as the measurement scale, in which 1 represents "strongly disagree", 2 represents "disagree", 3 represents "somewhat disagree", 4 represents "somewhat agree", 5 represents "agree", and 6 represents "strongly agree" (Vagias, 2006). By using an even number, it can avoid having respondents from being neutral (Chomeya, 2010).

### Inferential Analysis

PLS-SEM is used to maximize the dependent variables' explained variance by adopting an ordinary least squares estimation method. PLS-SEM is a nonparametric method; hence, it makes no distributional assumption which is considered to be exploratory (Ravand, 2016). To ensure that there is an accurate model, an inner model analysis is conducted. An inner model analysis can be seen through $R^2$ (R Square), which is categorized as 0.67, 0.33, and 0.19, which rates endogenous variables as substantial, moderate, and weak. R square values represent the total number of variances in endogenous variables that represent the structural model itself (Hsiao, 2017).

In the hypothesis testing, t-statistic and p-value were used to measure whether the variables were accepted/rejected. A p-value less than 0.05 ($p<0.05$) also needed to be fulfilled in order for a variable to be accepted.

## RESULTS AND DISCUSSION

### Pretest Study

In order for the study be measured accordingly, a pre-test study was conducted with 30 random respondents. The results of the pre-test led to 7 low loading factor questions which later were deleted in this study. There were two questions deleted from the low efficiency variable, which were: "It would cost me a lot of money to switch from a debit/ credit card to E-wallet" and "It would take me a lot of effort to switch from a debit/ credit card to E-wallet". There were three questions extracted from the personal experience variable which stated: "Nowadays, I still use a debit/ credit card", "I know how to use a debit/ credit card", and "I understand how a debit/ credit card works when completing a transaction". There was one question deleted from the alternative attractiveness variable which was "A debit/ credit card payment isn't convenient (has a physical form). Last but not least, there was one question deleted from the switching intention variable which was: "The likelihood of me switching from a debit/ credit card to E-wallet is high".

## Validity and Reliability Test

### Reliability Test

In consonance with previous research (Li, 2018), SmartPLS Version 3.0 was employed to perform the PLS analyses. In addition, bootstrapping was employed with 500 sub-samples to assess the significance of the indicators and path coefficients. In this study, the reliability and validity of all the constructs were determined. It also conducted an exploratory factor analysis to ensure high loadings of the hypothesized factors and low loadings of the cross-loadings in the data set. The standards for what makes a "good" α coefficient are entirely discretionary and depend on your theoretical knowledge of the scale in the questions. α coefficients that are less than 0.5 are more likely to be unacceptable (Goforth, 2015). In this research, the Cronbach's alpha of each multi-item was over 0.5 with the number of actual behavior (0.851); switching intention (0.846); alternative attractiveness (0.706); low efficiency (0.897); personal experience (0.519); security (0.824); social image (0.889); and switching cost (0.792) that were demonstrated to be above the standard of moderate reliability Hinton *et al.* (2004). The composite reliability measures were all above 0.8 with the number of actual behavior (0.893); switching intention (0.897); alternative attractiveness (0.819); low efficiency (0.833); personal experience (0.806); security (0.877); social image (0.919); and switching cost (0.854), which were higher than the recommended 0.7 which indicates that the internal consistency is adequate (Bagozzi & Yi, 1988; I-Cheng, Chuang-Chun, & Kuanchin, 2014). Thus, all of the variables meet the minimum requirements of both Cronbach's Alpha and the Composite Reliability test. It implies that all of the variables and constructs are reliable.

*Table 2. Cronbach's alpha and composite reliability*

| | Composite Reliability | Cronbach's Alpha |
|---|---|---|
| AB | 0.893 | 0.851 |
| ALTERNATIVE ATTRACTIVENESS | 0.819 | 0.706 |
| LOW EFFICIENCY | 0.833 | 0.897 |
| PERSONAL EXPERIENCE | 0.806 | 0.519 |
| SECURITY | 0.877 | 0.824 |
| SOCIAL IMAGE | 0.919 | 0.889 |
| SWITCHING COST | 0.854 | 0.792 |
| SWITCHING INTENTION | 0.897 | 0.846 |

*Table 3. Outer loadings of variables*

| | AB | ALTERNATIVE ATTRACTIVENESS | LOW EFFICIENCY | PERSONAL EXPERIENCE | SECURITY | SOCIAL IMAGE | SWITCHING COST | SWITCHING INTENTION |
|---|---|---|---|---|---|---|---|---|
| AA1 | | 0.710 | | | | | | |
| AA3 | | 0.716 | | | | | | |
| AA4 | | 0.755 | | | | | | |
| AA5 | | 0.733 | | | | | | |
| AB1 | 0.780 | | | | | | | |
| AB2 | 0.796 | | | | | | | |
| AB3 | 0.821 | | | | | | | |
| AB4 | 0.812 | | | | | | | |
| AB5 | 0.748 | | | | | | | |
| LE3 | | | 0.602 | | | | | |
| LE4 | | | 0.994 | | | | | |
| LE5 | | | 0.744 | | | | | |
| PE1 | | | | 0.815 | | | | |
| PE2 | | | | 0.828 | | | | |
| S1 | | | | | 0.825 | | | |
| S2 | | | | | 0.601 | | | |
| S3 | | | | | 0.816 | | | |
| S4 | | | | | 0.794 | | | |
| S5 | | | | | 0.781 | | | |
| SC1 | | | | | | | 0.751 | |
| SC2 | | | | | | | 0.632 | |
| SC3 | | | | | | | 0.723 | |
| SC4 | | | | | | | 0.744 | |
| SC5 | | | | | | | 0.817 | |
| SI1 | | | | | | 0.791 | | |
| SI2 | | | | | | 0.819 | | |
| SI3 | | | | | | 0.877 | | |
| SI4 | | | | | | 0.791 | | |
| SI5 | | | | | | 0.883 | | |
| SWI1 | | | | | | | | 0.809 |
| SWI3 | | | | | | | | 0.847 |
| SWI4 | | | | | | | | 0.810 |
| SWI5 | | | | | | | | 0.843 |

## Validity Test

The data was processed using SmartPLS 3.0 software to calculate and analyze the

*Table 4. Average variance extracted (AVE)*

| | AVE |
|---|---|
| AB | 0.627 |
| ALTERNATIVE ATTRACTIVENESS | 0.531 |
| LOW EFFICIENCY | 0.634 |
| PERSONAL EXPERIENCE | 0.675 |
| SECURITY | 0.590 |
| SOCIAL IMAGE | 0.694 |
| SWITCHING COST | 0.541 |
| SWITCH ING INTENTION | 0.685 |

PLS algorithm. The results are shown in the following table.

The results shown in Table 4 detail that the outer loadings of the variables and each construct are more than 0.5 in value. Then it was determined that all of the variables passed one of the requirements for discriminant validity. The highest outer loading is LE4 with the number of 0.994, while the lowest is S2 with 0.601 outer loadings.

Convergent validity is demonstrated because the AVE values for all constructs are above the suggested number which is 0.5, according to Fornell and Larcker (2013) as cited in (I-Cheng, Chuang-Chun, & Kuanchin, 2014). The discriminant validity was assessed using Fornell and Larcker's (2013) procedure.

Discriminant validity is the concept of a test to ensure whether the concept or measurement differs in every variable. For this purpose, this research used the cross loading of the data shown in Table 5. The data shows that all of the cross loading values and each variable of their own construct have the highest loading factors compared to other variables. The results also indicate the square roots of AVE are all higher than all of the inter-construct correlations, so that the discriminant validity is achieved.

## Respondents' Profiles

This research focused on the switching intention of debit/ credit card users towards E-wallet as an alternative payment system. Hence, this research targeted people who had completed a purchase using E-wallet such as OVO, Go-pay, T-cash, etc., more than one time to purchase something other than transportation. In addition, the respondents had to have a minimum purchase of 1 time a month. The questionnaire was filled in by 413 respondents with 409 respondents passing the screening questions and data from 357 respondents was used for the analysis.

*Table 5. Cross loading of variables*

| | AB | ALTERNATIVE ATTRACTIVENESS | LOW EFFICIENCY | PERSONAL EXPERIENCE | SECURITY | SOCIAL IMAGE | SWITCHING COST | SWITCHING INTENTION |
|---|---|---|---|---|---|---|---|---|
| **AA1** | 0.496 | **0.710** | 0.053 | 0.483 | 0.536 | 0.454 | 0.491 | 0.518 |
| **AA3** | 0.491 | **0.716** | -0.190 | 0.262 | 0.445 | 0.425 | 0.405 | 0.497 |
| **AA4** | 0.582 | **0.755** | -0.067 | 0.268 | 0.508 | 0.507 | 0.462 | 0.583 |
| **AA5** | 0.524 | **0.733** | -0.013 | 0.263 | 0.522 | 0.412 | 0.374 | 0.504 |
| **AB1** | **0.780** | 0.604 | 0.047 | 0.445 | 0.575 | 0.540 | 0.584 | 0.636 |
| **AB2** | **0.796** | 0.527 | 0.033 | 0.162 | 0.451 | 0.534 | 0.447 | 0.606 |
| **AB3** | **0.821** | 0.603 | -0.110 | 0.279 | 0.603 | 0.628 | 0.433 | 0.711 |
| **AB4** | **0.812** | 0.544 | -0.076 | 0.256 | 0.582 | 0.580 | 0.487 | 0.651 |
| **AB5** | **0.748** | 0.574 | 0.013 | 0.254 | 0.457 | 0.455 | 0.406 | 0.516 |
| **LE3** | 0.016 | 0.060 | **0.602** | 0.151 | -0.048 | -0.025 | 0.042 | 0.016 |
| **LE4** | -0.024 | -0.058 | **0.994** | 0.074 | -0.176 | -0.097 | -0.042 | -0.108 |
| **LE5** | 0.037 | 0.013 | **0.744** | 0.109 | -0.032 | -0.002 | 0.021 | -0.001 |
| **PE1** | 0.276 | 0.341 | 0.040 | **0.815** | 0.392 | 0.355 | 0.399 | 0.354 |
| **PE2** | 0.305 | 0.375 | 0.055 | **0.828** | 0.426 | 0.352 | 0.397 | 0.365 |
| **S1** | 0.620 | 0.619 | -0.059 | 0.486 | **0.825** | 0.567 | 0.534 | 0.656 |
| **S2** | 0.346 | 0.452 | -0.045 | 0.350 | **0.601** | 0.395 | 0.297 | 0.390 |
| **S3** | 0.557 | 0.560 | -0.174 | 0.380 | **0.816** | 0.536 | 0.407 | 0.587 |
| **S4** | 0.482 | 0.477 | -0.203 | 0.338 | **0.794** | 0.599 | 0.385 | 0.614 |
| **S5** | 0.565 | 0.535 | -0.215 | 0.358 | **0.781** | 0.500 | 0.401 | 0.575 |
| **SC1** | 0.406 | 0.438 | 0.020 | 0.443 | 0.397 | 0.342 | **0.751** | 0.405 |
| **SC2** | 0.333 | 0.391 | 0.052 | 0.300 | 0.239 | 0.218 | **0.632** | 0.260 |
| **SC3** | 0.468 | 0.466 | -0.086 | 0.334 | 0.355 | 0.328 | **0.723** | 0.421 |
| **SC4** | 0.456 | 0.419 | -0.080 | 0.305 | 0.428 | 0.465 | **0.744** | 0.537 |
| **SC5** | 0.498 | 0.483 | -0.049 | 0.405 | 0.481 | 0.561 | **0.817** | 0.574 |
| **SI1** | 0.536 | 0.446 | 0.005 | 0.417 | 0.503 | **0.791** | 0.471 | 0.558 |
| **SI2** | 0.588 | 0.468 | -0.006 | 0.330 | 0.496 | **0.819** | 0.433 | 0.588 |
| **SI3** | 0.619 | 0.559 | -0.168 | 0.343 | 0.649 | **0.877** | 0.467 | 0.695 |
| **SI4** | 0.562 | 0.556 | -0.091 | 0.315 | 0.509 | **0.791** | 0.472 | 0.614 |
| **SI5** | 0.596 | 0.541 | -0.137 | 0.393 | 0.661 | **0.883** | 0.461 | **0.727** |
| **SWI1** | 0.642 | 0.629 | -0.010 | 0.499 | 0.636 | 0.607 | 0.605 | **0.809** |
| **SWI3** | 0.682 | 0.591 | -0.099 | 0.329 | 0.567 | 0.637 | 0.487 | **0.847** |
| **SWI4** | 0.699 | 0.600 | -0.129 | 0.295 | 0.633 | 0.617 | 0.496 | **0.810** |
| **SWI5** | 0.604 | 0.573 | -0.162 | 0.327 | 0.631 | 0.685 | 0.482 | **0.843** |

## Respondents by Gender

Based on Figure 2, the gender majority of the respondents in this research was males with 51.8% equal to 185 male respondents. On the other hand, female respondents equaled 48.2% or 172 respondents in total.

## Respondents by Age

As shown in Figure 3, the majority of the respondents was dominated by people in the range of 21-25 years old with the number of 70.9% which equaled 253 respondents. It was followed by the 15-20 age group with the number of 23.2% which equaled 83 respondents. Then the respondents between the ages of 26-30 had 4% in number which equaled 14 respondents. Lastly, it was followed by respondents with ages above 30 which were 1.7% which equaled 6 respondents and 0.3% or 1 respondent under the age of 15.

## Respondents by Occupation

As the data from Figure 4 shows, the respondents generally came from students with the number of 72% which was equal to 257 respondents. It was followed by employees with the number of 24.1% which was equal to 86 respondents. In the third place, it was 3.1% which was equal to 11 respondents who did not work. Last but not least, there was 0.8% equal to 3 respondents who chose another option.

*Figure 2. Respondents' gender*

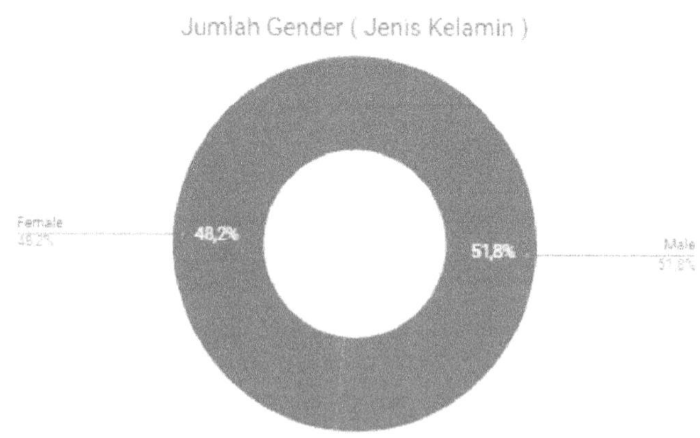

*Figure 3. Respondents' Age*

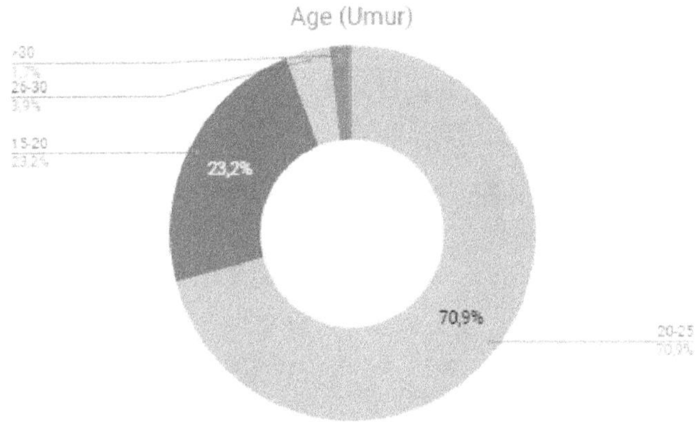


## Respondents by Educational Level

As seen from Figure 5, there was 89.9% which was equal to 321 respondents who were Undergraduate Degree holders. Second, senior high school students or below had the number of 8.4% which was equal to 30 respondents. Lastly, Master's Degree or above had the lesser number of 1.7% or equal to 6 respondents.

*Figure 4. Respondents' occupation*

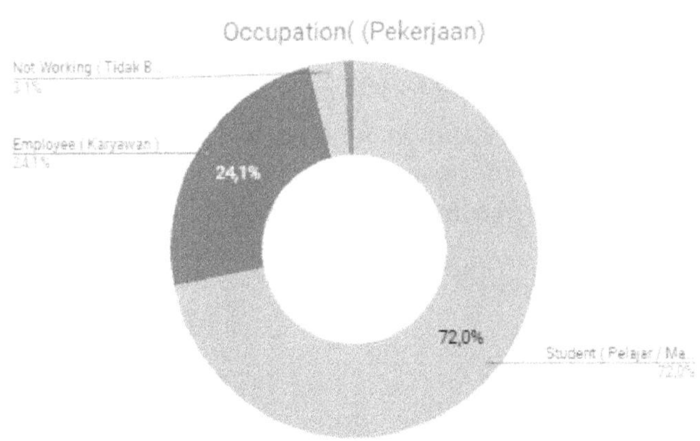

*Figure 5. Respondents' educational level*

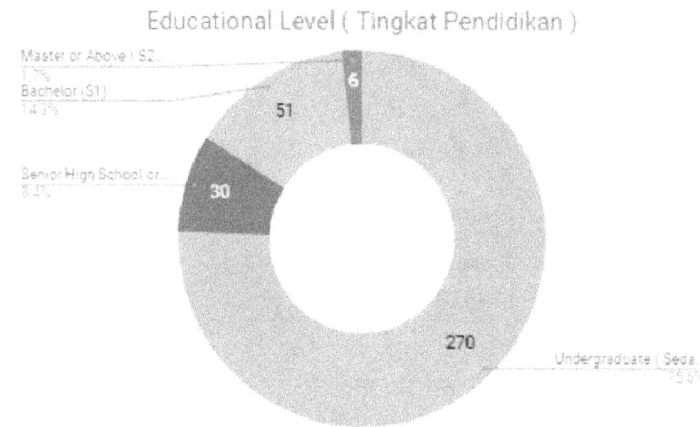


## Respondents by Monthly Income

As shown in Figure 6, the highest monthly income level came to IDR 3,000,001 – IDR 5,000,000 with the number of 41.5% people which was equal to 148 respondents. It was followed by IDR 1,500,000 – IDR 3,000,000 with 35.6% which was equal to 127 respondents. In the third place, there was 14.6% which was equal to 52 respondents that had an income level above IDR 5,000,000. Last but not least, there was 8.4% which was equal to 30 respondents who had an income level below IDR 1,500,000.

## Respondents by E-wallet Usage Frequency

As shown in Figure 7, the majority of the respondents was dominated by people in the range of 5-8 times a month with the number of 42.6% which was equal to 152 respondents. It was followed by the section of more than 8 times a month with the number of 29.1% which was equal to 104 respondents. Then the respondents that had used E-wallet between 2-4 times a month had a number of 24.1% which was equal to 86 respondents. Lastly, it was followed by respondents with a purchase of only once a month with a percentage of 4.2% which was equal to 15 respondents.

*Figure 6. Respondents' monthly income level*

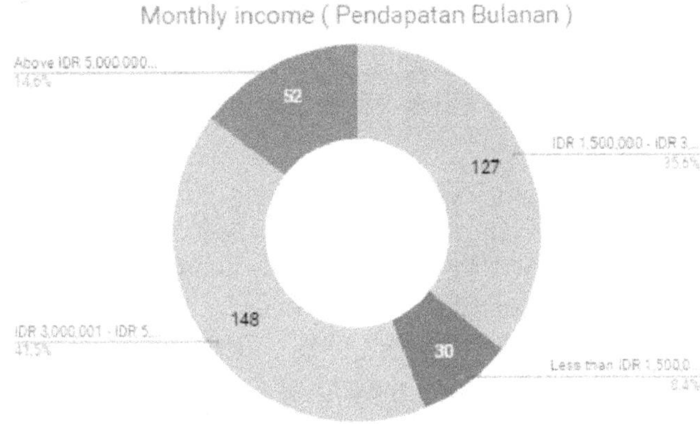


## Descriptive Analysis

### Low Efficiency

(LE3) "Using a debit/ credit card hasn't provided a genuine service" had a number of (m= 2.89; SD= 1.48). (LE4) "I don't think the fees given from a debit/ credit card are worth the services that I am paying for (monthly fee)" had a number of

*Figure 7. Respondents' e-wallet usage frequency*

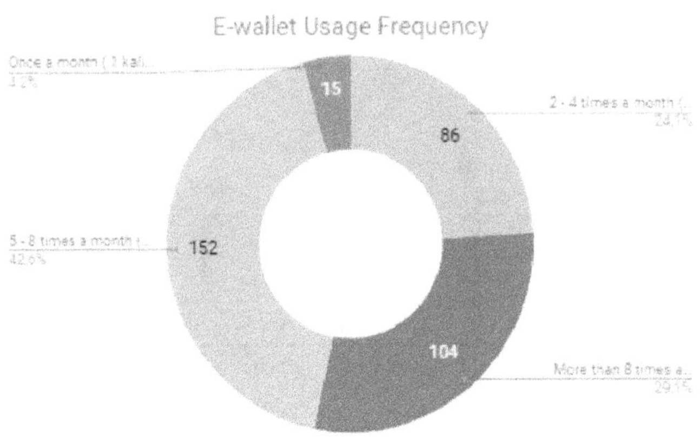

(m= 3.01; SD= 1.48). (LE5) "Overall debit/ credit card services don't help make my purchases more efficient" had a number of (m= 2.75; SD= 1.57).

As seen in Table 6, the LE3, LE4, and LE5 means were lower than 3.02, which was in the capacity of "slightly disagree". The standard deviations of LE3, LE4, and LE5 were more than 1.00, which can be considered as polarized.

## Switching Cost

(SC1) "It would not cost me a lot of money to switch from a debit/ credit card to E-wallet" had a number of (m= 5.24; SD= 0.95). (SC2) "It would not take a lot of effort to switch from a debit/ credit card to E-wallet" had a number of (m= 5.00; SD= 0.78). (SC3) "It would not take me a lot of time to switch from a debit/ credit card to E-wallet" had a number of (m= 5.27; SD= 0.86). (SC4) "There'll be no problem if I change from a debit/ credit card to E-wallet" had a number of (m= 5.00; SD= 0.96). (SC5) "I feel it's necessary to change from a debit/ credit card to E-wallet because I will not lose the benefits given by a debit/ credit card" had a number of (m= 5.05; SD= 1.00).

As seen in Table 6, SC1, SC2, SC3, SC4, and SC5 had means higher than 5.00, which are in the capacity of "agree". The standard deviations of SC1, SC2, SC3, and SC4 were less than 1.00, which can be considered as low polarized. This result is contrary with the standard deviation of SC 5, which is more than 1.00 and can be considered as polarized.

*Table 6. Low efficiency data summary*

| | N | Minimum | Maximum | Mean | Std. Deviation |
|---|---|---|---|---|---|
| LE3 | 357 | 1.00 | 6.00 | 2.8936 | 1.48585 |
| LE4 | 357 | 1.00 | 6.00 | 3.0112 | 1.48395 |
| LE5 | 357 | 1.00 | 6.00 | 2.7507 | 1.57830 |

*Table 7. Switching cost data summary*

| | N | Minimum | Maximum | Mean | Std. Deviation |
|---|---|---|---|---|---|
| SC1 | 357 | 1.00 | 6.00 | 5.2437 | .95957 |
| SC2 | 357 | 1.00 | 6.00 | 5.0000 | .78253 |
| SC3 | 357 | 2.00 | 6.00 | 5.2745 | .86618 |
| SC4 | 357 | 1.00 | 6.00 | 5.0028 | .95840 |
| SC5 | 357 | 1.00 | 6.00 | 5.0504 | 1.00433 |

## Personal Experience

(PE1) "In the past, I usually shopped using a debit/ credit card" had a number of (m= 5.20; SD= 1.01). (PE2) "In the past, I seldom changed my payment method" had a number of (m= 4.98; SD= 0.82).

As seen in Table 7, PE1 has a mean higher than 5.00, which is in the capacity of "agree". Then PE2 has a mean slightly below 5.00, which is still in the capacity of "agree". The standard deviation of PE2 is less than 1.00, which can be considered as low polarized. This result is contrary with the standard deviation of PE1, which is more than 1.00 and can be considered as polarized.

## Social Image

(SI1) "The people in my environment who use E-wallet usually are more educated than those who do not use it" had a number of (m= 5.04; SD= 1.13). (SI2) "The people in my environment who use E-wallet have a superior profile" had a number of (m= 4.78; SD= 1.07). (SI3) "Using E-wallet is a social status symbol in my environment" had a number of (m= 4.88; SD= 1.33). (SI4) "I prefer using E-wallet when I am with a group of friends" had a number of (m= 5.00; SD= 1.08). (SI5) "I use E-wallet because it also makes a good impression on others" had a number of (m= 4.87; SD= 1.31).

As seen in Table 8, SI1 and SI4 have means higher than 5.00, which are in the capacity of "agree". Then SI2, SI3, and SI5 have means slightly below 5.00, which are still in the capacity of "agree". The standard deviations of SI1, SI2, SI3, S4, and SI5 were higher than 1.00, which can be considered as polarized.

## Security

(S1) "I think E-wallet has a better mechanism to ensure the safe transmission of its users' information" had a number of (m= 5.09; SD= 1.02). (S2) "I don't feel discomfort when performing a transaction using E-wallet" had a number of (m= 5.01; SD= 0.81). (S3) "I know that my financial information is secure when I use E-wallet" had a number of (m= 5.17; SD= 0.92). (S4) "I don't worry that someone

*Table 8. Personal experience data summary*

| | N | Minimum | Maximum | Mean | Std. Deviation |
|---|---|---|---|---|---|
| PE1 | 357 | 1.00 | 6.00 | 5.2045 | 1.01957 |
| PE2 | 357 | 1.00 | 6.00 | 4.9804 | .82935 |

*Table 9. Social image data summary*

| | N | Minimum | Maximum | Mean | Std. Deviation |
|---|---|---|---|---|---|
| SI1 | 357 | 1.00 | 6.00 | 5.0448 | 1.13831 |
| SI2 | 357 | 1.00 | 6.00 | 4.7871 | 1.07555 |
| SI3 | 357 | 1.00 | 6.00 | 4.8852 | 1.33481 |
| SI4 | 357 | 1.00 | 6.00 | 5.0000 | 1.08359 |
| SI5 | 357 | 1.00 | 6.00 | 4.8739 | 1.31470 |

can access my financial information when I use E-wallet" had a number of (m= 4.90; SD= 1.20). (S5) "Overall E-wallet is a safe payment method" had a number of (m= 5.17; SD= 0.84).

As seen in Table 9, S1, S2, S3, and S5 have means higher than 5.00, which are in the capacity of "agree". Then S4 has a mean slightly below 5.00, which is still in the capacity of "agree". The standard deviations of S2, S3, and S5 were less than 1.00, which can be considered as low polarized. This result contrasts with the standard deviation of S1 and S4 which is more than 1.00 and can be considered as polarized.

## Alternative Attractiveness

(AA1) "E-wallet provides better services" had a number of (m= 5.28; SD= 0.81). (AA3) "E-wallet has more benefits than a debit/ credit card" had a number of (m= 5.31; SD= .82). (AA4) "I would feel more satisfied with the services of E-wallet than I am with a debit/ credit card" had a number of (m= 5.17; SD= 0.77). (AA5) "All considered, E-wallet would be less costly than a debit/ credit card" had a number of (m= 5.31; SD= 0.78).

As seen in Table 10, AA1, AA3, AA4, and AA5 have means higher than 5.00, which are in the capacity of "agree". The standard deviations of AA1, AA3, AA4, and AA5 are less than 1.00, which can be considered as low polarized.

*Table 10. Security data summary*

| | N | Minimum | Maximum | | Mean | Std. Deviation |
|---|---|---|---|---|---|---|
| S1 | 357 | 1.00 | 6.00 | | 5.0980 | 1.02438 |
| S2 | 357 | 1.00 | 6.00 | | 5.0168 | .81403 |
| S3 | 357 | 1.00 | 6.00 | | 5.1765 | .92086 |
| S4 | 357 | 1.00 | 6.00 | | 4.9020 | 1.20808 |
| S5 | 357 | 1.00 | 6.00 | | 5.1793 | .84555 |

## Switching Intention

(SWI1) "I am considering to switch from a debit/ credit card to E-wallet" had a number of (m= 5.19; SD= 0.93). (SWI3) "I intend to use E-wallet as my priority payment method" had a number of (m= 5.15; SD= 1.01). (SWI4) "I am determined to switch from a debit/ credit card to E-wallet" had a number of (m= 5.08; SD= 0.92). (SWI5) "I would not continue using a debit/ credit card if it's possible" had a number of (m= 4.93; SD= 1.29).

As seen in Table 11, SW1, SWI3, and SWI4 have means higher than 5.00, which are in the capacity of "agree". Then SWI5 has a mean slightly below 5.00, which is still in the capacity of "agree". The standard deviations of SWI1 and SWI4 were less than 1.00, which can be considered as low polarized. This result is contrary with the standard deviations of SWI3 and SWI5, which are more than 1.00 and can be considered as polarized.

## Actual Behavior

(AB1) "I gradually use E-wallet from time to time" had a number of (m= 5.20; SD= 0.92). (AB2) "I use E-wallet more than a debit/ credit card" had a number of (m= 5.01; SD= 0.99). (AB3) "I rarely use a debit/ credit card since I started using E-wallet" had a number of (m= 5.12; SD= 1.12). (AB4) "I do most of my transactions through E-wallet rather than a debit/ credit card" had a number of (m=

*Table 11. Alternative attractiveness data summary*

| | N | Minimum | Maximum | Mean | Std. Deviation |
|---|---|---|---|---|---|
| AA1 | 357 | 2.00 | 6.00 | 5.2801 | .81418 |
| AA3 | 357 | 2.00 | 6.00 | 5.3165 | .82997 |
| AA4 | 357 | 2.00 | 6.00 | 5.1765 | .77511 |
| AA5 | 357 | 2.00 | 6.00 | 5.3137 | .78049 |

*Table 12. Switching intention data summary*

| | N | Minimum | Maximum | Mean | Std. Deviation |
|---|---|---|---|---|---|
| SWI1 | 357 | 1.00 | 6.00 | 5.1905 | .93172 |
| SWI3 | 357 | 1.00 | 6.00 | 5.1541 | 1.01190 |
| SWI4 | 357 | 1.00 | 6.00 | 5.0840 | .92936 |
| SWI5 | 357 | 1.00 | 6.00 | 4.9300 | 1.29308 |

*Table 13. Actual behavior data summary*

| | N | Minimum | Maximum | Mean | Std. Deviation |
|---|---|---|---|---|---|
| AB1 | 357 | 1.00 | 6.00 | 5.2017 | .92329 |
| AB2 | 357 | 1.00 | 6.00 | 5.0112 | .99147 |
| AB3 | 357 | 1.00 | 6.00 | 5.1232 | 1.12751 |
| AB4 | 357 | 1.00 | 6.00 | 5.0952 | 1.02327 |
| AB5 | 357 | 1.00 | 6.00 | 5.2997 | .84622 |

5.09; SD= 1.02). (AB5) "I use E-wallet to help me on a daily basis" had a number of (m= 5.29; SD= 0.84).

As seen in Table 12, AB1, AB2, AB3, AB4, and AB5 have means higher than 5.00, which are in the capacity of "agree". The standard deviations of AB1, AB2, and AB5 were less than 1.00, which can be considered as low polarized. This result is contrary with the standard deviations of AB3 and AB4 which are more than 1.00 and can be considered as polarized.

## Inferential Analysis

### Good Fit Model

The Good Fit Model in SmartPLS 3.0 is measured by using SRMR (Standard Root Mean Residual). SRMR is the difference of the observed correlation and the model in the correlation matrix. It allows the researcher to assess the average magnitude of discrepancies between observed and expected correlations as one of the absolute measures of model fit criteria. Henseler et al. (2014) introduced SRMR as one of the goodness of fit measures for PLS-SEM to avoid model misspecification. With an SRMR value of 0.069, which is a value that is less than 0.08, it is considered a good fit.

### Hypothesis Testing

It was based on the minimum value of the t-table which was 1.96 and a p-value less than 0.05 ($p<0.05$). The following table shows the p-value of each variable.

H1: "Low efficiency influences switching intention" is not supported with a p-value of 0.836. H2: "Switching cost influences switching intention" is supported with a p-value of 0.001. H3: "Personal experience influences switching intention" is not supported with a p-value of 0.511. H4: "Social image influences switching intention" is significant since the score of the p-value is 0.000. H5: "Perceived

*Table 14. P-value*

| | Original Sample (O) | Sample Mean (M) | Standard Error (STERR) | T Statistics (\|O/STERR\|) | P Values |
|---|---|---|---|---|---|
| **ALTERNATIVE ATTRACTIVENESS -> SWITCHING INTENTION** | 0.231 | 0.232 | 0.065 | 3.575 | 0.000 |
| **LOW EFFICIENCY -> SWITCHING INTENTION** | -0.007 | -0.005 | 0.037 | 0.196 | 0.844 |
| **PERSONAL EXPERIENCE -> SWITCHING INTENTION** | -0.029 | -0.028 | 0.048 | 0.604 | 0.546 |
| **SECURITY -> SWITCHING INTENTION** | 0.261 | 0.264 | 0.071 | 3.675 | 0.000 |
| **SOCIAL IMAGE -> SWITCHING INTENTION** | 0.373 | 0.364 | 0.080 | 4.665 | 0.000 |
| **SWITCHING COST -> SWITCHING INTENTION** | 0.156 | 0.158 | 0.049 | 3.160 | 0.002 |
| **SWITCHING INTENTION -> AB** | 0.795 | 0.795 | 0.026 | 31.027 | 0.000 |

security influences switching intention" is supported since the score of the p-value is 0.000. H6: "Alternative attractiveness influences switching intention" is supported since the score of the p-value is 0.000. H7: "Switching intention influences actual behavior" is also supported with a p-value of 0.000.

A. Multiple R-square

The model was evaluated using the $R^2$ (R-square) that aimed to show the proportion of variance in the endogenous variables and is shown as a representative of the explanatory power of the structural model.

Table 15 shows the R-square of switching intention (0.735) and actual behavior (0.631). It describes that 73% of switching intention can be affected by low efficiency, personal experience, switching cost, social image, security, and alternative attractiveness. Thus, 63% of actual behavior can be affected by low efficiency, personal experience, switching cost, social image, security, alternative attractiveness, and switching intention.

*Table 15. R-square of switching intention and actual behavior*

| | Original Sample (O) |
|---|---|
| AB | 0.631 |
| SWITCHING INTENTION | 0.735 |

## Discussion

The first hypothesis which is: "Low efficiency influences switching intention" is rejected with a p-value of 0.836. The minimum score of the p-value is less than 0.05 ($p<0.05$). Both of the criteria do not meet the requirements, which leads to the rejected result. On the other hand, past research conducted by Li (2018) shows that low efficiency has a significant impact on switching intention.

To explore why there is no significant influence from low efficiency towards switching intention, this research conducted a focus group discussion with 6 respondents. Each of the respondents gave their own opinions on the question regarding debit/ credit card efficiency. The results had an unexpected answer. The first respondent stated that "I find that using a debit/ credit card is already efficient enough for this time being". The second respondent agreed and added "A debit/ credit card is easier to use since all of our money is already there, and there's no need to recharge the balance". The third respondent stated that "I think a debit/ credit card is already efficient and worth to pay for". The other three respondents agreed with the statements made by the first there with the following statement: "Yes, we agree and think that overall, a debit/ credit card is already sufficient for the time being". All of the respondents revealed that they perceived that a debit/ credit card is already efficient enough to support their usage until this time. They believe that the easiness of using E-wallet is not significant enough for them to believe that it is more efficient to use E-wallet rather than a debit/ credit card. Then to summarize, the efficiency alone will not lead the users' to switch from a debit/ credit card to E-wallet.

The second hypothesis which is: "Switching cost influences switching intention" is accepted since the score of the p-value is 0.001 to pass the minimum score of the p-value being less than 0.05 ($p<0.05$). Thus, there is a significant impact from switching cost towards switching intention. This aligns with past studies conducted by Ghasrodashti (2017) that show a noticeable influence of switching cost on switching intention. There was also a study conducted by I-Cheng *et al.* (2014) which shows that switching cost has a significant impact on switching intention.

The third hypothesis of "Personal experience influences switching intention" is rejected since the score of the p-value being 0.511 does not meet the minimum score of the p-value being less than 0.05 ($p<0.05$). The result of the study is the complete opposite of a past study by Li (2017) that showed "Personal experience has a significant impact on switching intention". Another study conducted by Hsieh *et al.* (2012) also revealed that there is a positive relation between personal experience as a mooring factor and switching intention.

To explore why there is no significant influence from personal experience towards switching intention, this research conducted a focus group discussion with six respondents. Each of the respondents gave their own opinions on the question regarding their past experiences in using a debit/ credit card. The first respondent stated: "I believe that my past experience using a debit/ credit card will not determine my intention to use E-wallet". It was followed by the second respondent who stated: "I will still use a debit/ credit card even though there's a new alternative". The 3rd and 4th respondents agreed with the statement of respondent 1. The last 2 appeared to have the same opinion that: "Their past experiences lead to a strong loyalty to debit/ credit card usage". It appears that all of the respondents agreed that even though they are using E-wallet, it is not because they have had a bad experience in using a debit/ credit card. They believe there are other factors that led to their decisions to switch, and even though they are using E-wallet, they will still use a debit/ credit card if it is necessary. Overall, personal experience will not affect users' decisions to switch from a debit/ credit card to E-wallet.

The fourth hypothesis of "Social image influences switching intention" is accepted since the score of the p-value is 0.000, which is considered satisfactory since the minimum score of the p-value is less than 0.05 ($p<0.05$). These findings align with a study conducted by Francisco *et al.* (2014) that demonstrated there is a positive relation between social image and switching intention based on the study. It is in contrast to a study by Han and Hyun (2012), which shows that there are no significant influences of social image towards switching intention.

The fifth hypothesis of "Perceived security influences switching intention" is accepted since the score of the p-value is 0.000. Thus, with the minimum score of the p-value less than 0.05 ($p<0.05$) being met, the result can be perceived as being significant. These findings are aligned with a study conducted by Lai (2015), who showed that there is a positive relation between security and switching intention.

The sixth hypothesis of "Alternative attractiveness influences switching intention" is accepted since the score of the p-value is 0.000. Thus, the minimum requirement for the score of the p-value being less than 0.05 ($p<0.05$) is met and the result can be perceived as significant. These findings align past research conducted by Ghasrodashti (2017), who revealed that there is a significant influence on alternative attractiveness towards switching intention. Another study by I-Cheng *et al.* (2014) also showed that there is a significant influence on alternative attractiveness towards switching intention.

"Switching intention influences actual behavior" is also accepted with the p-value being 0.000. Thus, the minimum requirement for the score of the p-value being less than 0.05 ($p<0.05$) is met and the result can be perceived as significant. These findings align with a study conducted by Li (2018), in that switching intention and actual behavior have a positive relationship. Another study conducted by Bansal *et al.* (2005) also proved that there is a positive relation between switching intention and actual behavior.

## SOLUTIONS AND RECOMMENDATIONS

Switching cost has a significant influence on switching intention. It shows that the provider should make sure that it is worth it to leave past services for a new one, even though they might lose some future benefits. Not only that, but the provider also needs to consider the cost to switch from past services. For example, in order to use T-cash, we need to buy the NFC chip first, which leads to a surprising high switching cost in terms of cost, effort, and time.

Social image also has a significant influence on switching intention. In this era, social image is one of the important variables to be taken into account since society has a strong impact on one's behavioral intention, and the provider should consider creating a good image both for society and peer influences. Security also needs to be considered, and the provider should advance and upgrade the line of security to make sure that consumers will continue to use the services and feel safe to use them.

Alternative attractiveness is one of the major variables that act as a pull factor on consumers' switching intention from a debit/ credit card to E-wallet. There are promotions, easy to use aspects, useful aspects, and others included in alternative attractiveness. As being the only pull factors in this research, attractiveness should be considered to be done accordingly to the extent of customers' needs. For example, providers do promotions of clothing merchants for Christmas, grocery promotions for Thanksgiving, etc.

## SUGGESTIONS FOR FUTURE RESEARCH

With this research, there are several findings which can be useful for future research. It shows what factors influence users' intention to switch from a debit/ credit card towards E-wallet.

It also shows that the additional factors, which are security and social image, affect users' decisions to switch in a payment system context.

1. For researchers, the study provides evidence on what elements have been demonstrated to have significant effects on switching intention and actual behavior.
2. Businesses can apply the results of this study to rearrange their strategies to be more efficient and attractive.
3. The government can also get insights into how people react towards E-wallet as a new payment method. Last but not least, the PPM method has been recognized to be significant in terms of measuring the willingness of users' switching behavior and their actual usage of post switching.

## CONCLUSION

The results of this research show that:

1) Low efficiency has no significant influence on switching intention.
2) Switching cost has a significant influence towards switching intention.
3) Personal experience has no significant influence towards switching intention.
4) Social image has a significant influence towards switching intention.
5) Security has a significant influence towards switching intention.
6) Alternative attractiveness has no significant influence towards switching intention.
7) Switching intention has a significant influence towards actual behavior.

Research on E-wallet as an alternative payment method is considered to be still in the early stages (Oliveira, Thomas, Baptista, & Campos, 2016). There are still a lot of elements to be explored. The variables in this framework only explain 73% of switching intention and 63% of actual behavior. This leads to a recommendation that future research needs to add more variables into the framework such as trust and satisfaction, as well as moderating variables like age, job, income level, and others in order to explain more of the users' intention to switch and their actual behavior.

## REFERENCES

Ajzen, I. (1991). The theory of planned behavior. Organizational Behavior and Human Decision Processes, 50(2), 179–211.

Bagozzi, R., & Yi, Y. (1988). On the evaluation of structural equation models. *Academy of Marking Science*, *16*, 76–94.

Bansal, H., Taylor, S., & James, Y. (2005). 'Migrating' to new service providers: Toward a unifying framework of consumers' switching behaviors. *Journal of the Academy of Marketing Science*, *33*(1), 96–115.

Bhattacherjee, A. (2001). Understanding information systems continuance: An expectation-confirmation model. *Management Information Systems Quarterly*, *25*(3), 351–370.

Calvo-Porral, C., & Lévy-Mangin, J.-P. (2015). Switching behavior and customer satisfaction in mobile services: Analyzing virtual and traditional operators. *Computers in Human Behavior*, *49*, 532–540.

Campo, K., & Breugelmans, E. (2015). Buying groceries in brick and click stores: Category allocation decisions and the moderating effect of online buying experience. *Journal of Interactive Marketing*, *31*, 63–78.

Casalo, L., Flavian, C., & Guinaliu, M. (2007). The role of security, privacy, usability and reputation in the development of online banking. Online Information Review, 31(5), 583-603.

Centeno, C. (2004). Adoption of internet services in the acceding and candidate countries,lessons from the internet banking case. Telematics and Informatics, 21(4), 411-417.

Cheng, Y.-M. (2014). Extending the expectation-confirmation model with quality and flow to explore nurses' continued blended e-learning intention. Information Technology & People, 27(3), 230-258.

Chomeya, R. (2010). Quality of Psychology Test Between Likert Scale 5 and 6 Points. *Journal of Social Sciences*, *6*(3), 399–403.

Chong, A. Y., Darmawan, N. O., & Lee, V. H. (2011). Determinants of 3G adoption in Malaysia: A structural analysis. *Journal of Computer Information Systems*, *51*, 71–80.

Chung, J. E., Stoel, L. X., & Ren, J. (2012). Predicting Chinese consumers' purchase intentions for imported soy-based dietary supplements. *British Food Journal, 114*(1), 143–161.

Clark, B. H., Abela, A. V., & Ambler, T. (2005). Organizational motivation, opportunity and ability to measure marketing performance. *Journal of Strategic Marketing, 13*(4), 241–259.

Daniel, D., Swartz, W., & Fermar, A. (2004). *Economics of a cashless society: An analysis of costs and benefits of payment instruments.* AEI-Brookings Joint Centre.

Davis, F. D. (1989). Perceived usefulness, perceived ease of use, and user acceptance of information technology. *Management Information Systems Quarterly, 13*(3), 319–340.

Dick, A., & Basu, K. (1994). Customer loyalty: Toward an integrated conceptual framework. *Journal of the Academy of Marketing Science, 22*, 99–113.

Dinakaran, R. (2016). *Why e-wallet wins over debit or credit cards.* Retrieved from https://www.thehindubusinessline.com/news/variety/why-ewallet-wins-over-debit-or-credit-cards/article9334954.ece

Ellahi, A., & Bokhari, R. H. (2013). Key quality factors affecting users' perception of social networking websites. *Journal of Retailing and Consumer Services, 20*, 120–129.

FactsofIndonesia. (2017, March 4). *Largest Cities in Indonesia – Development and Growth.* Retrieved from FactsofIndonesia.com: https://factsofindonesia.com/largest-cities-in-indonesia

Farah, M. F. (2017). Application of the theory of planned behavior to customer switching intentions in the context of bank consolidations. *International Journal of Bank Marketing, 5*(1), 147–172.

Fenollar, P., & Cuestas, P. J. (2010). Cómo es el innovador? Rasgos de la personalidad como antecedentes. *XXII Congreso Nacional de Marketing.*

Francisco, L.-C., Juan, S.-F., & Francisco, M.-L. (2014). Antecedents of the adoption of the new mobile payment systems. *Computers in Human Behavior*, 464–478.

Ghasrodashti, E. K. (2017). *Explaining brand switching behavior using pull–push–mooring theory and the theory of reasoned action.* Academic Press.

Goffman, E. (1967). *Interaction ritual.* New York: Pantheon.

Goforth, C. (2015). *Research Data Services + Science.* University of Virginia Library.

Han, H. (2012). Image Congruence and Relationship Quality in Predicting Switching Intention. *Journal of Hospitality & Tourism Research (Washington, D.C.), 37*(3), 303–329.

Han, H., Kim, W., & Hyun, S. S. (2011). Switching intention model development: Role of service performances, customer satisfaction, and switching barriers in the hotel industry. *International Journal of Hospitality Management, 30*(3), 619–629.

Henseler, J., & Dijkstra, T. K. S. M., Ringle, C. M., Diamantopoulos, A., & Straub, D. W. (2014). Common Beliefs and Reality about Partial Least Squares: Comments on Rönkkö & Evermann. *Organizational Research Methods, 17*(2), 182–209.

Hinton, P. R., Brownlow, C., McMurray, I., & Cozens, B. (2004). *SPSS explained.* London: Routledge, Taylor and Francis Group.

Höltl, A., & Trommer, S. (2013). Driver Assistance Systems for Transport System Efficiency: Influencing Factors on User Acceptance. *Journal of Intelligent Transportation Systems: Technology, Planning, and Operations, 17*(3), 245–254.

Hou, A. C., Chern, C. C., Chen, H. G., & Chen, Y. C. (2011). Migrating to a new virtual world: Exploring MMORPG switching through human migration theory. *Computers in Human Behavior, 27*(5), 1892–1903.

Hsiao, K.-L. (2017). What drives smartwatch adoption intention? Comparing Apple and non-Apple watches. *Library Hi Tech, 35*(1), 186–206.

Hsieh, J.-K., & Hsieh, Y.-C. C. H.-C., & Feng, Y.-C. (2012). Post-adoption switching behavior for online service substitutes: A perspective of the push–pull–mooring framework. *Computers in Human Behavior, 28*, 1912–1920.

Humphrey, D. (2004). Replacement of Cash by Cards in U.S. Consumer payments. *Journal of Economics and Business, 56*, 211–225.

I-Cheng, C., Chuang-Chun, L., & Kuanchin, C. (2014). The push, pull and mooring effects in virtual migration for social networking sites. *Information Systems Journal, 24*, 323–346.

Jung, J., & Han, H. (2015). Travelers' switching behavior in the airline industry from the perspective of the push-pull-mooring framework. *Journal of Tourism Management, 59*, 139-153.

Kim, B. (2012). The diffusion of mobile data services and applications: exploring the role of habit and its antecedents. Telecommunication Policy, 36(1), 69–81.

Kim, B., & Kang, M. (2016). How user loyalty and nonconscious inertia influence the continued use of mobile communications platforms. *International Journal of Mobile Communications*, *14*(4), 387.

Kim, G. S. B., & Lee, H. (2006). A study of factors that affect user intentions toward email service switching. *Information & Management*, *43*, 884–893.

Kotler, P., & Andreasen, A. (1996). *Strategic marketing for nonprofit organizations* (5th ed.). Englewood Cliffs, NJ: Prentice-Hall.

Lai, J.-Y., & Wang, J. (2015). Switching attitudes of Taiwanese middle-aged and elderly patients toward cloud healthcare services: An exploratory study. *Technological Forecasting and Social Change*, *92*, 155–167.

Lam, S. Y. (2004). Customer value, satisfaction, loyalty, and switching costs: An illustration from a business-to business service context. *Journal of the Academy of Marketing Science*, *32*(3), 293–311.

Lamberte, M. (n.d.). Efficiency and Implications for the Conduct of Monetary Policy. *The Phillipine Payment Systems*.

Lee, E. S. (1966). A theory of immigration. *Demography*, *3*(1), 47–57.

Lee, M. (2009). Factors influencing the adoption of internet banking: An integration of TAM and TPB with perceived risk and perceived benefit. *Electronic Commerce Research and Applications*, *8*(3), 130–141.

Legters, B. (2013). *The evolution of payments*. Retrieved from FIS Payments Leader: https://www.paymentsleader.com/the-evolution-of-payments/

Li, C. Y. (2018). The Power of a Thumbs-Up: Will E-commerce Switch to Social Commerce? *Information & Management*, *55*(3), 340–357.

Liljander, V., & Tore, S. (1995). The Nature of Customer Satisfaction in Service In Advances in Services Marketing and Management (vol. 4). London: JAI.

Lin, C., Wu, S., & Tsai, R. (2005). Integrating perceived playfulness into expectation-confirmation model for web portal context. Information & Management, 42(5), 683-693.

Liu, C.-H., & Technology, N. C.-Y. (2015). The impact on switching intention of E-trading systems for the securities industry. *Global Journal of Business Research*, *9*(1), 75–88.

Nakajima, M. (2012, February 20). The Evolution of Payment Systems. *The European Financial Review, 11.*

National Bank of Serbia. (n.d.). *Payment System.* Retrieved from NBS: https://www.nbs.rs/internet/english/35/index.html

Odior, E., & Banuso, F. (2012). Cashless banking in Nigeria: Challenges, benefits and policy. *European Scientific Journal*, *8*(12), 289–315.

Oei, S. W. (2016). *Pengaruh Kesesuaian Tugas-Teknologi, Kualitas Informasi, Dan Kualitas Sistem Terhadap Kinerja Individu Pengguna Sistem Erp.* Retrieved from http://repository.unhas.ac.id/handle/123456789/21393

Oliveira, T., Thomas, M., Baptista, G., & Campos, F. (2016). Mobile payment: Understanding the determinants of customer adoption and intention to recommend the technology. *Computers in Human Behavior, 61*, 404–414.

Oliver, R. (1997). *Satisfaction: A Behavioral Perspective on the Consumer.* McGrawhill.

Olsina Santos, L. (1999). Website quality evaluation method: A case study on museums, Los Angeles, California, USA. *ICSE99 2nd Workshop on Software Engineering over the Internet.*

Pal, A., Herath, T., De, R., & Rao, H. R. (2018). Factors facilitating Adoption of Mobile Payment. *PACIS 2018 Proceedings, 337.*

Patterson, P., & Smith, T. (2003). A cross-cultural study of switching barriers and propensity to stay with service providers. *Journal of Retailing*, *79*, 107–120.

Ping, R. Jr. (1993). The effects of satisfaction and structural constraints on retailer exiting, voice, loyalty, opportunism, and neglect. *Journal of Retailing*, *69*(Fall), 320–352.

Ravand, H. (2016). Partial Least Squares Structural Equation Modeling with R. *Practical Assessment, Research & Evaluation*, *21*(11), 1531–7714.

Rivest, R. L. (1997). Perspectives on Financial Cryptography. In *International Conference on Financial Cryptography* (pp. 145-149). Berlin: Springer.

Rusbult, C. (1980). Commitment and satisfaction in romantic associations a test of the investment model. *Journal of Experimental Social Psychology*, *16*, 172–186.

Ryu, H.-S. (2018). What makes users willing or hesitant to use Fintech?: the moderating effect of user type. Industrial Management & Data Systems, 118(3), 541-569.

Sari, R. D. (2015). *Pengaruh Penggunaan Celebrity Endorser Terhadap Keputusan Pembelian Konsumen Secra Online Pada Media Sosial Instagram Di Bandar Lampung.* Retrieved from http://digilib.unila.ac.id/7165/1/ABSTRAK.pdf

Sharma, N., & Patterson, P. G. (2000). Switching costs, alternative attractiveness and experience as moderators of relationship commitment in professional, consumer services. *International Journal of Service Industry Management, 11*(5), 470–490.

Shen, J. (2012). Social comparison, social presence, and enjoyment in the acceptance of social shopping websites. *Journal of Electronic Commerce Research, 13*(3), 198–212.

Shen, Q.-L., & Li, Y.-T. (2010). Explore antecedent factors of switching costs and intentions and their impact on customer loyalty. *The 13th Conference on Interdisciplinary and Multifunctional Business Management, 13*, 271-288.

Shih, Y.-Y., & Fang, K. (2006). Effects of network quality attributes on customer adoption intentions of internet banking. Total Quality Management & Business Excellence, 17(1), 61-77.

Song, K. F. (2007). Telepresence and fantasy in online apparel shopping experience. *Journal of Fashion Marketing and Management: An International Journal, 11*(4), 553–570.

Stewart, K. (1998). An exploration of customer exit in retail banking. *International Journal of Bank Marketing, 16*, 6–14.

Sugiyono. (2015). *Statistik Nonparametris untuk penelitian.* Bandung: Alfabeta.

Susanto, A., Chang, Y., & Ha, Y. (2016). Determinants of continuance intention to use the smartphone banking services. Industrial Management & Data Systems, 116(3), 508 - 525.

Taherdoost, H. (2016). *Validity and Reliability of the Research Instrument; How to Test the Validation of a Questionnaire/Survey in a Research.* SSRN Electronic Journal. doi:10.2139srn.3205040

Tan, M., & Teo, T. (2000). Factors influencing the adoption of internet banking. *Journal of the AIS, 1*(5), 1-40.

Umar, A., & Baharun, R. (2017). The mediating relationship of customer satisfaction between brand trust, brand social responsibility image wiith moderating role of switching cost. *Universiti Teknologi Malaysia, 23*, 9020–9025.

Venkatesh, V. (2008). Technology acceptance model 3 and a research agenda on interventions. *Decision Sciences*, *39*(2), 273–315.

Venkatesh, V., Morris, M. G., & Davis, G. B. (2003). User acceptance of information technology: Toward a unified view. *Management Information Systems Quarterly*, *27*(2), 425–478.

Wu, Y.-L., & Tao, Y.-H. L. C.-P., Wang, S.-Y., & Chiu, C.-Y. (2014). User-switching behavior in social network sites: A model perspective with drill-down analyses. *Computers in Human Behavior*, *33*, 92–103.

Yoon, S.-J. (2012). A social network approach to the influences of shopping experiences on e-wom. *Journal of Electronic Commerce Research*, *13*(3), 213–223.

Zhou, T. (2016). Examining user switch between mobile stores: A push-pull-mooring. *perspective. Information Resources Management Journal*, *29*(2), 1–13.

# Chapter 8
# The Phenomenon of Blockchain Technology and the Future of Self-Stabilising and Self-Adaptive Systems of Systems

**Ben van Lier**
https://orcid.org/0000-0001-6377-1767
*Centric, The Netherlands*

## ABSTRACT

*Thinking in terms of blockchain technology has developed rapidly over the past decade, with the focus primarily being on all the new possibilities created by this essentially technology-driven development. At its core, blockchain technology consists of a combination of reliable communication mechanisms between systems and the performance of joint decision-making processes by these systems. Consensus algorithms enable combinations of hardware, algorithms, and software (i.e., cyber-physical systems) to operate in networks and engage in secure intercommunication and interaction in these networks. Thanks to intercommunication, cyber-physical systems are able to reach consensus on decisions on information transactions to perform. This, in turn, empowers interconnected systems to increase their level of autonomy and adapt their individual and joint operations to changes that emerge from within them or from their environment.*

DOI: 10.4018/978-1-7998-1879-3.ch008

## INTRODUCTION

In 2008, Satoshi Nakamoto published a paper entitled 'Bitcoin: A Peer-to-Peer Electronic Cash System.' Nakamoto's paper gave birth to the bitcoin cryptocurrency and the network within which bitcoins are produced and bitcoin transactions are performed. For the purposes of my essay, the manifestation of bitcoin as a cryptocurrency is irrelevant, as I will primarily go into the technology behind bitcoin and explore the historical backgrounds to this technology. In his paper, Nakamoto shares that there were two important reasons that drove him to publish his paper. Firstly, there was the given that "*commerce on the Internet has come to rely almost exclusively on financial institutions serving as trusted third parties to process electronic payments*" (Nakomoto, 2008, p. 1). Nakamoto highlights that online commerce, both goods and services, has become an important factor in society. Commerce is now largely taking place on the Internet, which is a network of interconnected electronic systems that, in turn, are made up of combinations of hardware, algorithms and software. Commerce through a combination of networked electronic systems, i.e. generally online, takes places between electronic systems owned and operated by persons and companies (online stores, marketplaces, platforms, etc.) in the form of data or information transactions. These data and information transactions between parties are processed and goods are transferred after the financial side of the transaction has been completed through a reliable third party, i.e. a financial institution such as a bank. According to Nakamoto, banks' growing role in these transactions is having too much of an influence on the costs involved. He argues that what is needed for the financial settlement of the transaction is "*an electronic payment system based on cryptographic proof instead of trust, allowing any two willing parties to transact directly with each other without the need for a trusted third party*" (Nakomoto, 2008, p. 1). Nakamoto assumes that such an electronic payment system needs to be available to all systems operating online to facilitate any commercial activity on the Internet. In designing his system, Nakamoto had to take into account that combinations of random systems that want to engage in commercial transactions with each other are, in fact, unknown to each other and return to being strangers to each other as soon as a transaction has been completed. On this point of the randomness of nodes, Nakamoto (2008) states the following in his paper: "*Nodes can leave and rejoin the network at will, accepting the proof-of-work chain as proof of what happened while they were gone.*" (p. 8). For the purposes of this essay, what I want to pick up on from Nakamoto's train of thought is his assumption of participating systems accepting what has taken place previously in the network before proceeding to take part in new transactions. Nakamoto (2008) also says about these systems that "*they vote with their CPU power, expressing their acceptance of valid blocks by working on extending them and rejecting invalid blocks by refusing to work on them"* (p. 8). This voting on data and information transactions between electronic systems' processing units was, when Nakamoto wrote his paper, already a known phenomenon used mainly in the development and application of distributed systems. The execution of voting rounds enables interconnected electronic systems to jointly reach a decision on transactions to perform. In the final sentence of his paper, Nakamoto (2008) says that "*any needed rules and incentives can be enforced with this consensus mechanism*" (p. 8), closing his paper with the introduction of a consensus mechanism for decision-making by distributed systems in the processing of transactions. For Nakamoto, there can be no mistake about it that these consensus mechanisms can be used for much more than only the electronic payment system he envisioned. In this essay, I will first touch on the phenomenon of technology from a philosophical perspective, followed by an analysis of how technology is currently driving the rapid development of the (Industrial) Internet of Things and cyber-physical systems. Next, the focus will shift to where these consensus mechanisms originally came from and what possibilities they create in terms of facilitating decision-making processes between interconnected electronic systems. The concept of the consensus mechanism will be analysed here based on the inception and development of the Paxos algorithm, which the US National Institute of Standards and Technology (NIST, 2018) considers the cornerstone of consensus algorithms. And finally, I will address the question whether interconnected cyber-physical systems making decisions by consensus can help shape capabilities such as self-stabilisation and self-adaptation in the functioning of these systems of systems. This essay will close with conclusions and recommendations for further research.

## PHENOMENOLOGY

In the early 20th century, German philosopher Edmund Husserl (2014) wrote the following line: "*An individual object is not merely in general an individual one, a This here! A 'once and only' object.*" (p. 11). Expressing one of the key ideas of phenomenology, he claims that an object can be a catch-all for many different but interconnected manifestations. In Husserl's theory, an object of research always comes in different manifestations, whereby the current manifestation is shaped based on a core of what is actually displayed combined with the way in which the object is perceived, comprehended and interpreted at that moment in time, or, in Husserl's (2014) words, "*by a horizon of (both) 'what is co-given,' but not genuinely and a more or less vague indeterminacy*" (p. 11). As a phenomenon, blockchain technology and the way in which its manifestation is perceived, comprehended and interpreted can be analysed and described by looking at the history of this manifestation and the choices made within in. The choices made by designers such as Nakamoto largely determine the essence of what we now know as blockchain technology. According to Husserl (2014): "*Essence initially designated what is to be found in the being that is proper to an individual itself as its what*" (p. 11). Finding the essence of a phenomenon, as described by Husserl, is mainly about looking for and putting into words what could be the essence of, in this case, the phenomenon of blockchain technology. This essence can, in Husserl's view, only show itself in our perception and our experience. An analysis and description of a phenomenon and its manifestations has, in Husserl's words (2014), its legitimacy in itself. Heidegger defines phenomenon as something that shows itself, something that seems, or something that appears. Heidegger (1999) urged those conducting research into a phenomenon as for instance blockchain technology to always bear in mind the given that "*something can show itself as what it, in itself, is not*" (p. 51). The way you approach a phenomenon can, therefore, according to Heidegger, be considered an exceptional encounter with something that is. Research into and encounters with existing or new phenomena, such as blockchain technology, should, in Heidegger's view, be focused on (learning to) understand the specific object of research, and what is said or written about it can, in turn, be used to demonstrate or explain the manifestation or functioning of the object. How the object functions is, according to Heidegger, not a stand-alone given, but should instead be researched based on the mutual connections and relationships that make the object what it is or determine how it manifests itself to us as an interconnected manifestation. Heidegger adds that it is important to focus such research into an object on the nature of the object's being based on a predefined demarcation of what makes the object a whole of interconnected parts. Phenomenological research into technology-based objects is important, according to Heidegger, because the being of technology will only manifest itself in its presence in reality. Heidegger (1977): "*Technology comes to presence in the realm where revealing and unconcealment take place, where aletheia, truth happens*" (p. 13). Heidegger (1977) believed it is important to explore and describe technology to identify the manifestation of its essence because "*we must take that challenging that sets upon man to order the real as standing-reserve in accordance with the way in which it shows itself. That challenging gathers man into ordering. This gathering concentrates man upon ordering the real as standing-reserve*" (p. 19). In Heidegger's theory, the essence of technology lies in what he calls '*enframing*'. Technology has a tendency, according to Heidegger (1977), to penetrate our existence and, at the same, change our existence as humans through the enframing, leading him to conclude that "*the question as to how we arrive at a relationship to the essence of technology asked in this way, always comes too late*" (p 25). In being too late, we, as humans, are first going to have to adapt to the essence of the manifestation of technology and then look for new possibilities or manifestations of it from there. Technology is, in Heidegger's mind, not something that can by definition be described as human, while it is certainly not purely technological either. For Heidegger (1968), the essence of technology lies primarily "*in what from the beginning and before all else gives food for thought*" (p. 22). The first possible question that we can raise following Nakamoto's observation of online commerce is, considering today's technological developments, what other technological manifestations would be compatible with consensus algorithms?

## CYBER-PHYSICAL SYSTEMS OF SYSTEMS

Physical objects, such as cars, trucks, planes, solar panels, wind mills, MRI scans, drones, etc., but also television sets, washing machines, smart grids or smart factories are increasingly integrated and interconnected in networks. These are some examples of a fast development that are leading the development of a global (Industrial) Internet of Things in which more and more devices are connected in these networks in the form of cyber-physical systems. Interconnecting devices in networks enables such objects to interact and communicate by exchanging and sharing information. Interconnection of devices in networks, such as the Internet, makes it possible for these objects to not only communicate and interact with other objects, but also to communicate and interact with people who are also connected to these networks through their desktop and laptop PCs, smartphones, wearables, mobile medical applications, etc. The physical and stand-alone object is slowly but surely developing into a networked cyber-physical system. In the view of Baheti and Gill (2011), a cyber-physical system refers to a new generation of systems with "*integrated computational and physical capabilities that can interact with humans through many new modalities. The ability to interact with, and expand the capabilities of, the physical world through computation, communication and control is a key enabler for future technology developments*" (p. 1). According to the US National Institute of Standards and Technology (NIST) (2015), these new and interconnected and integrated systems can fill new social needs "*to improve quality of life and enable technological advances in critical areas, such as personalized healthcare, emergency response, traffic flow management, smart manufacturing, defense homeland security, and energy*" (p 1). Poovendran (2010) argues that the development, production and application of cyber-physical systems in our day-to-day work and living environment is bound to change how we view modern physical systems (televisions, planes, cars). Poovendran furthermore claims that the basic principle right from square one in the development of cyber-physical systems has to be "*to include human, infrastructure, and platform in a system-of-systems, creating a uniquely large scope and context in which the system behavior must be predictable and provable*" (p. 1364). Van Lier (2018a, 2018b) claims that based on the possibilities offered by cyber-physical systems, new environments and concepts develop, such as the internet of things, mobile health care, Industrial Internet/advanced manufacturing, etc. The essence of these concepts is subsequently constituted by the fact that people and cyber-physical systems are increasingly interconnected in networks as a natural given and based on equality, and are able to communicate and interact in these networks by exchanging and sharing data and information. Hissam et al. have branded these networks containing both people and cyber-physical systems 'socio-adaptive systems'. In the view of Hissam et al. (2013), socio-adaptive systems are "*systems in which human and computational elements interact as peers. The behavior of the systems arises from the properties of both types of elements and the nature of their collective reaction to changes in their environment, the mission they support, and the availability of resources they use*" (p. vii). The socio-adaptive system containing a wide range of people or groups of people and cyber-physical systems can also be referred to as a system of systems. The Trans-Atlantic Research and Education Agenda (2013) defines system of systems as: "*an integration of a finite number of constituent systems which are independent and operable, and which are networked together for a period of time to achieve a certain higher goal*" (p. 11). According to NIST (2015), a system of systems exists when it has at least three of the following five characteristics: "*operational and managerial independence, geographic distribution, emergent behavior, and evolutionary development*" (p. 17). Cyber-physical systems of systems are, in the definition used by NIST, characterised by interaction with their environment, which generally also includes people. A roadmap compiled by the European cyber-physical systems-of-systems project defines cyber-physical systems of systems (CPSoS) (2015) as follows: "*The concept of Systems of Systems (SoS) has been developed to characterize large, distributed systems that consist of interacting and networked, but partially autonomous, elements and can show emergent behaviour.*" (p. 5) Together, the variety of components that make up CPSoS fulfil a specific need, provide a specific service or produce specific products. Although tasks and functions can be performed separately by components, the overall functioning of the system as a whole depends on the orchestration of the various components. The authors of the roadmap also claim that "*physical size or geographic distribution of the system are not essential factors to make it a system of systems, but rather is its complexity*" (p. 7). It is also this complexity that makes it impossible to run or control from a central perspective cyber-physical systems of systems that are made up of autonomous components such as people or cyber-physical systems as they perform their time-independent and location-independent actions or activities. The European roadmap considers autonomy to be the presence of local conditionalities or priorities that cannot be executed or controlled from the system as a whole. "*Rather, incentives or constraints are given to the subsystem control in order to make it contribute to the global system targets*" (p. 8). In the opinion of the roadmap authors, this kind of autonomy can lead to self-organising systems. The new combinations of hardware, algorithms and software in the manifestation of cyber-physical systems enable these systems to engage in intercommunication and interaction, and to self-organise. The resulting new possibilities also empower these systems to jointly reach consensus on joint decisions concerning the transfer of data and information to other systems within the network, regardless of whether or not these also operate jointly. The possibility of joint decision-making through consensus opens up an entire realm of new possibilities, according to Nakamoto (2008), since "*any needed rules and incentives can be enforced with this consensus mechanism.*" (p. 8) The next section will analyse the development of consensus algorithms based on the development of the Paxos algorithm, which NIST considers the original consensus algorithm.

## Self-Stabilising Systems

Ashby (1954) considered it a given that two or multiple systems can be interconnected. By creating interconnections, different systems acquire an ability to communicate and interact, thus forming a new whole. Creating stability of the whole is what Ashby considers a kind of 'adaptive behaviour', enabled by the interconnections between the various systems. Ashby believed that "*change of stability can only be due to change of value of a parameter, and change of value of a parameter causes a change in stability*" (p. 79). In the early 1970s, Edsger W. Dijkstra (1974) raised the question whether self-stability as a feature of a whole made up of interconnected systems could be interesting on a scale "*from a world-wide network to a common bus control*" (p.1). Dijkstra defines self-stability in systems in this context as follows: "*We call the system self-stabilising if and only if, regardless of the initial state and regardless of the privilege selected each time for the next move, at least one privilege will always be present and the system is guaranteed to find itself in a legitimate state after a finite number of moves*" (p. 2). That same year, Lamport (1974) added the following to Dijkstra's definition: "*Self-stabilizing systems represent ones which are self-correcting: even if they reach an incorrect state through some transient malfunction, they will eventually resume correct operation*" (p. 1). The nuance that Lamport added to the definition opens up the possibility of having networked machines communicate as neighbours as soon as a specific connection has been established between the neighbours. Lamport thus laid the basis for thinking about distributed operating systems that are interconnected in networks, which leads him to state the following: "*a distributed system consists of a collection of distinct processes which are spatially separated, and which communicate with one another by exchanging messages. A network of interconnected computers, such as the ARPA net, is a distributed system*" (Lamport, 1978, p. 558). Reliable exchange of messages makes sure that the functioning of the whole of interconnected systems cannot be disrupted by changes to one or several of the connected systems. Lamport ultimately included the principles of self-stabilisation in the PAXOS algorithm, paving the way for the development of a self-stabilising whole of interconnected systems that, based on collective decision-making and information-sharing, are able to keep functioning, even in the face of changes from outside the system in one or multiple connected systems. Developing and implementing a reliable communication system for the execution of information transactions between separate, autonomous and distributed systems will have to meet a number of general conditions. The system will have to be robust or fault-tolerant, meaning that the system must always be able to keep running and keep enabling information transactions between systems in a reliable manner. Within the Industrial Internet Reference Architecture (2015) fault tolerance is described as follows: "*the ability of the connectivity framework to ensure that redundant connectivity endpoints are properly managed, and appropriate failover mechanisms are in place when an endpoint or a connection is lost.*" (p. 75) When distributed systems as van Lier states (2018c) want to conduct direct transactions with each other, consensus will have to be reached between the systems involved on whether they can jointly accept and process a reliable transaction based on the information received from the environment. It must be possible for every autonomous and distributed system to record every agreed transaction so that the origins of the information transaction can always be traced without the information having to be available in a central location. Finally, there has to be a protocol in place that specifies all conditionalities for consensus on decisions and distributed recording of these decisions. In the following, these four aspects will be worked out based on research by Lamport. First, Lamport assumes that every process is capable of sending these messages directly to other processes and of receiving similar messages directly from other processes. The ability to send and receive messages between different processes requires distributed algorithms that ensure that each process follows similar rules for the sending and receiving of messages, meaning that there is no longer a need for centralised synchronisation or storage of these messages. Lamport (1978) states that "*this approach can be generalized to implement any desired synchronization for such a distributed multiprocess system*" (p. 562). With such a direct way of sending and receiving messages between processes he does, however, specify a further condition in that application of distributed algorithms requires all other processes involved to participate actively. Active participation is possible, Lamport (1978) explains, when all processes "*know all the commands issued by other processes, so that the failure of a single process will make it impossible for any other process to execute State machine commands, thereby halting the system*" (p. 562). Communication processes' interconnections with and dependency on random and distributed systems means that a system of systems must be able to keep functioning without problems in one or multiple separate systems or components of systems leading to the system of systems malfunctioning or not functioning at all. This means, in Lamport's view, that we have to think about fault-tolerant systems. He considers the concept of a disruption of one or multiple processes within a system meaningless without a notion of time. He therefore states that "*we can only tell that a computer system has failed ("crashed") when we have been waiting too long for a response*" (Lamport, 1978b, p. 96). Another condition that has to be met to make fault-tolerant systems possible is that "*each machine must maintain its own copy of the user machine state*" (p. 109). In Lamport's view, communication between systems that function as part of a whole can be considered secure when it is impossible, or at least difficult, to disrupt the required communication between the systems through, for example, unauthorised activity or by spreading information that has not been approved beforehand. For a combination of distributed systems to ultimately be able to jointly form a fault-tolerant system, Pease, Shostak and Lamport (1980) claim that what is needed is an ability to absorb the effects of faulty functioning or non-functioning of distributed systems by using "*voting schemes involving more than one round of information exchange; such schemes might force faulty processors to reveal themselves as faulty or at least to behave consistently enough with respect to the non-faulty processors to allow the latter to reach an exact agreement*" (p. 228). Lamport et al. assumes that distributed systems will have to be able to reach consensus on transactions unaided when distributed algorithms can be developed that can regulate the consistency of these voting schemes. One of the basic conditions for these kinds of algorithms is that they must work on the basis of 3 systems + 1, i.e. a minimum of four systems. In his opinion, the ability to continuously maintain an interactive form of consistency between separate systems is a fundamental precondition for the design and development of distributed systems, where executive control is also distributed. To be able to solve the problem of interactive consistency, they work out a solution to what is known as the Byzantine Generals' Problem (BGP) (1982). According to Lamport (1983), a solution to the BGP implies "*obtaining agreement among a collection of processes some of which may be faulty*" (p. 668). A distributed algorithm that is intended to solve the BGP will have to take as its starting point the signed and therefore reliable messages that systems can send to each other, the contents of which are based on the contents of previously received messages. An algorithm that intends to solve the BGP will be functional when the principle of 3m+1 is combined with the principle that a minimum of 6 messages + 2 messages, i.e. a minimum of 5 messages, are used to reach consensus between parts of the system as a whole on the activities or transactions they are to perform jointly. Once consensus has been reached on the execution of activities or transactions, the components involved in the decision-making will themselves record this decision in their own and therefore distributed ledger. In 1990, Leslie Lamport (1998) submitted a research article to the Association for Computing Machinery (ACM). The article, entitled 'The Part-Time Parliament', sat there for eight years before finally being approved for publication and centres around what is considered one of the more obscure algorithms in distributed computing. In this article, Lamport describes the workings of a parliament in an ancient civilization, using this description as the basis for a decision-making algorithm that is focused on reaching consensus between the part-time members of this parliament, who are not all able to attend parliamentary meetings at the same time to take part in decision-making procedures. Lamport uses the story of this parliament and its members as a metaphor for the consensus random parts of a system have to reach in joint decision-making. He describes in great detail how such a protocol for consensus and decision-making would have to be designed. He captures this protocol in a detailed algorithm, with which consensus, decision-making and recording of transactions to be carried out can be realised between the entities. The key requirements behind this algorithm are, firstly, fundamental trust between the entities involved and, secondly, consistency where "*each Paxon legislator maintained a ledger in which he recorded the numbered sequence of decrees that were passed*" (Lamport, 1998, p. 2). Key conditions for the use of these individual ledgers used by individual systems are described in what is known as the Paxos protocol, including the condition that each decision be recorded using indelible ink so that recorded decisions cannot be changed at a later stage. The Paxos protocol is focused primarily on achieving consistency in recording decisions in the respective distributed ledgers to prevent saving of contradictory information. The Paxos protocol also contains, among other things, rules to ensure that decision-making procedures are initiated and ballots are conducted, rules on quorums for these ballots and how to reach consensus between separate systems on decisions to be made. Furthermore, the protocol provides rules on the manner in which the decision made is to be recorded in the respective ledgers. Once a decision has been recorded by all involved in their own distributed ledger and can no longer be changed, this decision can be considered to be a shared block that appears in all distributed ledgers. This block is the basis for subsequent decision-making procedures and connected decisions on whether or not blocks can be considered to be a chain of decisions, i.e. a blockchain. The Paxos protocol therefore seems to provide nearly all characteristics needed for fault-tolerant and distributed exchanging and sharing of information between separate systems, as it functions on the basis of a mutually agreed joint protocol for consensus and decision-making. In a cyber-physical system of systems, consensus would then enable decision-making on joint activities or transactions by random systems. Decisions made are recorded securely in distributed ledgers. The total of distributed ledgers consequently offers every individual component an up-to-date overview of joint decisions, meaning that these components no longer have to rely on centralised storage of decisions by a trusted third party.

Lamport's groundwork paves the way for a form of self-stabilisation between interconnected systems based on decision-making by consensus. These decisions enable distributed systems to incorporate into their functioning any changes emerging from their own operations or that are introduced from outside the system, i.e. to adapt their individual or joint operations to these changes. This development towards self-adaptive systems of systems will be further elaborated upon in the final part of this essay.

## Self-Adaptive Systems

Autonomy, as van Lier (2018a) states, can be defined as self-government or self-regulating and the functioning of systems increases as they become better able to adapt to changes emerging from within them or from their environment. This capability of autonomous systems is also referred to as self-adaptiveness, which will be detailed in the following. Based on a form of awareness of itself as a whole, a system is able to adapt to changes emerging from within itself or from its surroundings. This ability can be considered the system's adaptive capability. An autonomously operating system's ability to adapt to changes or change its behaviour without human intervention is what we refer to as self-adaptation. The development of this ability to self-adapt to changes arising from a system's environment is an initial and necessary precondition for the autonomous operation of any random system. According to a report published by the United States Department of Defense (DoD) (2012), "*autonomy is a capability (or a set of capabilities) that enables a particular action of a system to be automatic or, within programmed boundaries, self-governing*" (p. 1). This report was published back in 2012, when the Department of Defense still believed that all autonomous systems are in one way or another under the responsibility of human operators. The report does recognise, however, that alongside these human operators, there are algorithms and software that regulate the behaviour of autonomous or semi-autonomous systems. According to Mitchell (2005), the term algorithm generally refers to "*steps by which an input is transformed to an output*" (p. 129). Steiner (2012) defines the core of an algorithm as "*a set of instructions to be carried out perfunctorily to achieve an ideal result. Information goes into a given algorithm, answer comes out*" (p. 54). Steiner argues that the value of an algorithm is determined by the speed at which it can perform the requested tasks. According to him, this speed is determined largely by the hardware on which the algorithms operate. According to the US DoD in 2012, autonomous systems still use algorithms and software in which humans have specified boundaries within which the autonomous system operates autonomously, is able to make decisions, or execute delegated actions. In the above description, the autonomy of a system is more than an intrinsic property of an isolated and unmanned system. The autonomy of a system should, according to the DoD, also be considered an outcome of a process of collaboration between human and system(s), both in the development and in the execution of tasks or actions. The new combination of object (hardware), rules based on which this object operates (algorithms) and the way in which the object performs its tasks (software) is what determines, together with humans, the boundaries of autonomy within which the system is able and allowed to operate independently. This new combination of hardware, algorithms, software and humans controls the execution of tasks and actions by the object or objects. The US DoD added in 2012 that the increasing complexity of interconnected humans, algorithms, software and hardware creates a great variety of challenges, both in the area of interaction between interconnected systems in dynamic environments and in collaboration between human and system. The biggest challenge for increasing the autonomy of the system in its functioning, however, requires a shift of attention from the system hardware to the development and application of algorithms and software that make this increase in autonomy possible. In 2016, the DoD observed that the development of system autonomy had up to then produced a result that ensued from the transfer of competencies from human to system to enable the system to perform actions independently within predefined boundaries. The restriction imposed by these boundaries basically curtails or even eliminates the system's possibilities of operating outside these boundaries, thus also constituting a restriction of the system's autonomy. To be able to operate with a far-reaching level of autonomy, the US DoD (2016) argues that "*a system must have the capability to independently compose and select among different courses of action to accomplish goals based on its knowledge and understanding of the world itself and the situation*" (p. 4). Further development of system autonomy therefore requires such autonomy to be embedded in an increasing number of algorithms and interconnected software entities. To inspire stakeholder confidence in decisions made by systems individually or jointly, the question of how to regulate this decision-making process must be addressed at an early stage in the design process for these procedures. Getting designers and stakeholders together at an early stage to have them come up with possible conditions that this decision-making has to meet will create the possibility to apply "*adequate indicator capabilities so that inevitable context-based variations in operational trustworthiness can be assessed and dealt with at run-time*" (p. 14). According to Scharre et al. (2015), an essential dimension of autonomous systems is thus increasingly created by the level of complexity of the system itself and the environment within which the system has to operate. Scharre: "*Complexity matters because it affects the human operators' ability to predict the behavior of the system.*" (p. 11). The complexity created by interconnections, intercommunication and interaction within the system and between the system and other objects in its environment will reduce the transparency of the system's functioning, making it harder for human stakeholders to fathom the system's operations. The result of such increasing complexity could, according to Scharre, be that: "*predicting the system's behavior, particularly when operating in complex and unstructured real-world environments can be more challenging*" (p. 11). Work on the algorithms and software that are needed has been ongoing for years. Laddaga (1999) argues, among other things, that for a system to be self-adaptive, it needs to be capable of self-evaluation of "*its own behavior and changes in behavior when the evaluation indicates that it is not accomplishing what the software is intended to do, or when better functionality or performance is possible*" (p. 27). In this same context, Laddaga argues that for algorithms and software to be able to give an autonomous system the ability to self-adapt, the software needs to be able to implement any changes on the fly. Laddaga and Robertson (2004) conclude that this basic premise means that "*we design and code an application as a control system. The runtime software is treated like a factory, with inputs and outputs, and a monitoring and control facility that manages the factory*" (p. 1). The algorithms and software that autonomous systems need should, according to Laddaga and Robertson, consist of parts that jointly control and monitor the whole. Self-adaptive software will play a major role in the development of all kinds of embedded software for use in areas such as robotics, manufacturing, aerospace, self-driving cars and sensor systems. Laddaga and Robertson: "*As such, self-adaptive software is an ideal framework for building pervasive computing systems.*" (p. 2) In the view of Shen et al. (2004) control systems consist of at least two components, i.e. a controlled object and a controller. They believe that the controller takes care of implementing changes in the object's behaviour "*by delivering control signals (called control inputs) which force the controlled object to achieve a desired goal (called set point)*" (p. 1). Salehie and Tahvildari (2009) argue that when you take a system of software components that are able to regulate themselves and the behaviour of other systems as the starting point, you need interoperability of information between these parts. They state the following: "*Interoperability is always a concern in distributed complex systems for maintaining data and behavior integrity across all constituent elements and subsystems.*" (p. 50) As result of the fact that, according to Brun et al., algorithms and software have "*become the bricks and mortar of many complex systems (i.e. systems composed of interconnected parts that as a whole exhibits one or more properties (behaviors among the possible properties) not obvious from the properties of the individual parts)*" (p. 16), algorithms and software are now basically de facto essential factors in the development of self-adaptive systems. Brun et al. claim that self-adaptive systems that both operate in a distributed manner and work together differentiate themselves through their self-organising capability. These systems use their self-organising capability to jointly perform activities on a local level while adhering to simple rules. This led Brun et al. to the following argument: "*The global behavior of the system emerges from these local interactions. It is difficult to deduce properties of the global system by analyzing only the local properties of its parts. Such systems do not necessarily use internal representations of global properties or goals; they are often inspired by biological or sociological phenomena.*" (p. 50) Interconnected systems' self-adaptation capabilities increase the autonomy in the functioning of the whole as a cyber-physical system of systems, and such growing autonomy will produce an inevitable shift in responsibility from humans to the whole made up of combinations of hardware, algorithms and software that does so much of the work for us. As responsibility shifts from humans to technology, the question that arises is the one I asked at the start of this essay, i.e. whether new forms of technology automatically force us to think about the meaning and essence of current manifestations of technology.

## CONCLUSION

In the past decade, the paper published by Nakamoto ultimately led to the development of the bitcoin network. Seemingly new possibilities were created based on networked systems made up of new combinations of hardware, algorithms and software that lead to new and, until then, unknown possibilities. The development of this network of distributed systems that are able to jointly make decisions based on consensus is currently very much in the public eye across the globe and has triggered an unprecedented hype in recent years. Still, this hype is primarily about the seemingly revolutionary possibilities that are offered by this new technology, while the origins of the underlying combinations of hardware, algorithms and software on which blockchain technology is built are largely overlooked. Research into the essence of this technology based on the principles of phenomenology shows that the origins of these new combinations of technologies date as far back as the mid-20th century, when such combinations of technologies were used for industrial purposes. Developments in our current time, such as the (Industrial) Internet of Things, where more and more physical systems are interconnected in networks in combination with algorithms and software, seem to offer huge opportunity to harness capabilities such as the reliable communication and joint decision-making that is made possible by blockchain technology. Given that cyber-physical systems are combinations of hardware, algorithms and software that have a capability to communicate and interact in networks, they are able to jointly make consensus-based decisions. Joint decision-making and the interconnected systems' ability to adapt to changes from within or from outside the whole of interconnected systems are enabling the whole and its constituent parts to continuously self-stabilise or self-adapt to these changes. This kind of self-stabilisation or self-adaptation can shape the phenomenon of homeostasis for electronic ecosystems, as described by Van Lier (2018a). The latter point calls for further research, analysis and experimentation with new and current combinations of hardware, algorithms and software that were basically already formulated and established in the previous century.

## REFERENCES


Ashby, R. W. (1954). *Design for a Brain*. New York: John Wiley & Sons Inc. doi:10.5962/bhl.title.6969

Baheti, R., & Gill, H. (2011). Cyber-Physical Systems. In T. Samad & A. M. Anaswamy (Eds.), The Impact of Control Technology. Academic Press.

Brun, Y., di Marzo Serugendo, G., & Gacek, C. (2009). Engineering Self-Adaptive Systems. Through Feedback Loops. In *Self-Adaptive Systems* (pp. 48–70). Heidelberg, Germany: Springer-Verlag Berlin. doi:10.1007/978-3-642-02161-9_3

Defence Science Board. (2012). Task force report: The role of autonomy in dod systems. US Department of Defense, Office of the Under Secretary of Defence for Acquisition, Technology and Logistics.

Defence Science Board. (2016). Summer study on autonomy. US Department of Defense. Office of the Under Secretary of Defence for Acquisition, Technology and Logistics.

Dijkstra, E. W. (1974). Self-Stabilizing Systems in Spite of Distributed Control. *Commun. ACM, 17*(11), 643-644.

Heidegger, M. (1977). *The Question Concerning Technology and Other Essays*. New York: Harper & Row Publishers.

Heidegger, M. (1999). Zijn en Tijd [Being and Time]. Nijmegen: Uitgeverij SUN.

Heidegger, M. (2004). *What is Called Thinking?* (J. G. Gray, Trans.). Harpers Perennial.

Hissam, S., Klein, M., & Moreno, G. A. (2013). *Socio-Adaptive Systems Challenge Problem*. Workshop report. Carnegie Mellon Software Engineering Institute Chicago. CMU/SEI-2013-SR-010.

Husserl, E. (2014). Ideas I. Ideas for a Pure Phenomenology and Phenomenological Philosophy (D. O. Dahlstrom, Trans.). Indianapolis, IN: Hackett Publishing Company.

Industrial Internet Reference Architecture (2015) tech-arch.001, 2015-06-04 Version 1.7

Laddaga, R. (1999). Creating Robust Software Through Self-Adaptation. *IEEE Intelligent Systems & their Applications*, *14*(3), 26–29. doi:10.1109/MIS.1999.769879

Laddaga, R., & Robertson, P. (2004). Self-Adaptive Software: A Position Paper. *ACM Transactions on Autonomous and Adaptive Systems*, *4*(2), 14.

Lamport, L. (1974). *On Self-Stabilizing Systems*. Massachusetts Computer Associates Inc.

Lamport, L. (1978). The Implementation of Reliable Distributed Multiprocess Systems. *Computer Networks*, *2*(2), 95–114. doi:10.1016/0376-5075(78)90045-4

Lamport, L. (1978a). The Implementation of Reliable Distributed Multiprocess Systems. *Computer Networks*, *2*(2), 95–114. doi:10.1016/0376-5075(78)90045-4

Lamport, L. (1978b, July). Time, Clocks and the Ordering of Events in a Distributed System. *Communications of the ACM*, *21*(7), 558–565. doi:10.1145/359545.359563

Lamport, L. (1983, July). The weak byzantine generals problem. *Journal of the Association for Computing Machinery*, *30*(3), 668–767. doi:10.1145/2402.322398

Lamport, L. (1998, May). The Part-Time Parliament. *ACM Transactions on Computer Systems*, *16*(2), 133–169. doi:10.1145/279227.279229

Lamport, L. (2001). Paxos made simple. *ACM Sigact News*, *32*(4), 18–25.

Lamport, L., Shostak, R., & Pease, M. (1982, July). The Byzantine Generals' Problem. *ACM Transactions on Programming Languages and Systems*, *4*(3), 382–401. doi:10.1145/357172.357176

Mitchell, M. (2005). Self-Awareness and Control in Decentralized Systems. In *Spring Symposium on Metacognition in Computation*. Menlo Park, CA: AAAI Press.

Nakamoto, S. (2008). *Bitcoin: A Peer-to-Peer Electronic Cash System*. Academic Press.

NIST. (2015). *Preliminary Discussion Draft Framework for Cyber-Physical Systems*. NIST.

NIST 2015 76. US National Institute of Standards and Technology (2016) *Framework for Cyber-Physical Systems*, Release 1.0 May 2016

Pease, M., Shostak, R., & Lamport, L. (1980). Reaching agreement in the presence of faults. *Journal of the Association for Computing Machinery*, *27*(2), 228-234.

Poovendran, R. (2010). Cyber-Physical Systems: Close Encounters Between Two Parallel Worlds. *Proceedings of the IEEE*, *98*(8), 1363–1366. doi:10.1109/JPROC.2010.2050377

Salehie, M., & Tahvildari, L. (2009). Self-Adaptive Software: Landscape and Research Challenges. *ACM Transactions on Autonomous and Adaptive Systems*, *4*(2), 1–42. doi:10.1145/1516533.1516538

Scharre, P. D. (2015). The Opportunity and Challenge of Autonomous Systems. In Autonomous Systems. Issues for Defence Policymakers. NATO Headquarters Supreme Allied Command. Allied Command Transformation.

Shen, J., Wang, Q., & Mei, H. (2004). Self-Adaptive Software: Cybernetic Perspective and an Application Server Supported Framework. *Proceedings of the 28th Annual International Computer Software and Applications Conference*.

Steiner, C. (2013). *Automate this: how algorithms took over our markets, our jobs, and the world*. New York: Penguin Group.

Towards a European Roadmap on Research and Innovation in Engineering and Management of Cyber-Physical Systems of Systems. (2015). *Cyber-Physical Systems of Systems: Research and Innovation Priorities*. Initial Document for Public Consultation.

*Trans-Atlantic Research and Education Agenda in Systems of Systems (T-AREA-SoS)*. (2013). Loughborough University. Document no. TAREA-PU-WP2-D-LU-9.

van Lier, B. (2018a). *Thinking about ecologies of autonomous cyber-physical systems and their ethics. Inaugural Lecture*. Rotterdam University of Applied Sciences.

van Lier, B. (2018b). Cyber-Physical Systems of Systems and Complexity Science: The Whole is More than the Sum of Individual and Autonomous Cyber-Physical Systems. *Cybernetics and Systems*, *49*(7-8), 7–8, 538–565. doi:10.1080/01969722.2018.1552907

van Lier, B. (2018c). Blockchain Technology: The autonomy and Self-Organisation of Cyber-Physical Systems. In H. Treiblmaier & R. Beck (Eds.), *Business Transformation Through Blockchain*. Cham, Switzerland: Palgrave McMillan.

Yaga, D., Mell, P., Roby, N., & Scarfone, K. (2018). *Blockchain Technology Overview. Draft report NISTIR 8202, January 2018*. National Institute of Standards and Technology.

# Section 3

# Trust, Equity, and Security

# Chapter 9
# Consumer Protection in Online Shopping:
## An Investigation of Virtual Credit Card Awareness in Turkey

**Emre Yildirim**
https://orcid.org/0000-0001-8703-7743
*Sakarya University, Turkey*

## ABSTRACT

*The main purposes of this research are (1) to reveal the virtual credit card (VCC) awareness of online shoppers and (2) to prepare a to do list for managers and relevant institutions to provide a more secured internet shopping process by enhancing the trust perception of consumers. To this end, a structured questionnaire consists of three sections was designed and conducted during the period April-May 2018 in Turkey. The findings achieved show that online shoppers mostly attach importance to the recognition of the website, which means they seek trust in online shopping process. On the other hand, they mostly use credit and debit cards, which may make consumers encounter serious fraud issues. VCC usage is only 4.9% in general although VCC awareness is 55.1%. This low level of VCC awareness is associated with education level and the information provided by the financial banks.*

DOI: 10.4018/978-1-7998-1879-3.ch009

## 1. INTRODUCTION

Today's increasing competitive market forces companies to keep pace with the technology, new trends and innovative changes to capture customer value (Berman, 2012). These attempts and investments are not a preference for companies, but a must when the consumers' power in terms of limitless choice among the variety of brands is considered. The main purpose is to be able to up to date; in other words, companies have to create new ways of doing business. Today, they succeed this by benefiting from the digital transformation. They can boost their sales, productivity and profits as well by serving more and more customers beyond the traditional ways (Matt et al., 2015). In this manner, every company wants to cross the limits via digitalization, but this attempt heavily depends on the Information Technologies (IT) ability and the internet infrastructure.

Considering the internet infrastructure worldwide, Asia is the continent where the most internet users live with the rate of 49% according to the global internet usage statistics of Internet World Stats by 2018 (internetworldstats.com, 2018). On the second position, Europe comes with 16.8% and Africa follows with 11% on the third position. These statistics show that almost half of the internet users live in Asia. However, a different picture appears when the number of internet users is proportioned to the total population of the continents, also called as internet penetration. In this context, North America has the first place with the penetration rate of 95%, Europe comes second with 85.2% and Australia/Oceania is the third with 68.9% according to the new picture. These internet penetration rates indicate that the individuals at North America and Europe use the internet with a ratio over 80%.

Europe statistics by 2018 imply that the internet penetration throughout the Europe is increasing rapidly. According to the report prepared by Ecommerce Foundation, Sweden has the highest penetration rate with 99.7% while Switzerland is at the second position with 98.3% and Iceland is the third with the ratio of 98% (Ecommerce Foundation, 2018). On the other hand, Ukraine is at the bottom of the list with 66% when Turkey has a ratio of 68.4%, which means Turkey's penetration is one of the lowest ratios among European countries. This state of Turkey indicates that the country is supposed to cover a lot of ground to catch up with the level of the developed countries in Europe where the internet penetration ratio is remarkable over 85%.

The other issue closely related to internet penetration is online shopping which has a high importance in today's economic system. In this context, considering the European region statistics again, the most online shoppers live in the Netherlands and Switzerland with the ratio of 82% according to the report prepared by Ecommerce Foundation. Sweden is the second with 81% when Denmark and Luxembourg follow it with 80% (Ecommerce Foundation, 2018). As understood from here, internet penetration and online shopping differ widely across the EU. The highest internet penetration rates do not mean a highest online shopping volume. Then, the question, "why does not high penetration rates guarantee high online shopping volumes?" comes to the mind.

In response to this question, the background of online shopping has to be initially comprehended. Online shopping is a system that creates closer relationships by benefiting virtual infrastructure of the internet based on the exchange of products and money among producers and consumers, reduces the costs, and gives opportunity to consumers to reach the products/services easily and independently of time and space (Chang & Chen, 2008; Lee et al., 2010; Gomez-Herrera et al., 2014). Online shopping has many advantages for both producers and consumers. It makes easier the way of doing business via digitalization; it provides a limitless choice of products/services and comparison of them to get with a better price (Mokhtarian, 2003; Liu & Forsythe, 2010).

However, there is a heavy cost coming with this online system. As Yıldırım (2013) mentioned, the digitalization of the transactions between producers and consumers makes the virtual environment uncertain and risky. The reasons resulting from lack of face-to-face communication, not being able to touch and see the products physically (unobservability), the problems associated with delivery and payment transactions, and theft of personal and/or financial information pose an obstacle to the development of online shopping. In addition to these factors, cultural values may sometimes create barriers as well (Kueh & Voon, 2007). Consumers tend to avoid online shopping especially when they have high levels of uncertainty avoidance scores. They feel uncomfortable and do not prefer to purchase online because of the factors associated with uncertainty such as information asymmetry, and the possibility of third parties' abusive intentions like information theft and financial frauds (Patterson, 2012; Petrova & Valles, 2012; Yıldırım & Barutçu, 2016).

In this manner, the importance of being secured financially stands out among online shoppers in Turkey where the uncertainty avoidance score is 85% (Hofstede Insights, 2019). Although there are some tools such as anti-virus software, security certificates (SET, SSL etc.) and secured http (Diker & Varol, 2013) to prevent the online frauds, none of them is found completely enough. Accordingly, other tools were needed, especially in the payment process, to make the price-sensitive consumers pay and purchase online (Valls et al., 2011). One of the tools developed to provide payment security is called Virtual Credit Card (VCC). This system works as a virtual reflection of the physical credit card without having a real limit. It is not physically available, and has a different card number. The most important feature of it is to allow the owner to make the limit zero when not in use (Oz, 2002; Erdağ & Batuman, 2006; Kaya, 2009). In this regard, VCC is thought to be a promising system in terms of consumer protection in online shopping process when considering the low level of Turkey's online shopping turnover and high level of uncertainty avoidance score.

In the light of all these information, it can be said that today's main problems are (1) how to provide a secured online shopping process to consumers, and (2) how to make them aware of the systems that provide security. Within this direction, this study is designed to (1) reveal the awareness of consumers through VCC and to (2) prepare a to do list for shareholders such as consumers, consumer agencies and other relevant institutions of the government and organizations based on the results to enhance the awareness of VCC. Providing a more secured internet shopping issue is thought to be significantly important in Turkey where the price-sensitive, uncertain-avoidant and conventionally purchasing consumers live.

## 2. BACKGROUND

### 2.1. Current State of Online Shopping in Turkey

As mentioned above, high levels of internet penetration rates do not guarantee the high volumes of online shopping. In the meantime, it is getting worse if the penetration level is low. Turkey is almost in the middle position with the ratio of 53% (Montenegro is the last with 13%), although it is one of the countries that have lowest internet penetration rates. Despite Turkey has shown a great progress when compared to last year's ratio, which is 29% (Ecommerce Foundation, 2017), there is still much to reach the level of the developed countries. The B2C turnover of Turkey, which is €12.5 billion, is also an evidence that supports the current state of online shopping where the UK is the first with 177.7, France with 93.2 and Germany with €93 billion. According to the report named as 'Household Information Technology Usage Report' released by TSI (Turkish Statistical Institute, 2018), the ratio of purchasing goods and services from the internet has increased from 11.8% to 29.3% in the last five years. Although this ratio increased approximately 4.5 point when compared to the previous year, online shoppers have to be encouraged to spend more in online shopping in order to reach a greater turnover.

The reasons why the online shopping turnover of Turkey is low can be explained with some reasons: low ratio of internet access, also called as internet penetration with 68.4%, the insufficient technological infrastructure, the problems caused by the e-retailer or e-seller and online shopping process, and other reasons such as trust issue and lack of experience of usage towards IT. Considering the ratio of purchasing goods and services from the internet, which is 29.3% by 2018 as mentioned before, also indicates the consumers who do not shop online. Stated in other words, the ratio of consumers who do not prefer to shop online can be expressed as 70.7%, which is significantly high. The reasons why consumers do not prefer to shop online in Turkey are explained in the report of TSI (2015) related to this topic. According to this report, the first reason not to shop online, which has a ratio of 81.2%, stems from the traditional shopping habits such as purchasing the products in person, like to see or touch the products and being loyal to the physical store. The latter is 'payment security and privacy concerns' with 44.2% and the third reason is distrust towards the process of receiving or returning goods with the ratio of 27%. As seen from here, excluding the first reason which is associated with traditionalism and regular practicing towards purchasing the products by seeing and touching more, the second most important factor that generates barriers towards developing the online shopping volume by enhancing the risk perceptions of consumers is payment security and privacy concerns. This concern leads consumer to feel financially insecure.

The privacy concerns of consumers in online shopping process generally stems from fraud and financial information theft that frequently occur. The state of unobservability and information asymmetry by the structure of the internet, which is attractive to malevolent third parties, enhances the risk perception of consumers (Tan & Thoen, 2010; Liu & Wei, 2003). The JavaScript codes uploaded to e-commerce sites by these malevolent individuals (Constantin, 2016), fake web sites, security vulnerability, virus attacks etc. are the some kinds of methods used for fraud (Aslanyürek, 2016). Preventing these kinds of fraud methods via websites and providing a more secured online shopping process are the businesses of the managers of e-commerce sites that directly interact with the online shoppers. These kinds of websites try to prevent the information leakage in the virtual environment by using tools such as anti-virus software, security certificates (SET, SSL etc.) and secured http (Diker & Varol, 2013). Besides the software security, some systems that guarantee payment process are also used. For instance, online shoppers have the chance of paying at the door concurrently with the delivery. However, consumers generally prefer to use credit cards due to the advantages such as the ease of use and installment option (Zengin & Güngördü, 2013; Kahraman, 2016) in online shopping. The statistics of 'Online Card Payment Transactions' that were released by The Interbank Card Center (ICC, 2018) in Turkey also support this tendency of online shoppers. According to the statistics of the last three years, domestic and international use of domestic card transactions amounted to 68.8 million Turkish Liras (TL) in 2016, TL 98.2 million in 2017 and TL 95.8 million in the first three quarter of 2018 that indicates a rising graph over years.

The rise of credit card usage in online shopping over years reveals that consumers have to be protected more at the point of online payment. In the contrary case, consumers may be vulnerable and encounter problems such as fraud and financial information theft. These kinds of problems may negatively influence the attitudes of consumers towards the next online payment preference via credit cards. Regarding this issue, the problems that consumers encounter in online shopping were announced by TSI (2018) in the 'Household Information Technology Usage Report'. Considering the results of the report, 21.5% of the consumers who shop online encountered at least one problem in the online shopping process by 2017. The most encountered problem is the delivery of wrong or damaged goods/services with the ratio of 49.1%. After that, the delayed delivery comes with the ratio of 46.5% and technical failure during the ordering or payment process follows it with 20.1%. In addition to these, fraud issue that contains the misuse of credit card information has a ratio of 15%. This ratio shows significant increase when compared to last years, which is 10.6%. This rise displays that online shoppers are now under financial fraud risk so much that they cannot be overlooked.

## 2.2. Online Payment Methods and Virtual Credit Card (VCC)

Consumers cannot give up using credit cards and debit cards in online shopping due to some advantages they offer such as the ease of use, installment options etc., and this increases the volume of the use of cards in online shopping. This intense card usage, which creates a very attractive environment for third parties, caused an increase in the number of card fraud cases based on stealing card information and this situation made the banks involved in the procedure as well. Banks assure the payment procedure via the system called as 3D Secure by sending a confirmation code to consumers' cellphones (Kaya, 2009). Since there appears the law "Debit Cards and Credit Cards Law" no. 5464, which holds the banks responsible for providing system security, and does not make the card holders responsible for the losses that may occur through the theft of card information (Uçar, 2015). In other words, the unjust treatments, which are likely to occur through the theft of card information in online shopping, have to be undertaken by the financial banks. However, not all online sellers use this system enhancing the payment security yet because it is not mandatory.

All these previously mentioned issues show that the security of payment procedure in internet shopping cannot be fully satisfied with the efforts of a single party. In other words, the way to increase security in payment procedure is to use more than one security technique at the same time. In fact, the security of the online payment process on the websites, which have security certificates, cannot be fully assured by the security of the banks with 3D secure. On the other side, the banks do not want to suffer from the weighty responsibility that was imposed by the Law no. 5464. With this purpose, banks serve another secured payment system called as 'Virtual Credit Card (VCC)' to their customers in order to secure themselves and respond to the requests due to the aforementioned advantages.

VCC is connected with the digital information of the physical credit card, and it is not physically available as a form of plastic card (Oz, 2002). In other words, VCC has no chip or magnetic strip. It can only be viewed online and there is no risk of loss or theft because the cardholder does not need to carry on it (Kaya, 2009). VCC corresponds to the same account number as the physical credit card of the customer; however, it has a different card number, expiration date and CVC code (Erdağ & Batuman, 2006). VCC can only be used for the purchases and other transactions made from the internet. The limit of VCC can be reset when not in use and it can be increased just before online shopping (Konuk & Yıldırım, 2012). Thus, it cannot be used in the physical shopping environments such as stores. Through the VCC, it is aimed to decrease and block the online card frauds based on credit card information theft in online shopping (Özgüven, 2011). The ability of consumers to determine the credit card limit as they wish in online shopping and to reset it when not in use minimizes the loss that may occur even if the card information is stolen by third parties. In such a case, it will be impossible to make a purchase with a zero card limit.

Although secure payment can be provided by VCC, it is seen that there are a lot of credit card fraud cases carried out with the stolen card information nowadays. It is thought that this is associated with the lowness of VCC awareness, and low usage as well. In other words, greater awareness in VCC leads to greater usage of VCC in online shopping by more consumers. Otherwise, low level of awareness will cause financial frauds to increase over time. In a study carried out by Konuk and Yıldırım (2012) at Sakarya province of Turkey, it was concluded that 63% of the participants who had the majority of education level as high school with 56.5% did not hear about the VCC before. In other saying, the ratio of the VCC awareness is only 37%. In addition to this, the ratio of VCC usage is only 26.8% inside this group. With reference to this point, it can be assumed that it is not enough to have only awareness about the VCC because according to results of the same research, the participants claimed that the banks did not provide adequate information about the VCC.

As can be seen from the results of the research of Konuk and Yıldırım (2012), low level of VCC awareness may cause low level of VCC usage. From different point of view, the way to increase the level of VCC usage is to raise the awareness of consumers about VCC. At this point, the research called as "Turkey Internet Measurement Research" done by IAB Turkey (2012) has rich content about the internet usage of Turkish consumers and VCC usage as well. According to the results of this research that represents the general internet profile of Turkish consumers who are primary school and high school graduate with a clear majority of 60%, VCC usage ratio is only 12.16% among the other online payment methods. Considering the payment methods, credit card usage leads the way with the ratio of 77.28%. On the other hand, VCC usage comes after the cash on delivery method (19.31%) and money order/electronic funds transfer (EFT) method (16.92%) as the fourth with this ratio. Another research conducted by Çetin and Irmak (2014) lays out that VCC usage is 28.1% among the other payment methods. This result is much higher than the results of IAB Turkey, but still not enough. One of the reasons to reach a higher VCC usage ratio is possibly the education level of the participants who have PhD degree in majority with the ratio of 66.2% in this research.

Consequently, the level of VCC awareness among Turkish consumers is seen low, and this negativity may bring significant losses in the future. The current condition of the level of VCC awareness clearly states that VCC should be used in online shopping more. The usage of VCC by more consumers through providing the high awareness will also decrease the pressure of the Law no. 5464 on financial banks, and prepare a more secured shopping process as well by benefiting the advantages of physical credit cards. Starting from this point, it can be put forward that making the consumers use VCC in online shopping process can be provided by raising the level of VCC awareness.

Considering the literature as a whole, this research is expected to answer the following questions:

1) Which factors do consumers consider in online shopping?
2) Which payment methods do consumers generally prefer and use in online shopping?
3) What are the reasons to prefer the relevant payment method?
4) What are the strengths and weaknesses of online payment methods?
5) Do consumers have any knowledge/awareness about VCC?
6) Which factors are important to be aware about VCC?

## 3. METHODOLOGY

The main purposes of this research are (1) to reveal the Virtual Credit Card (VCC) awareness of online shoppers and (2) to prepare a to do list for managers and relevant institutions to provide a more secured internet shopping process by enhancing the trust perception of consumers. In this general framework, in order to minimize the fraud issues especially during the online payment process and to protect the consumers from serious financial losses in online shopping procedure:

**Purpose 1**: To reveal the awareness of consumers through VCC, and to identify possible factors that reduce and/or block level of VCC awareness. In this context, it is aimed to;

- Determine the payment methods used by consumers in online shopping.

- Reveal the levels of VCC awareness within these payment methods.
- Reach the main reasons by identifying the factors that reduce and/or block level of VCC awareness.
- Develop the possible factors that have the potential to enhance the awareness after identifying them.

**Purpose 2**: To prepare a to do list based on the results of the 'purpose 1' and to inform the shareholders such as consumers, consumer agencies and other relevant institutions of the government and organizations to enhance the awareness of VCC level to provide a more secured internet shopping process. In this context, it is aimed to;

- Provide information what VCC is, how it is used and what factors to be considered when using it.
- Raise the awareness of the public through VCC.
- Minimize the fraud issues that consumers may encounter during online payment process.

To this end, firstly, a couple of interviews with two academics and two experts from the field were performed to make revisions and to have a better understanding about the questions, which were planned to use in the survey. Some of the questions were gathered from the previous study of Konuk and Yıldırım (2012), and the interviewers determined the rest. After that, a structured questionnaire consists of three sections including the questions to measure the (1) online payment systems, (2) virtual card awareness and (3) demographic characteristics of the participants that were designed according to the findings of the interviews. The questionnaires were conducted via face-to-face interviews during April-May 2018 in Turkey and 325 questionnaires were gained via convenience sampling. After the elimination, 314 questionnaires were analyzed. This research employs the non-parametric statistical tests in order to reach the findings of VCC awareness. The reason to benefit from the exploratory research is to reveal the deeper insights about VCC when considering the lack of the VCC studies in the literature.

## 4. FINDINGS

Table 1 shows the demographic characteristics of the participants.

In this context, females are in majority with the ratio of 56.1%, and single participants are with the ratio of 58.9% as well. Considering the education level that is very important for this study, the participants who are high-school graduate are in majority with 48.7% that represents almost the half. Subsequently, graduate participants come with 30.6% and upper secondary comes after it with the ratio of

*Table 1. Demographics*

| | | f | % |
|---|---|---|---|
| Gender | Male | 138 | 43.9 |
| | Female | 176 | 56.1 |
| Marital Status | Married | 129 | 41.1 |
| | Single | 185 | 58.9 |
| Education Level | Primary-school | 16 | 5.1 |
| | High-school | 153 | 48.7 |
| | Upper secondary | 36 | 11.5 |
| | Graduate | 96 | 30.6 |
| | Postgraduate | 13 | 4.1 |
| Occupation | Civil servant | 57 | 18.2 |
| | Private sector | 126 | 40.1 |
| | Self-employment | 20 | 6.4 |
| | Retired | 9 | 2.9 |
| | No work | 102 | 32.5 |
| Personal Income | Under 1500 TL | 75 | 23.9 |
| | 1501-3000 TL | 131 | 41.7 |
| | 3001-4500 TL | 48 | 15.3 |
| | 4501-6000 TL | 13 | 4.1 |
| | 6001 + | 8 | 2.5 |
| | No income | 39 | 12.4 |

11.5%. The ratio of the participants from primary school is significantly low with 5.1%. Regarding the occupation, the participants are mostly from private sector with the ratio of 40.1%, and the participants who have no work come after it with 32.5%. Lastly, the income level shows that the major group is 1501-3000 TL with the ratio of 41.7%.

Table 2 represents the factors that are considered primarily in online shopping process. As can be seen from the table above, the most important factor that online shoppers attach is the recognition of the website with the ratio of 18.1%. Guarantee, return and exchange policies provided by the website come as the second one with 14%, and customer reviews appear as the third one with the ratio of 11.1%. Besides, payment options come as 4$^{th}$ with 10.9. Regarding the table as a whole, it can be stated that online shoppers mostly attach importance to feel secure in online shopping. They prefer to purchase from the well-known websites, they want to see other customers' experiences about the website, and they want to see the security

*Table 2. The factors considered primarily in online shopping*

| | f | % |
|---|---|---|
| Recognition of the website | 226 | 18.1 |
| Guarantee, return and exchange policies provided by the website | 175 | 14.0 |
| Customer reviews on the website | 139 | 11.1 |
| Payment options provided by the website | 136 | 10.9 |
| Variety of products provided by the website | 131 | 10.5 |
| Security tools (SET, SSL protocols etc.) provided by the website | 128 | 10.2 |
| Product delivery time of the website | 122 | 9.8 |
| Ease of use of the website | 110 | 8.8 |
| Cargo company that contracted with the website | 83 | 6.6 |

tools such as SET and SSL protocols not to be exposed to fraud issues. At the same time, they request different payment options; because, one option may not fit the customers' needs. When the website provides only one payment method, customers who are very sensitive to security may not purchase from there.

Figure 1 displays the bank choices of consumers in online shopping. As can be seen from the figure, the most preferred bank is Garanti Bank with the ratio of 23.2%, which represents almost one quarter of the participants. Ziraat Bank comes with 21% right after Garanti. These two banks lead the way among all the banks in this case. Additionally, Ziraat Bank and Garanti Bank are the most valuable brands

*Figure 1. Preferred banks in online shopping*

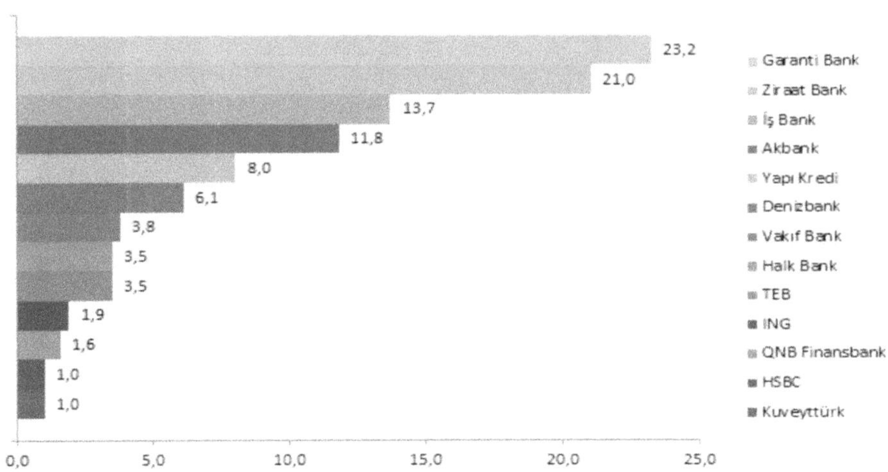

among the banks, respectively (Brand Finance, 2019). On the other hand, HSBC, QNB Finansbank, ING and TEB do not have many consumers and not widely preferred as others have.

Figure 2 shows the payment methods that are generally preferred and recently used by online shoppers. Considering the first graph, credit card is the most preferred payment method with the ratio of 52.4%, which represents half of the whole. Secondly, debit card comes with 23.9% and in the third place; there is cash on delivery method with 13.5%. On the other hand, VCC choice is only 4.9% (only preferred by 22 participants) which shows a significant lowness. Regarding the second graph, it is seen that the levels of all payment methods except credit card are decreasing. This indicates that online shoppers cannot give up using credit cards in online shopping due to the aforementioned advantages although this payment method has several risks. Therefore, the importance of VCC comes into prominence again. However, the usage of VCC is only 3.5% as the last online shopping payment method. This may stem from the low level of VCC awareness as mentioned before.

Right after the Figure 2, Figure 3 represents the reasons why online shoppers prefer the relevant payment method in their last online purchase. In this context, the first reason to use the relevant payment method is the ease of use with the ratio of 39.1% in majority. This finding is mostly associated with the physical credit cards. The second reason is the payment security provided with 22.8% when installment option is the third with 17.9%. The other reasons do not come into prominence with low ratios as these first three do. Moving from here, online shoppers mostly seek easiness, security and installment options in the purchase process.

*Figure 2. Generally preferred and recently used payment methods in online shopping*

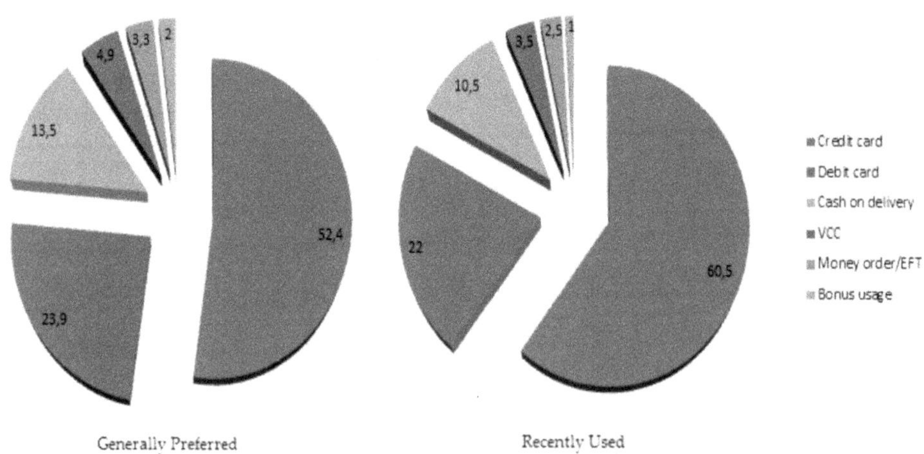

*Figure 3. Reasons to prefer the recently used payment method*

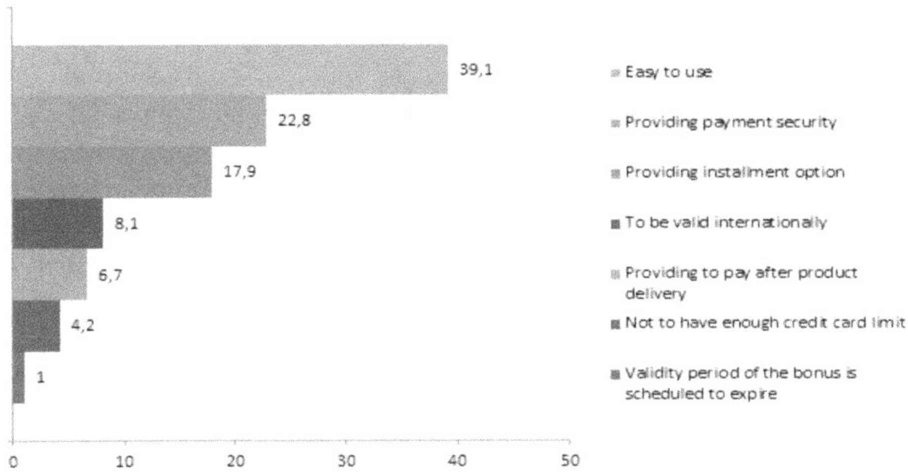


Table 3 was generated in order to reveal the reasons associated with the payment methods to use in online shopping. With reference to this point, the reason 'easy to use' is mostly associated with credit card (67.5%) and debit cards (25%). This is not a surprise, because their usage is so simple by taking them out from the wallet/purse and typing the numbers on them into the online purchase screen. On the other hand, VCC is not found to be easy to use by online shoppers because of the working principle of it. Consumers have to log in their bank accounts, determine the limit and then they can purchase the order. This process prevents consumers

*Table 3. Crosstabs of the recently payment method and the reason to prefer*

| | Credit Card | Debit Card | VCC | Money Order/ EFT | Cash on Delivery | Bonus Usage | Total |
|---|---|---|---|---|---|---|---|
| Easy to use | 135 | 50 | 1 | 3 | 9 | 2 | 200 |
| Providing payment security | 65 | 21 | 10 | 4 | 22 | 1 | 123 |
| Providing installment option | 77 | 12 | 3 | - | 1 | - | 93 |
| To be valid internationally | 27 | 11 | 2 | 1 | 1 | 1 | 43 |
| Providing to pay after delivery | 14 | 4 | - | 2 | 14 | 1 | 35 |
| Not to have enough credit card limit | 13 | 5 | - | 2 | 1 | 1 | 22 |
| Validity period of the bonus | 3 | 2 | - | - | - | - | 5 |
| Total | 334 | 105 | 16 | 12 | 48 | 6 | |

from using VCC. The interesting point here is that 'providing payment security' is again mostly associated with credit cards (52.8%) although it has many fraud risks in online shopping. Consumers may not be aware of the risks they may encounter on the internet. 'Cash on delivery' is the second with 17.8% and debit card using is the third with the ratio of 17% although it has the same risks with credit cards. Considering VCC, the ratio is low with 8.1%, but the ratio is 62.5% when a self-evaluation is made. In other words, consumers who prefer VCC in their last online shopping use VCC because it provides payment security. Moving from here, VCC is also preferred because of the reasons which are providing installment option, to be valid internationally and easy to use, respectively.

Figure 4 shows the level of VCC awareness of online shoppers and sources of this awareness. In this context, consumers are mostly aware of VCC with the ratio of 55.1%. On the other hand, the level of consumers who do not know VCC is not low enough to be ignored with the ratio of 44.9%. Considering the sources, online shoppers mostly hear this payment method from financial banks with 28%, internet & social media with 23.3% and friends with 22.9%, respectively. In addition to these, family members (12.7%) and TV (11.6%) are important sources about VCC awareness as well. From this point forth, it can be stated that VCC has an average awareness level, which cannot be seen as enough for consumer protection in online shopping. Moreover, the share of financial banks, which have the first responsibility in case of fraud cases, is considerably low although it is the first source of information. Additionally, the level of internet & social media has to be increased by benefiting from the power of it to make the consumers be aware of VCC.

*Figure 4. Awareness of VCC and VCC hearing sources*

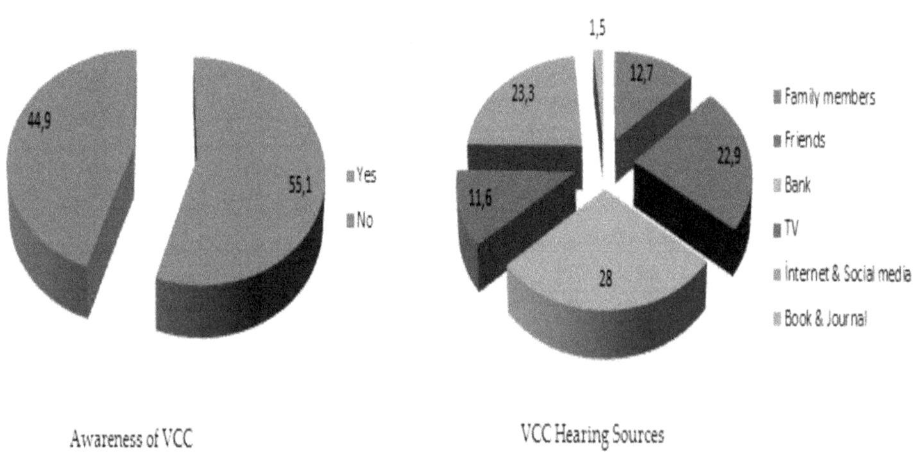

Considering the abovementioned graphs in Figure 5, the participants were asked to write the possible reasons why they have not heard about VCC so far, and whether they knew from which sources they can get information about VCC. In this context, the most encountered reason, why online shoppers have not heard about VCC, is the lack of interest with the ratio of 34.2% in majority. In other words, consumers are not interested in VCC, because they mostly attach importance to the ease of use of the payment method as mentioned in Figure 3. The second important reason is the lack of exploration with 23.3% and the third one is the lack of knowledge with 20.5%, which are also associated with consumers' interest. Besides, financial banks are also responsible for the low awareness, because they do not make promotions (16.4%) to give information about VCC. The second graph supports this finding as well. Regarding the second graph, most of the consumers with 86.3% do not know where they can get information about the VCC. These findings may be seen as the reasons why VCC awareness is not enough among online shoppers.

Figure 6 represents the awareness of online shoppers about VCC provided by their banks and whether adequate information is provided. Moving from here, consumers are mostly not aware of their bank's VCC application with the ratio of 64.3%. On the other hand, consumers who are aware of VCC state that the bank did not provide adequate information, which means consumers learn about the VCC from other sources aforementioned before in Figure 4. Therefore, it can be re-expressed that the banks do not carry out their duties although their role is quite important for consumer protection in online shopping. This can be seen in detail below in Table 4.

*Figure 5. Reasons not to hear about VCC and awareness of information source to be informed*

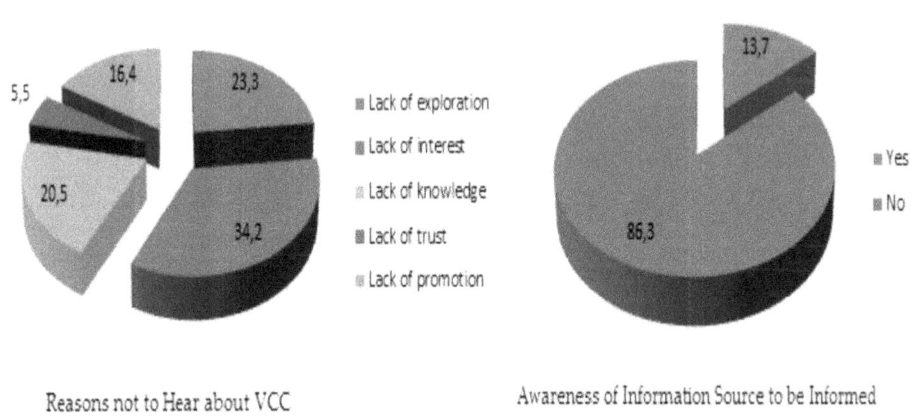

*Figure 6. Awareness of the chosen bank's VCC application and providing adequate information*

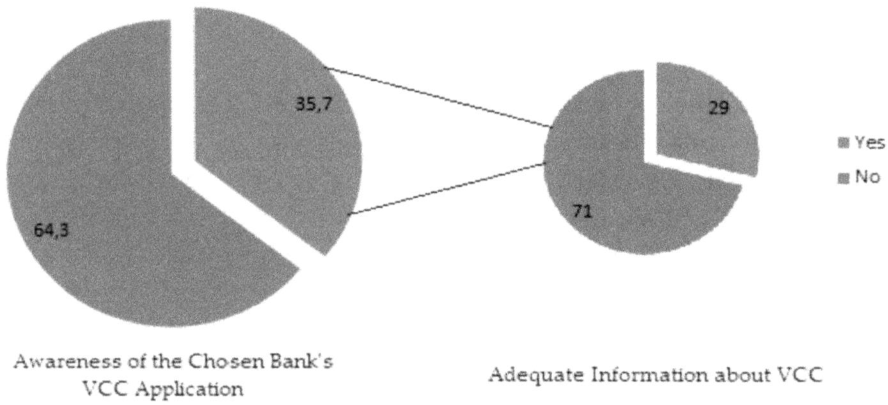


Table 4 shows the crosstabs through awareness of the chosen bank's VCC application and whether adequate information is provided. In this context, online shoppers who are aware of their bank's VCC application also state that provided information about VCC is not enough (76.7%). In parallel, consumers who are not aware of the VCC application utter that no adequate information was provided (97.5%). In this regard, it also means that only 2.5% part of them was equipped with adequate information about VCC by their banks. It seems that there is a relationship between these two variables; therefore, chi-square analysis was conducted to test this relationship. As seen from Table 4, the significant level of chi-square analysis is under 0,001, which means this relationship is significantly supported by the data.

*Table 4. Chi-square analysis through awareness of the chosen bank's VCC application and adequate information provided*

| | | **Adequate Information Provided?** | | | **Chi-square (sig.)** |
|---|---|---|---|---|---|
| | | **Yes** | **No** | **Total** | |
| Awareness of the Chosen Bank's VCC Application | Yes | 86 | 26 | 112 | ,000*** |
| | No | 5 | 197 | 202 | |
| | Total | 91 | 223 | | |

***sig.<0,001

In other words, as the banks provide adequate information for their customers, VCC awareness of the customer raises.

Table 5 indicates the crosstabs through awareness of VCC and education level. Considering the table, most of the online shoppers, who are primary and high school graduates, are not aware of VCC. More precisely, considering these education levels, 62.5% of primary school graduate consumers and 54.2% of high school graduate consumers are away from VCC awareness. On the other hand, a significant breaking point is seen with upper-secondary graduate consumers. 72.2% of these consumers have VCC awareness when graduate ones are at 73.2% and postgraduate ones are at 61.5% ratio. Moving from here, the group with the highest VCC awareness level is the consumers who have graduate level. It seems that there is a relationship between these two variables; therefore, chi-square analysis was conducted to test this relationship. As can be seen from Table 5, the significant level of chi-square analysis is under 0,01, which means this relationship is significantly supported by the data. That is to say, as education level increases, VCC awareness also increases.

## CONCLUSION

Today's consumer purchase preference is mostly shaped around online shopping due to the developments in the Internet and Information Technologies. This virtual environment has its own rules and way of working. It has several advantages, but also disadvantages such as intangibility, fraud issues and so on. These issues make consumers more sensitive in online shopping procedure especially at the point of payment. It enhances the risk perceptions of consumers while decreasing the online purchase volumes for the danger of encountering serious financial problems.

*Table 5. Chi-square analysis through education level and awareness of VCC*

| | | Awareness of VCC | | | Chi-square (sig.) |
|---|---|---|---|---|---|
| | | Yes | No | Total | |
| Education Level | Primary-school | 6 | 10 | 16 | ,003** |
| | High-school | 70 | 83 | 153 | |
| | Upper secondary | 26 | 10 | 36 | |
| | Graduate | 63 | 33 | 86 | |
| | Postgraduate | 8 | 5 | 13 | |
| | Total | 173 | 141 | | |

**sig.<0,01

Therefore, the payment security has to be provided by relevant institutions and organizations.

In this research, it is concluded that online shoppers primarily consider the recognition of the website at the beginning of online shopping process. In other words, they do not prefer the websites that they are not acquainted with because they seek trust. Besides, consumers in online shopping also see the clarity of the policies that the website has and other customers' reviews as important factors. All these factors can be considered as the effort of the consumers to seek trust. Furthermore, they mostly prefer credit and debit cards as online payment methods. They prefer these payment methods due to some advantages, especially the ease of use during the process. Moreover, in the recent online shopping they used these payment methods, particularly the credit cards increasingly. However, these preferences as online payment methods throw online shoppers into the fire. As previously mentioned, they seek trust in online shopping but they do not know how to achieve this via payment methods exactly, because they have a misunderstanding about credit and debit cards provide payment security as well.

In parallel, consumers do not mostly prefer VCC in their online shopping process, although the level of VCC awareness is surprisingly about 55%. More than half of online shoppers are aware of VCC, but fewer of them know its benefits. Accordingly, VCC awareness is not low enough to be ignored with 45%, because the ratio of consumers who use this payment method is only 4.9%. This may stem from the insufficient efforts of financial banks as an information source although they have the first responsibility in case fraud issues occur. Internet & social media and friends have almost the same percentage with banks, which means consumers seek information about VCC outdoor since banks do not do their part.

On the other hand, online shoppers are as guilty as financial banks, because they mostly attach importance to the ease of use of the payment methods. They seek trust but not interested in VCC; they have lack of knowledge and lack of exploration. In other words, it can be stated that online shoppers seek payment security in wrong places. The low level of awareness of information source that consumers can get information also supports this. The significantly high percentage with 86% shows that online shoppers are not aware of the information sources to be informed about the VCC. In addition to this, most consumers also do not know whether there is a VCC application in their chosen banks, because their banks did not provide adequate information about VCC. These results can be considered as the reasons why VCC awareness is not enough among online shoppers. As seen, this is a reciprocal procedure; therefore, consumers and banks have close responsibilities to decrease the level of online payment fraud issues. Moreover, education level is also important, because consumers who have low level of education do not have adequate information about

VCC. Therefore, such problems can be solved via training programs provided by the relevant institutions.

This research also has some limitations. First of all, it has been tried to reveal the awareness of VCC. To this end, the questionnaire was designed in terms of descriptive statistics due to the lack of VCC studies in the literature. There are not any questions associated with the use of VCC, so further researches can be conducted in the light of the results of this research. The other thing is sampling method. Convenience sampling was used to gather data due to the lack of time and effort, and the study was carried out almost in two months. It can be extended over a period of time to enrich the results.

## MANAGERIAL IMPLICATIONS

Considering the weighty responsibility imposed by the Law no. 5464 to financial banks, it can be stated that the only way for banks to wash their hands is to provide payment security for consumers in online shopping. According to the obtained results of the present research, there is a method providing payment security and half of the consumers know it. However, the results show that fewer consumers, who are aware of VCC, use this method. This proves the informing issue. Therefore, top organization of the financial banks called "Banking Regulation and Supervision Agency" should consider the results of this study, and make the banks give broad information about the VCC application. The banks can decrease the level of fraud issues all together and even prevent them. By doing this, they can create a trustworthy online shopping and protect themselves from the weighty responsibility of Law no. 5464.

Another relevant organization called "Information and Communication Technologies Authority" should also consider the results. There is a misunderstanding about VCC and consumers generally do not prefer to use it because of lack of interest, exploration and knowledge. In parallel, consumers mostly attach importance to the ease of use of the payment methods. From this point of view, regarding the relationship between VCC awareness and education level, when the usage of VCC is facilitated via technological investments, consumers may be willing to use it more. In addition to this, the results provide important feedbacks to "Consumer Problems Association". If an increase can be achieved in VCC awareness and usage as well, the complaints stemming from the fraud issues in online shopping process can be decreased. This positive development will reduce the workload of "Consumer Arbitration Committees" which is one of the stakeholders in resolving complaints.

Consequently, the main purpose of both the institutions and the government is to make online shoppers use VCC in their shopping process. Almost there is no exact way to prevent financial frauds in the virtual environment, but it is possible to avoid.

However, this is only possible when consumers know how to protect themselves against malevolent third parties. In this manner, VCC awareness can increase the use of VCC, and VCC usage can enhance consumers' trust perception especially at the point of online payment. Ultimately, all these developments contribute to the online shopping turnover of Turkey.

## REFERENCES

Aslanyürek, M. (2016). İnternet ve sosyal medya kullanıcılarının internet güvenliği ve çevrimiçi gizlilik ile ilgili kanaatleri ve farkındalıkları. *Maltepe Üniversitesi İletişim Fakültesi Dergisi*, *3*(1), 80–106.

Berman, S. J. (2012). Digital transformation: Opportunities to create new business models. *Strategy and Leadership*, *40*(2), 16–24. doi:10.1108/10878571211209314

Brand Finance. (2019). *Banking 500 2019*. Retrieved from https://brandirectory.com/rankings/banking-500-2019

Çetin, H., & Irmak, S. (2014). Elektronik alışverişte akademisyenlerin güvenlik ve risk algılarının belirlenmesi. *Yönetim Bilimleri Dergisi*, *12*(24), 275–294.

Chang, H. H., & Chen, S. W. (2008). The impact of online store environment cues on purchase intention: Trust and perceived risk as a mediator. *Online Information Review*, *32*(6), 818–841. doi:10.1108/14684520810923953

Constantin, L. (2016). *Thousands of online shops compromised for credit card theft*. Retrieved from https://www.computerworld.com/article/3131085/security/thousands-of-online-shops-compromised-for-credit-card-theft.html

Diker, A. & Varol, A. (2013). E-ticaret ve güvenlik. *1th International Syposium on Digital Forensics and Security Proceedings*, 29-33.

Ecommerce Foundation. (2017). *European ecommerce report light version*. Retrieved from https://www.eurocommerce.eu/media/142202/c_european_ecommerce_report_2017_v170623-published_28basic_29.pdf

Ecommerce Foundation. (2018). *European ecommerce report 2018 edition*. Retrieved from https://www.haendlerbund.de/de/downloads/ecommerce-europe/european-ecommerce-report-2018.pdf

Erdağ, N., & Batuman, E. (2006). *Elektronik ticaret el kitabı*. İstanbul: Arıkan Yayınları.

Gomez-Herrera, E., Martens, B., & Turlea, G. (2014). The drivers and impediments for cross-border e-commerce in the EU. *Information Economics and Policy*, *28*, 83–96. doi:10.1016/j.infoecopol.2014.05.002

Hofstede Insights. (2019). *Values of Turkey's cultural dimensions*. Retrieved from https://www.hofstede-insights.com/product/compare-countries/

ICC. (2018). *Online card payment transactions*. Retrieved from: https://bkm.com.tr/internetten-yapilan-kartli-odeme-islemleri/

Internet World Stats. (2018). *World internet user statistics*. Retrieved from http://www.internetworldstats.com/stats.htm

Kaya, F. (2009). *Türkiye'de kredi kartı uygulaması*. İstanbul: Graphis Matbaa.

Konuk, F. A., & Yıldırım, E. (2012). Sanal kart bilinirliliği üzerine bir araştırma: Sakarya ilinde bir uygulama. *11. National Business Congress Proceedings*, 85-87.

Kueh, K., & Voon, B. H. (2007). Culture and service quality expectations: Evidence from generation y cosumers in Malaysia. *Managing Service Quality*, *17*(6), 656–680. doi:10.1108/09604520710834993

Lee, H. H., Kim, J., & Fiore, A. M. (2010). Affective and cognitive online shopping experience: Effects of image interactivity technology and experimenting with appearance. *Clothing & Textiles Research Journal*, *28*(2), 140–154. doi:10.1177/0887302X09341586

Liu, C., & Forsythe, S. (2010). Sustaining online shopping: Moderating role of online shopping motives. *Journal of Internet Commerce*, *9*(2), 83–103. doi:10.1080/15332861.2010.503848

Liu, X., & Wei, K. K. (2003). An empirical study of product differences in consumers' e-commerce adoption behavior. *Electronic Commerce Research and Applications*, *2*(3), 229–239. doi:10.1016/S1567-4223(03)00027-9

Matt, C., Hess, T., & Benlian, A. (2015). Digital transformation strategies. *Business & Information Systems Engineering*, *57*(5), 339–343. doi:10.100712599-015-0401-5

Mokhtarian, P. L. (2003). A conceptual analysis of the transportation impacts of B2C e-commerce. *Transportation*, *31*(3), 257–284. doi:10.1023/B:PORT.0000025428.64128.d3

Oz, E. (2002). *Foundations of e-commerce*. Prentice Hall.

Özgüven, N. (2011). Tüketicilerin online alışverişe karşı tutumları ile demografik özellikleri arasındaki ilişkinin analizi. *KMÜ Sosyal ve Ekonomik Araştırmalar Dergisi, 13*(21), 47–54.

Patterson, A. (2012). Social-networks of the world, unite and take over: A meta-introspective perspective on the Facebook brand. *Journal of Business Research, 65*(4), 527–534. doi:10.1016/j.jbusres.2011.02.032

Petrova, A., & Valles, C. (2012). *Perceived risks as adoption barriers of Facebook commerce: Exploring Facebook users' perception* (Master's Thesis). Jönköping University.

Tan, Y. H., & Thoen, W. (2010). A logical model of trust in e-commerce. *Electronic Markets, 10*(4), 258–263.

TSI. (2015). *Household information technology usage report.* Retrieved from http://www.tuik.gov.tr/PreHaberBultenleri.do?id=18660

TSI. (2018). *Household information technology usage report.* Retrieved from http://www.tuik.gov.tr/PreTablo.do?alt_id=1028

Turkey, I. A. B. (2012). *Türkiye internet ölçümleme araştırması.* Retrieved from http://www.iab-turkiye.org/arastirma-raporlari.php

Uçar, A. (n.d.). *Bankacılık sektöründe bilişim suçları.* Retrieved from https://books.google.com.tr/books?id=Ka-fCQAAQBAJ&printsec=frontcover&hl=tr&source=gbs_ge_summary_r&cad=0#v=onepage&q&f=false

Valls, J. F., Andrade, M. J., & Arribas, R. (2011). Consumer attitudes towards brands in times of great price sensitivity: Four case studies. *Innovative Marketing, 7*(2), 60–70.

Yıldırım, E. (2013). The effects of user comments on e-trust: An application on consumer electronics. *Journal of Economics. Business and Management, 1*(4), 360–364.

Yıldırım, E., & Barutçu, M. T. (2016). How uncertainty avoidance, power distance and indulgence affect social commerce expenditure? An investigation based on Facebook. *International Journal of Science Culture and Sport, 4*(4), 403–421. doi:10.14486/IntJSCS593

## ADDITIONAL READING

Çetin, H., & Irmak, S. (2014). Elektronik alışverişte akademisyenlerin güvenlik ve risk algılarının belirlenmesi. *Yönetim Bilimleri Dergisi*, *12*(24), 275–294.

Constantin, L. (2016). *Thousands of online shops compromised for credit card theft*. Retrieved from https://www.computerworld.com/article/3131085/security/thousands-of-online-shops-compromised-for-credit-card-theft.html

Konuk, F. A., & Yıldırım, E. (2012). Sanal kart bilinirliliği üzerine bir araştırma: Sakarya ilinde bir uygulama. *11. National Business Congress Proceedings*, 85-87.

Özgüven, N. (2011). Tüketicilerin online alışverişe karşı tutumları ile demografik özellikleri arasındaki ilişkinin analizi. *KMÜ Sosyal ve Ekonomik Araştırmalar Dergisi*, *13*(21), 47–54.

## KEY TERMS AND DEFINITIONS

**Bonus:** Extra money which is too little and uploaded to the credit and debit cards through the shopping amount.

**Cash on Delivery:** The system enables to pay after taking the products.

**Installment:** The payment option that enables consumers to pay in parts.

**Online Shopper:** Consumers who purchase from the internet.

**VCC Awareness:** The knowledge level of consumers about virtual credit card.

**Virtual Credit Card:** The virtual version of the physical credit card with different numbers.

# Chapter 10
# A Systematic Review of the Use of Social Media as a Recruitment Tool on the Social Ethical and Privacy Context

**Ghadah Althawwad**
*University of Ottawa, Canada*

## ABSTRACT

*The influence of social media such as Facebook, LinkedIn, and Twitter plays an increasingly influential role in the daily lives of people. Despite the rise of interest in this topic, the research discussing the ethical concerns of using social media for recruitment purposes remains in exploratory stages. This chapter provides a systematic review of recent research that was published from 2012 to 2018 and focused on ethical issues related to the use of social media for recruitment purposes. The techno-ethical lens, which studies the impact of technology on ethics, was used to explore the social and ethical aspects of how recruiters use social media for recruitment purposes.*

## INTRODUCTION

In recent years, social media has been used gradually more and more as a new e-recruitment method. According to the Society for Human Resource Management Report (2015), 81% of screened corporations used social media for recruitment purposes. Moreover, 71% of organizations used social media to find employees with specific qualifications. Importantly, a great number of corporations used social media

DOI: 10.4018/978-1-7998-1879-3.ch010

to exclude some applicants from the job competition. It was estimated that 36% of these corporations excluded job candidates because of the type of information they used in their public social media profiles (Society for Human Resource Management Report, 2015).

Ethical concerns have been raised about privacy and the level of acceptance and relevance for managers to seeking information regarding a candidate's personal life. Ethical issues, such as invasion of privacy and discrimination, are possible risks associated with the rise of social media and, ultimately, the decision by a hiring manager as to whether to invite a person for an interview and/or hire him or her. The question of whether or not candidates should have a right to know if recruiters are utilizing social media sites to screen out applicants has become relevant over the past decade. With the open nature of social media platforms, viewing nearly anyone's online profile, at any time, is the norm of our time (Karlen, 2014; Schneider, Goffin, & Daljeet, 2015; Schroeder & Lile, 2016). To address the current gap in research that addresses the ethical implications related to the use of social media platforms for recruitment purposes, this paper identifies it as critical to address the following research question on the matter: "What is the current state of knowledge concerning the use of social media as a recruitment tool in society from an ethical perspective?"

## BACKGROUND

Nowadays, employers can collect jobseekers' personal and professional information beyond the information traditionally contained in the applicant's resumé or CV, from social media platforms. Utilization of social media in the employment process is relatively new, which means both jobseekers and employers are not fully aware of the potential ethical or legal risks associated with these activities. However, there remains a lack of clear standards and policies to regulate the use of social media for recruitment practices. This study will use a technoethical inquiry approach to explore and evaluate the ethical uses surrounding the reliance on online social media platforms by hiring managers to screen individual jobseekers and society (Luppicini, 2010). The limited knowledge about users' rights is one of the biggest challenges that jobseekers and human resources employees face when they turn to social media.

A technoethical lens helps to synthesize the ethical issues related to the use of social media in recruitment practices. According to Luppicini (2010), techno-ethics can "advance technological relationships of humans with a focus on ethical implications for human life, social norms and values, education, work, politics, law, and ecological impact" (p. 40). There are several academic papers that use technoethical inquiry principles to explore ethical issues related to the use of advanced technologies in society, such as the use of commercial drones and ethical hacking (Abu-Shaqra & Luppicini, 2018; Luppicini & So, 2016).

## METHODOLOGY

The main goal of this paper is to review key literature on the use of social media in the hiring process and assess the current status of how this practice interacts with the principles of ethics in technology. Therefore, the study uses a systematic review process to identify literature that, from both a social and ethical perspective, addresses the utilization of social media for recruitment purposes. This systematic review is grounded in the review approach developed by Gough, Oliver and Thomas (2013). Accordingly, the review, in summary, was conducted in three stages:

1. **Review Question and Search Strategy**: Determine the question and design search strategy.
2. **Description of Study and Quality Assessment**: Establish the inclusion and exclusion criteria for selecting the papers most relevant to the research question.
3. **Synthesis**: Collect the findings from different studies and write research result.

### Search Methods

The review started by searching for the key terms related to the ethical issues surrounding the use of social media as a recruitment tool. Data collection procedures were applied to four electronic e-business databases: ABI/INFORM Global, SCOPUS and Web of Science. The systematic review scope included academic scholarly papers only. The inclusion criteria should meet the following criteria:

- Published papers between the date of 2012-2018.
- **Source Type**: Peer-reviewed books and scholarly journals.
- **Language**: English.

After reading the titles and abstracts of the retrieved papers, the full-text papers that were read were meeting the following criteria: papers dealt with the use of social media for recruitment purposes and surrounding ethical issues All the papers that did not meet the inclusion criteria were excluded. Additionally, exclusion criteria covered the papers that discussed software programs used for recruitment purposes, trade journals, conference papers, notes, and editorials.

## Search Strategy

Targeted search strategies were developed for the selected databases. In order to refine the preliminary results to the most relevant papers, the search was limited to inclusion criteria. The search query that was used in the three databases was: (employer*) AND ("social media" OR "social networking" OR Facebook OR LinkedIn OR twitter) AND (recruit*) AND (ethic* OR privacy)

After removing duplicate papers from the three databases and screening the search results against inclusion and exclusion criteria, a total of 32 papers were retrieved and reviewed. Figure 1 shows the selection process.

## Data Extraction

A template was used to extract data from the papers that met the inclusion and exclusion criteria. The data extracted included title, date, author(s), and the main themes. Table 1 is presented in the Appendix and summarizes the retrieved studies associated with the key themes and their frequencies of occurrence. Graphical percentage of the key themes related to retrieved papers are shown in Figure 2.

*Figure 1. Selection process*

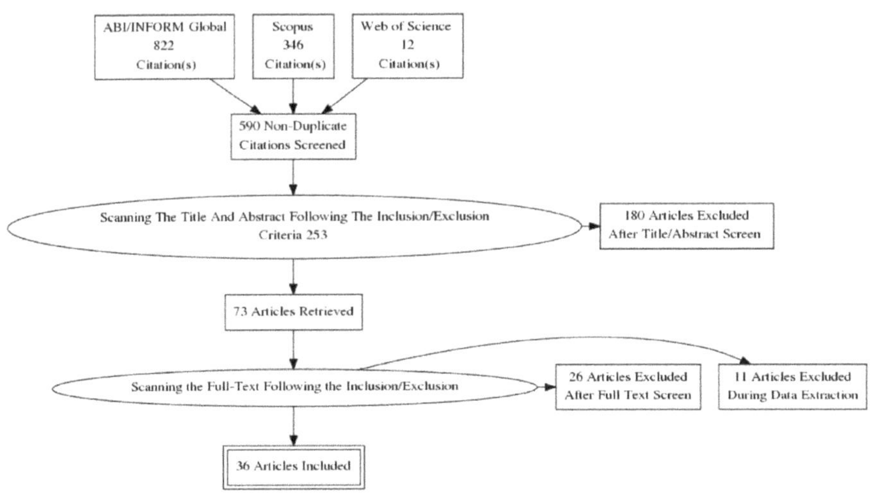

## FINDINGS AND ANALYSIS

General findings revealed different social and ethical issues, including advantages, disadvantages, gender, industry type, privacy, discrimination and legal. Table 2 illustrates the list of different constructs discussed in the retrieved studies, as well as the frequency of occurrence of each theme. Figure 3 provides a graphical representation of different constructs discussed in the retrieved studies, showing the frequency of occurrence of each theme. All of the retrieved papers discussed at least one type of social media platform, both non-professional and professional platforms. Only two papers focused solely on the professional social media "LinkedIn." The rest of the studies discussed the use of both non-professional and professional types of social media from various perspectives, which will be discussed later in this paper.

The dominant theme occurring across the reviewed literature was concerned with the implications of using social media as a job recruitment tool. All of the retrieved papers addressed online recruitment practices from either the point of view of employers or jobseekers. The many advantages associated with using social media as a recruitment tool were the most frequently reported on theme, and was discussed in 75% of the reviewed papers. At the other end of the spectrum, the disadvantages associated with using social media during the hiring process were discussed in 50% of the total papers. The influence of gender was less frequently discussed, only 12.5% of the papers reviewed. Matters that related to privacy and discriminations were mentioned only in 15% of retrieved papers, whereas 18% of the papers dealt with legal problems.

## ANALYSIS OF KEY THEMES

The objective of this literature review was to identify various ethical issues arising from using social media for recruitment purposes. The identified themes have been grouped into four main groups: 1). Social media types; 2). Social media implications; 3). Ethical considerations; 4). Influence of gender and industry type.

## SOCIAL MEDIA TYPES

The most popular type of social media used by jobseekers and employer for recruitment purposes was Linkedin. Facebook and Twitter are considered non-professional platforms, designed to connect people virtually, emphasising social rather than professional connectivity.

Zide, Elman, and Shahani-Denning (2014) defined professional social media as a platform that helps users "present their professional credentials" (p. 24), while non-

*Table 1. Summary of the retrieved studies associated with the key themes*

| Articles | Authors | Year | Social Media Type | | Advantages | Disadvantages | Gender | Industry Type | Privacy | Discriminations | Legal |
|---|---|---|---|---|---|---|---|---|---|---|---|
| | | | Professional | Non professional | | | | | | | |
| 1 | Aguado D., Rico R., Rubio V.J., Fernández L. | 2016 | x | x | x | | | | | | |
| 2 | Allden, N., & Harris, L. | 2013 | x | x | x | x | | | | | |
| 3 | Batagan, L., & Boja, C. | 2015 | x | x | x | x | | | | x | |
| 4 | Berkelaar B.L., Buzzanell P.M. | 2014 | | x | x | x | | | | | |
| 5 | Berkelaar, B. L., Scacco, J. M., & Birdsell, J. L | 2015 | x | x | x | | | | | | |
| 6 | Bohmova, L., & Pavlicek, A | 2015 | x | x | x | | | x | x | | |
| 7 | Bonsón E., Bednárová M. | 2013 | x | | | | | x | x | | |
| 8 | Broughton, A., Foley, B., Ledermaier, S., & Cox, A. | 2013 | x | x | | x | | | | | x |
| 9 | Capiluppi, A., Serebrenik, A., & Singer, L | 2012 | x | x | x | x | | | | | |
| 10 | Drouin, M., O'Connor, K. W., Schmidt, G. B., & Miller, D. A | 2015 | x | x | | x | | | | | x |
| 11 | El Ouirdi M., Pais I., Segers J., El Ouirdi A. | 2016 | x | x | x | x | | | x | | |
| 12 | El Ouirdi, M., Segers, J., El Ouirdi, A., & Pais, I | 2015 | x | x | x | x | x | | | | |
| 13 | Gibbs C., MacDonald F., MacKay K. | 2015 | x | x | x | | | | | | |
| 14 | Harrison J.A., Budworth M.-H. | 2015 | x | x | x | | | | | | |
| 15 | Hoek J., O'Kane P., McCracken M. | 2016 | x | x | | x | x | | | x | |
| 16 | Hurrell S.A., Scholarios D., Richards J. | 2017 | x | x | | x | | | | | |
| 17 | Jeske D., Shultz K.S. | 2016 | x | x | x | x | | | | x | |
| 18 | Joachim, J., Martin, M., Lange, H., Schneiker, A., & Dau, M. | 2018 | | x | x | | | | | | |
| 19 | Kashi K., Zheng C., Molineux J. | 2016 | x | x | x | x | | | | x | x |
| 20 | Micik, M., & Eger, L. | 2015 | x | x | x | | | | | | |
| 21 | Nikolaou I. | 2014 | x | x | x | | | | x | | x |

*continued on following page*

*Table 1. Continued*

| Articles | Authors | Year | Social Media Type | | Advantages | Disadvantages | Gender | Industry Type | Privacy | Discriminations | Legal |
|---|---|---|---|---|---|---|---|---|---|---|---|
| | | | Professional | Non professional | | | | | | | |
| 22 | Priyadarshini C., Kumar Y.L.N., Jha R.R. | 2017 | x | x | x | | x | | | | |
| 23 | Rosoiu, O., & Popescu, C | 2016 | x | x | x | | | | | | |
| 24 | Roulin N. | 2014 | x | x | | x | | | | | |
| 25 | Sameen, S. S. C | 2013 | x | x | x | | | | | | |
| 26 | Schneider T.J., Goffin R.D., Daljeet K.N. | 2015 | x | x | | x | | | | | |
| 27 | Schroeder, A. N., & Lile, C. R | 2016 | x | x | | x | | | | | x |
| 28 | Sivertzen A.-M., Nilsen E.R., Olafsen A.H. | 2013 | x | x | x | | | | | | |
| 29 | Slovensky R., Ross W.H. | 2012 | x | x | x | | | | | | x |
| 30 | Stoughton J.W., Thompson L.F., Meade A.W. | 2015 | x | x | x | | | | x | | |
| 31 | Theurer, C. P., Tumasjan, A., Welpe, I. M., & Lievens, F | 2018 | | x | x | | | | | | |
| 32 | Zide J., Elman B., Shahani-Denning C. | 2014 | x | | x | x | x | x | | x | |
| Frequencies | | | 29 | 30 | 24 | 16 | 4 | 3 | 5 | 5 | 6 |
| percentage | | | 90.00% | 93.70% | 75.00% | 50.00% | 12.50% | 9.30% | 15.00% | 15.00% | 18.00% |

professional social media platforms are used for social activities, such as connecting users to families and friends. The most effective social media sites for recruitment are LinkedIn (73%), and Facebook (14%), whereas Instagram, YouTube, and Twitter, make up less than 1% (SHRM, 2016).

Broughton, Higgins, Hicks, & Cox provided a concise and beneficial summary of the most popular social media platforms. LinkedIn which is a business-related social networking site primarily used for professional networking. Users usually a list of communication details of people whom they have relationship with to use to create a contact network in order to follow various companies and find jobs, individuals and groups or organisations. Facebook is another social networking service that provides its users personal profiles and let them add other users and switch messages. While Instagram delivers more information regarding the personal lifestyle, Facebook lets its users share information about the political, ethnic, and religious beliefs of an individual. In addition, Facebook users may meet common-interest groups such as workplace groups. Instagram is an electronically accessible mobile photo and video sharing and social networking service enabling Instagram users to take pictures and record videos to share them on a different social networking services like Facebook and Twitter. On other hand, Twitter enables its users to send out and read overtly visible messages named Tweets, which can be written up to 280 characters. YouTube, a video sharing website where its users upload, share and view videos involving film and television clips that enable its users to leave notes and exchange comments.

A recent study conducted by Hurrell, Scholarios, and Richards (2017) showed that young students raised concerns and displayed negative attitudes towards the possibility that hiring managers and recruiters might review their personal online profiles for reasons of making decisions on employment matters. However, the study also showed that respondents had a positive perception when the hiring managers only used professional sites, such as LinkedIn. The reason behind these positive perceptions is that professional social media focuses only on the professional aspects of the users (Zide et al., 2014), and should form the primary online basis to influence the decision over whether to hire someone.

## SOCIAL MEDIA IMPLICATIONS

Several studies have explored the important role that social media plays in finding and attracting the right job candidates (Allden & Harris, 2013; Berkelaar & Buzzanell, 2014; Broughton, Foley, Ledermaier, & Cox, 2013; Gibbs, MacDonald, & MacKay,

*Figure 2. Percentage of the key themes related to retrieved papers*

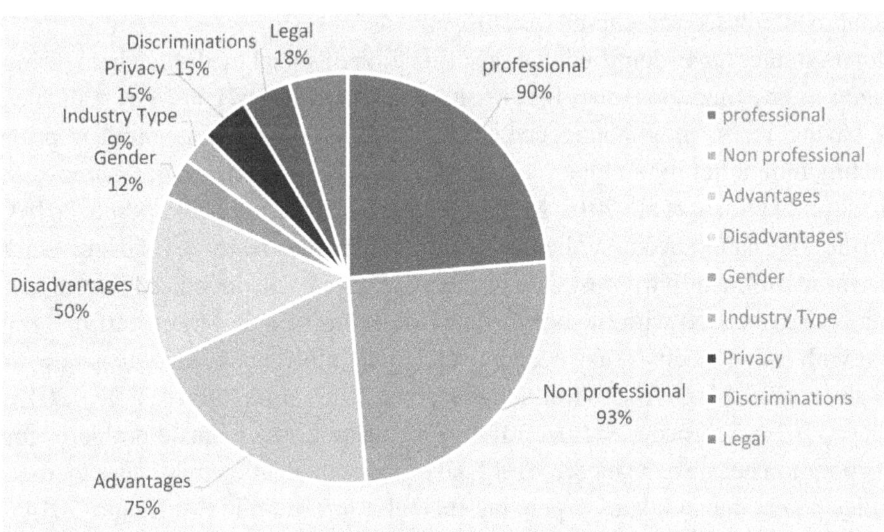

2015; Jeske & Shultz, 2015; Priyadarshini, Kumar, & Jha, 2017; Slovensky & Ross, 2012; Stoughton, Thompson, & Meade, 2015). Furthermore, several studies have described social media as the perfect tool to gain valuable information about applicants (Batagan & Boja, 2015; Bohmova & Pavlicek, 2015; Kashi, Zheng, & Molineux, 2016; Priyadarshini et al., 2017; Sameen, 2013). Social media can also help companies market their values and build a reputation as an organisation that people want to work for. Joachim, Martin, Lange, Schneiker, and Dau (2018), further found that an added advantage was that "social media offered companies an additional means of, not only addressing prospective employees, but also distinguishing themselves from their competitors" (p. 307).

Meanwhile, Hoek, O'Kane, and McCracken (2015) explained that reviewing online profiles through social media could "...offer cues as to what to ask the interviewee" (p. 9). Zide et al. (2014), had further suggested that the information posted on social media could be used as an indicator of a candidate's values and personality, because "the more information a candidate has on his or her profile (i.e. full descriptions of previous work experience), the more likely hiring professionals will be able to gain an understanding of the candidate's personality and behaviour" (p. 12).Another important advantage of using social media is that it allows for building a strong online presence (Theurer, Tumasjan, Welpe, & Lievens, 2018). Through social media, companies are able to expand their sphere of influence, increase their number of followers, and target new market segments (Bonsón & Bednárová, 2013). Similarly, jobseekers can take also benefit from the advantage of building their personal brand online. In their study, El Ouirdi, Pais, Segers, and El Ouirdi (2016) proposed that "individuals who engage in careful impression management and personal branding would be more likely to be assessed positively in social media-based screening" (p. 10).

Others studies have identified that the utilization of social media can help hiring managers to save time and money because it allows them to target the best candidates while having access to an abundance of free information about a candidate prior to meeting him or her for an interview (Berkelaar & Buzzanell, 2014; Bohmova & Pavlicek, 2015; Kashi et al., 2016; Micik & Eger, 2015; Rosouiu & Popescu, 2016).

On the other hand, several studies have examined the disadvantages of using social media for recruitment purposes. One of the main drawbacks identified relates to a hidden cost associated with the need to provide equipment and train staff on how to deal with issues, such as fake accounts and false information, as well as how to avoid any potential legal issues (Allden & Harris, 2013; Broughton & Foley, 2013; Nikolaou, 2014; Schroeder & Lile, 2016). Face accounts may create problems for the corporation because the incorrect or fake profile information may cause them to misjudge potential candidates depending this false information (McKenna, 2014).

For example, while LinkedIn are specifically created for potential employers to view, others social networking sites like Facebook has been designed for a more personal use. Davison, Maraist and Bing (2011) indicate that:

*Employers often assume that the information posted will be more accurate than what candidates provide in a cover letter ... because they do not expect the website to be viewed by employers (i.e., they have no motive to ''fake good'') ... Nonetheless, an individual who creates a webpage may be trying to impress friends and family and the type of distortion or ''faking'' may differ depending on the intended viewer. (p.156)*

In addition, employers might also reach to misinformation which again cannot be verified about a candidate on these social platforms, which creates a question of whether job-related features can be assessed accurately by using these platforms (Davison, Maraist & Bing, 2011).

A further disadvantage associated with recruiting through social media is the fact that many potentially-qualified employees risk being left out of the hiring process because of their poor online presence on social media (Vasilescu, Capiluppi & Serebrenik, 2013; Drouin, O'Connor, Schmidt, & Miller, 2015; El Ouirdi et al., 2015; Harrison & Budworth, 2015; Hoek et al., 2015; Zide et al., 2014). Further troublesome issues occur if an applicant is selected only due to their posts about a particular hobby or interest, which "can sway how attractive they are to recruiters because they may coincide with the recruiters' own hobbies" (Zide et al., 2014, p. 8). Therefore, using social media as the primary basis for decisions about who to hire may not be as effective for finding the most suited candidates, overall.

## ETHICAL CONSIDERATIONS

Nowadays, employers have the opportunity to collect jobseekers' personal and professional information beyond the conventional information entailed in a covering letter, CV, or resumé through the use of social media platforms. The literature review showed that some job applicants might consider that when hiring managers use the information available on social media it constitutes an invasion of their privacy (Priyadarshini et al., 2017). This premise for this sentiment was confirmed in a survey conducted by Stoughton et al. (2015) who found that "pre-employment social networking website screening increases applicants' perceptions of invasion of privacy"(p. 76). Furthermore, Berkelaar and Buzzanell (2014) indicated that employers use social media with the intention "to check on individuals if it believes that there is a specific issue that needs to be addressed" (p. 457). Although some social media platforms give the users an option to hide their information, Schroeder and Lile (2016) argued that the privacy options are often overlooked and may not prove to be effective in protecting all of a person's sensitive information" (p. 684). To avoid the problems associated with privacy invasion, some companies recently informed all job candidates that they would scan their social media profiles or that they would be asking them for their passwords (Kashi et al., 2016; Schneider et al., 2015). In their study on privacy issues in Northern American hotels, Gibbs et al. (2015) noted that by developing "standard guidelines related to the use of social media for areas like raising awareness, attracting applicants, and communicating with candidates, hotels could avoid the legal pitfalls that come with privacy concerns" (p. 181).

The literature review further highlighted the problem of discrimination, which may occur intentionally or unintentionally when employers use social media to inform the recruitment process. Jeske and Shultz (2016) questioned whether hiring managers are able to ignore information related to an applicant's ethnicity/ race, gender, health, disabilities, or sexual orientation during the hiring process. Broughton, Foley, Ledermaier, and Cox, (2013) demonstrated several cases of discrimination in organizations where an employee was dismissed on the grounds of lacking competency. The employee argued that they were dismissed due to certain political or religious beliefs that the employer could only have become aware of through scouring the employee's social media at an early stage of employment. On the other hand, positive discrimination may take place to for the underrepresented group by, for instance, giving first choice to Asian applicants over their white equivalents although this kind of positive discrimination is not legally permissible (Roberts, 2005). This kind of discrimination and other unethical actions can easily occur and hard to avoid because of the ease and availability of social media in the recruitment process whether jobseekers applying using traditional ways such as submit a resume or applying for a job online.

Another type of discrimination occurs in relation to whether a jobseeker is overall lacking or has a poor online presence. This fact risks that the jobseeker might automatically be excluded from being considered in the hiring process because employers prefer applicants they can already gauge knowledge of before they have even met them (Hoek et al., 2015; Zide et al., 2014). Should a candidate's name not appear in a Google search, this is most likely to put them at a disadvantage in the hiring process.

As the use of social media in the hiring process is a relatively new phenomenon, jobseekers and employers are yet to fully realize the potential legal risks associated with these so-called pre-screening activities (Broughton et al., 2013). A recent study by Schroeder and Lile (2016), examined Federal Court cases in the US related to the use of social media by human resources. It was found that in 17 out of 21 termination cases, the organizations involved had terminated employment based on the information that was available on an employee's social media profile during the probationary stage of hire. The authors noted that jobseekers are not aware that organizations use the online information about them. When Drouin et al. (2015) investigated the opinions of US students in relation to legal issues related to using social media for recruiting purposes, they found that one-third of their study participants held the view that "people should be able to post illegal material without the threat of losing their jobs" (p. 127).

Kashi et al. (2016) indicated that the social media policy in recruitment practices merely provide human resources professionals with the guidelines to meet their recruitment needs rather than ensure that jobseekers are being protected by various legal measures.

## INFLUENCE OF GENDER

Several studies have also dealt with the problem of how men and women are perceived differently on social media, which ultimately impacts their chances of gaining hire. Zide et al. (2014) found a significant difference between how both genders use social media. Men are more likely to display their personal and professional interests and exchange recommendations with male peers. Female employers, on the other hand, focused much more on professional content compared to men (El Ouirdi et al., 2016). Surprisingly, females who have professional presence on social media platforms are more likely to increase their employment opportunities compared to their male counterparts (El Ouirdi et al., 2016).

## INDUSTRY TYPE

Another influential factor in companies' use of social media that the literature review identifies relates to the type of industry in which they operate. This finding applies especially to professional social media platforms such as LinkedIn. Bonsón and Bednárová (2013) found that the highest level of activity was detected in the technology, telecommunications, travel, and leisure sectors, as well as in the oil and gas industries. Bohmova and Pavlicek (2015) found that the two types of jobs that were the most frequently contracted through social media were for roles in upper management and IT. In addition, Zide et al. (2014) also held that sales and marketing professionals have the greatest presence on professional social media sites, so they can "connect with as many people as possible, provide multiple methods for people to contact with them, [and] reveal personal information to people they do not know" (p. 13). Two additional studies also discussed the relationship between the size of a company and its use of social media. Hoek et al. (2015) found that using social media for recruitment purpose was more beneficial for small companies. A recent study by Gibbs (2015) found that midsize hotels are the most active users of social media when it comes to attracting jobseekers. However, Slovensky and Ross (2012) highlight the important role that social media can play in helping to scan jobseekers' background criminal activities on social media in certain sectors, such as the military, armed forces, national defence, financial sectors, and work involved vulnerable groups. In these types of sectors, added consideration and special attention of each candidate is required before making any hiring decisions. Recruits and jobseekers in these sectors may be more accepting of having their background screened on social media in addition to the mandatory criminal background checks.

## DISCUSSION

Several of the papers reviewed in this research emphasised the many ethical issues related to the use of social media for recruitment purposes. There is lack of clear standards, practices, and policies that regulate the use of social media for recruitment practices. The limited knowledge about users' rights is one of the biggest challenges that jobseekers and human resources departments face when they turn to social media. As applied in this paper, a technoethical lens helps to synthesize the ethical issues related to the use of social media in recruitment practices. This study will use technoethical inquiry principles adapted from Luppicini (2010) to assess the ethical uses of social media in recruitment practices, which include: theoretical perspectives, socio-cultural perspectives, economic perspectives, levels of influence, possible side effects, and finally, an overall assessment in terms of efficiency and fairness.

### Theoretical Perspectives

Different theories were reviewed in the literature to increase the understanding of how hiring managers use social media for recruitment purposes. Media's richness' theory was used in the literature review and analysis to assess the effectiveness of social media use for recruitment purposes.

The theory was first proposed by Daft and Lengel (1986), who defined media richness as the "ability of information to change understanding within a time interval. Communication transactions that can overcome different frames of reference or clarify ambiguous issues to change understanding in a timely manner are considered rich" (p. 560). The authors suggested that media could be considered rich when they are able to transfer information through the medium over a period of time to decrease both equivocality (i.e. when there is a need for more information) and uncertainty (i.e. when there is lack of clarity and chance for different interpretation). Daft and Lengel (1986) further described four criteria that determined the level of richness in the media: 1). Feedback speed within the media and between the users; 2). The range of multiplied cues, whether it is a written text or a non-verbal behaviour; 3). Personalization of the signal through the media and; 4). Transfer of natural language. Apers and Derous (2017) used media richness theory to test the level of accuracy in video résumés and paper-based résumés to evaluate applicants' personal characteristics and attractiveness. The authors suggested using both types of résumés to improve accurate judgment about applicants' information. The study performed by (Berkelaar et al., 2015) also showed that reputational consistency between online and offline information appeared vital to assessments of whether candidates were being genuine or not.

Despite the need to use both online and offline sources of information documented in the previous studies, social media allows recruiters to acquire rich information about a job candidate that is freely and conveniently available in the pursuit for the best candidates. One example is a feature on LinkedIn profiles, which enables recruiters to find different formats of information, such as written texts, blogs, publications, pictures, and videos. All of these additionally sources of information can help build an overview of the applicants' job- and non-job related activities. The rich diversity of online information is very beneficial to organizations for increasing the chances to create a personal connection with the candidate. Hoek et al. (2015) explained that reviewing different online profiles using social media "could offer clues as to what to ask the interviewee" (p. 47).

Indeed, social media platforms, such as Facebook, allow for video calls, which enable organizations to conduct virtual interviews. Meanwhile, the LinkedIn platform can be a better option for replacing the role of letters of recommendation, because it simplifies the connection and reaches a wider range of professionals that can establish the previous experience with the applicant in light of mutual connections (El Ouirdi et al., 2016; Van Hoye & Lievens, 2005). Therefore, social media platforms constitute the richest and most diverse possible medium that any organization that is hiring can use for recruitment purposes.

'Signalling theory,' as noted in the literature, was used to predict how social media was used for recruitment purposes. Bangerter, Adrian, Roulin, and König (2012) defined signalling theory as "the conditions under which the exchange of accurate information is possible among rational individuals with partly divergent interests." The authors described the process of how true information is transferred between the sender and the receiver, even if there is a conflict of interest between them. Moreover, the process helps the applicants and recruiters to specify the information needed to make decisions. The authors used signalling theory as a framework to explain and predict how candidates and recruiters used personal selection through social media.

There are two types of honest signals: the signal that is hard to deceive, such as intelligence quotient (IQ) tests, and signals that require money or time to acquire, such as an academic degree. Social media provides an opportunity for recruiters to find out which one of the honesty signals an applicant might possess. Applicants can use misleading signals to deceive the recruiters for the purpose of serving their own interests. Bangerter et al. (2012) predicted that jobseekers could manipulate the information posted online on order to attract the desired type of recruiter. As a result, the recruiter, upon realizing that the candidate has not been honest, will distrust the information posted online. This leads to an "evolution of a signaling system" (Bangerter et al., 2012, p. 728). Therefore, adopting the honest signal strategy can help recruiters to find the truthful information about a candidate's qualities.

Additionally, social media is a medium that helps recruiters acquire honest signals. Social media, especially the professional platforms such as LinkedIn, show the professional aspects related to the users' organizations, such as where they worked, the education institutions they attended, and also their mutual contacts, which allows site visitors to verify any information posted on the users' profiles. However, users can manipulate the information they decide to post and send deceptive signals with the intention of influencing other users (i.e. hiring managers, but also competitors in the field). Magnani (2007) emphasized the need to design new rules and regulations in order to avoid the ethical issues related to technology: "We can no longer apply old moral rules and old-fashioned arguments to beings that are simultaneously biological and virtual" (p. xiii). Therefore, software developers will need to create programs that will bridge this gap and prevent candidates from sending deceptive signals in the hopes of building a fake trust with hiring managers, and, instead, ensure that the social media platform is one of reliable and verifiable honest communications from both sides.

## Socio-Cultural Perspectives

Regarding the frequency of social media use, Trepte and Masur (2016) compared the different countries and found that more than 90% of all participants stated that they used social media (mostly Facebook) and more than 75% of them used video platforms like Youtube while around one third of them said that they used reading blogs. The researchers also established that more Chinese and Americans used microblogging platforms like Twitter. In contrast, European users involved more in the usage of Wikipedia. However, it was in this study noted that the frequency of social media use was quite similar in all countries and participants from various countries declared that they spent about one and a half hours on social media networks. In terms of microblogging use, it was fairly different among the countries: about 30 minutes in Germany and 76 minutes in the USA.

Castells (2009) argued that online communities, such as Facebook and LinkedIn, could be considered as real and unreal communities because of they are, "they do not follow the same patterns of communication and interaction as physical communities do. But they are not "unreal"; they work in a different plane of reality. They are interpersonal social networks" (p. 389).

Despite the ethical issues that arise because of using social media for recruitment purposes, it may still be beneficial for both recruiters and jobseekers. All organizations have their own unique cultures that shape their identity, values, goals, and work environments. Social media provides great opportunities for organizations to express their culture. This can help recruiters attract candidates that would be the best fit for the company and match their values and work culture. Broughton et al. (2013) stated that

*A potential candidate can gain quite a lot of information about a potential employer through its website and social media pages. This can help them to decide whether they like the general feel and culture of an organization. (p. 21)*

The authors use an example of the company 'Pets at Home' that uses Facebook to give jobseekers "a flavour of the company's culture and what it would be like to work there" (Broughton et al., 2013, p. 49). According to Slovensky and Ross (2012), collecting any information available on social media is necessary for some sectors and in "certain industries, such as military suppliers, banking, childcare, and private security firms, were hiring the wrong candidate may have serious consequences" (p. 5). Luppicini (2009) also holds that "a systems approach to social and ethical considerations rejects all ethical theories attempting to provide universal principles that apply to all societies all of the time" (p. 23). Therefore, socio-cultural perspectives recommend regulations and rules should be built in a way that is able to suit the different cultures that companies have come to adopt.

## Economic Perspectives

The level of ease and speed at which social media facilitates the collection of information can help job seekers and employers make better decisions. Bonsón and Bednárová (2013) found that "the crucial aspect of all social media strategies should be to increase general awareness of the company. Therefore, if a company wants to increase its visibility, using a particular social network may enable them to attract a higher number of followers" (p. 981). In addition, jobseekers can gain a better insight into the economic benefits to be expected from a job or working with a company by consulting the relevant social media platforms, such as Glassdoor. Harrison and Budworth (2015) stated that there is "a positive association between verbal impression management utterances displayed online and in hiring and salary recommendations. In addition, the use of professional images was also positively associated with these employment-related recommendations for job seekers" (p. 306).

On the other hand, there are some negative consequences with using social media for recruitment purposes. One of the negative outcomes, according to Broughton et al. (2013), is the time, money, and effort that it takes to manage social media sites in order to ensure that:

*enough staff time could be devoted to monitoring sites, and also in providing training for managers who would be operating aspects of the online content. The use of monitoring software also requires investment—in both equipment and qualified staff to operate it. (p. 26)*

Moreover, qualified candidates may miss out on employment opportunities as a result of mis-representations on social media sites. The positive and negative impact of using social media on the economy also needs to be addressed. The economic advantages of using social media for employment purposes are increasingly important to address levels of unemployment. However, the negative aspects should be considered as well. Policymakers, governments, employers, employees, and jobseekers, need to work together to increase the economic efficiency of using social media.

## Stakeholder Perspectives

It is essential for organizations to adopt a multi-stakeholder approach in the recruitment process, because stakeholders play different roles in increasing or decreasing opportunities and challenges for the recruitment and retention of employees in organizations. Therefore, organizations should not focus only on current employees and potential candidates, but on all stakeholders. Baruch and Altman (2016) hold that:

*Career sustainability can be achieved by responding to an ever-shifting environment imaginatively, constructively, and purposely. Employers and employees are no longer two sides of one coin. Rather, the coin is a single-faced playing field and all parties concerned are presented on it. (p. 18)*

Similarly, May (2013) asserted that organizations needed to understand and evaluate the perspectives of all stakeholders, which included "regulators, government agencies, and communities, among others. From this ethical point of view, an organization might emphasize the importance of dialogue, participation, and collaboration, in order to build and strengthen relationships" (p. 19).

Chou, Hammond and Johnson (2013) held that there were three key stakeholder groups that organizations should consider. The first stakeholder group is the employees who manage the online presence of organizations. The second group is the employers who use social media for hiring purposes. The third stakeholder group is those who manage the content and type of social media in terms of information accuracy and credibility. Chou et al. (2013) found that Facebook users tended to post positive information that might not reflect their accurate attitude about the current job. Moreover, Aguado et al. (2016) found that non-professional social media "are perceived by applicants as less useful as a valid assessment" (p. 188).

A fourth group of stakeholders has also been identified and is represented by lawmakers and prosecutors who establish new laws and regulations both, locally and globally. Davison, Bing, Kluemper, and Roth (2016) asserted, "Employers need to be particularly cautious when considering international hires. Other nations have even more restrictive privacy regulations, including the European Union" (p. 21).

Additionally, Zhao, Lampe, and Ellison (2016) proposed that there exists a fifth kind of stakeholder group. These stakeholders include software developers and designers who are keen to improve users' levels of satisfaction in order to attract more users and to compete with other websites.

Since there are many stakeholders involved in the management and use of social media, it is clear that the advantages stemming from the use of social media will not be exclusive to the organizations, but applies to all identified stakeholders (Lam, 2016).

## Levels of Influence and Possible Side Effects

The use of social media in the recruitment process and the overall influence that it creates on the process is not limited to only the relationship between jobseekers and employers. As discussed, it further extends to different societal levels through its economic, ethical, and legal aspects. Luppicini (2010) emphasized the importance of including different levels of influence when applying the technoethical inquiry approach. Although the use of social media as a recruitment tool has expanded globally, the possible ethical and legal side effects have not yet been fully identified and addressed. Luppicini and So (2016) emphasised that perhaps studying the use of social media on all aspects of societal undertakings needs to be done on a case-by-case basis so as to not threaten the wellbeing of the human environment.

Furthermore, Gibbs et al. (2015) found that many hotels in North American are not aware of the legal issues that can occur from using social media. They recommended a focus on "the development of standard guidelines related to the use of social media in areas, such as raising awareness, attracting applicants, and communicating with candidates, [so] hotels can avoid the legal pitfalls that come with privacy concerns (p. 181). A more recent study on legal trends of social media use in the USA, by

Schroeder and Lile (2016), noted, “Although some states provide limited protection for social media users, the legal ramifications for organizational social media use in employment decisions are unclear” (p. 32).

## Overall Assessment in Terms of Efficiency and Fairness

The vast growth in use of social media as a recruitment tool by jobseekers and hiring managers is a positive signal that proves the effectiveness of these sites. Lam (2016) highlighted the effectiveness of using social media for employment purposes by noting that the adoption of the benefits perceived by the interviewees vary from reduction in time-per-hire or cost-per-hire (i.e. efficiency gains) to more strategically oriented gains, such as creating more effective management of candidate relations and improvement in an employer’s brand, image, and reputation.

Recently, organizations have begun to take advantage of the rich and large amount of information available online in order to develop strategies to improve the retention of employees and attract the best fit among applicants (Aguado, Rico, Rubio, & Fernández,, 2016; Bangerter et al., 2012; Berkelaar & Buzzanell, 2014; El Ouirdi et al., 2016; Walker, Feild, Giles, Armenakis, & Bernerth, 2009; Zide et al., 2014). Walker et al. (2009) also highlighted the benefits associated with online recruitment compared to “traditional recruitment’s capacity to communicate a greater amount of information and reach more potential job seekers” (p. 1345).

Social media platforms can also be used to improve an organization’s image. Van Hoye and Lievens (2005) found that the negative reputation of organizations, which occurs from sources such as newspapers, could be mitigated and recovered by using different information sources. They suggested “exposing potential applicants to either recruitment advertising or positive word-of-mouth as a second information source after negative publicity” (p. 184), which can improve the perception of how attractive an organization is to work for.

However, the ethical and legal issues arising from using social media for recruitment are increasing as well. Black, Stone, and Johnson (2015) argued, “most organizations have not developed fair information management policies that may ameliorate applicants’ perceptions of invasion of privacy” (p. 145). Whereas Broughton et al. (2013) further highlighted issues that might affect the use of social media:

*When recruitment involves employers accessing highly sensitive and personal information not publicly available, the issues are no longer time and capital costs. Instead, the employer risks not only violating the privacy of applicants, but also potential lawsuits due to invasion of privacy and the fact that this is not a validated recruitment and hiring practice. (p. 45)*

Nevertheless, complete fairness in using social media as a recruitment tool is difficult to achieve. As Broughton et al. (2013) further explained:

*Equality and diversity practice tends to restrict hiring managers' access to demographic information about applicants, which is considered irrelevant to their ability to do the job and/or could lead to discrimination. Legislation in many countries highlights particular 'protected classes' who could be subject to this kind of discrimination, but other attributes—such as political affiliations or physical appearance—tend to fall outside of legally protected classes, but may still unfairly bias hiring decisions. (p. 31)*

Moreover ensuring the fairness and reducing ethical issues becomes harder to apply globally. For example, privacy invasion is one of the negative ethical implications of using this technology for employment purposes. However, the privacy perception is varied depending on how each culture interprets it. Ess and Hongladarom (2008) made a comparison between the Western and Buddhist perceptions of privacy. He found that privacy in Western culture is looked at from an individual perspective while Buddhists are concerned less about the individual, when it comes to privacy.

May (2013) asserted the importance of including and balancing different dimensions of ethical dilemmas for successful solutions. These dilemmas included:

*Justice versus mercy, individual versus community, cost versus quality, competition versus collaboration, flexibility versus structure, and long-term versus short-term, among others. Balancing these demands and managing among these values require, not only ethical vigilance, but also ethical insight, combining on-going analysis. (p. 26)*

In order to maximize the effectiveness and fairness of using social media in recruitment practices, organizational members, jobseekers, software designers, lawmakers, and policymakers should work together locally and globally to analyse and evaluate the positive and negative aspects of the entire process (Kashi et al., 2016). Walker et al. (2009) found that using a rich online content increases the attractiveness and credibility of organizations as well as improves the employees' understanding of performance feedback.

## LIMITATIONS

There are some threats that can limit the internal and external validity of the research presented in this review. Internal validity risk of bias is high because all retrieved papers were selected and reviewed by a single reviewer. To mitigate the threat, two strategies were used: consulting an e-business expert master librarian to assist with research string and results, and including multiple databases to cover the topic from different aspects. There are several external validity factors that prevent the review findings to be generalized outside the scope of the study. First, all studies were conducted in North America except for two studies, which were conducted in India and the Czech Republic. Thus, it should also be recognised that hiring practices vary from culture to culture, and so a longitudinal study to explore other factors that might influence the extent to which hiring practices use social media platforms, at what point in the hiring process does it comes in, and where it is used, if there is an objective and systematic weighted approach to scoring jobseekers' social media presence in the screening process.

Second, the retrieved papers did not cover topics such as the influence of education and economic level on utilizing social media for recruitment purposes. Therefore, the field is ripe for further research and will become this topic will become increasingly important over time as the rise of the use of social media for recruitment's hiring and vetting of personnel is growing alongside individuals' expanding digital footprints.

## FUTURE RESEARCH

The future of relying on social media platforms for decision-making and gaining an impression of candidates during the hiring process has been evaluated to have both benefits and risks associated with its practice. However, as the world is becoming more and more digitized and global, the trends are hardly going to go away. As limited regulations and policies regarding the practice and, in particular, on the issue of ethical use of social media by hiring managers, immediate research that is crucial to address this issue is required. If the world of professional social media is to turn from its current state of having features of the lawless 'wild-west' to a more coherent and useful tool, the starting point for next steps to follow is to undertake a policy review of Internet laws, rules, and regulations, in order to identify where and how certain countries may have legislated in this area, and identify examples of best practices. In the advent of cyber security concerns, it would seem that such future research is highly relevant to protect the privacy rights that most people enjoy in daily life to avoid having that privacy transposed universally onto the Internet.

## CONCLUSION AND RECOMMENDATIONS

The use of social networking platforms in recruitment and hiring processes has become popular amongst both employers and jobseekers. However, social media as an employment tool has both merits and demerits. Although social media platforms offer additional information about jobseekers to employers, using these platforms raises ethical questions, because the information contained on these platforms could invade a candidate's privacy and lead to discrimination, even if some sectors need to use social media to thoroughly investigate jobseekers' backgrounds relating to potential involvement in criminal activities. Moreover, employers should use social media for recruitment with caution, because they could be deceived by unqualified candidates, who create social media profiles that do not reflect their real skills and abilities. It is very important to set clear policies and regulations to protect social media users and society as a whole from any negative ethical implications that may occur from using this technology in recruitment practices.

With the increasing focus on where to draw the boundaries between societal life and digital societal life, cybercrime, security, and bullying appears to have made its move away from the streets and onto the internet. While this may be difficult to grasp for older generations, the implications that social media presence has on the younger generations are more prominent than ever, and will, in some form, influence their reputations and chances of landing jobs in the future. Therefore, in addition to legislation of proper conduct for companies and governments, teaching proper conduct on the Internet to its users now ideally sits on par with teaching mannerisms and how to be a good citizen. Thus, schooling and curriculum that focus on both how to protect ourselves in the Internet environment, but also how to be good digital citizens is necessary as we advance into the digital age.

## REFERENCES

Abu-Shaqra, B., & Luppicini, R. (2018). A technoethical study of ethical hacking communication and management within a Canadian university. In R. Luppicini (Ed.), *The changing scope of technoethics in contemporary society* (pp. 307–326). Hershey, PA: IGI Global. doi:10.4018/978-1-5225-5094-5.ch017

Aguado, D., Rico, R., Rubio, V. J., & Fernández, L. (2016). Applicant reactions to social network web use in personnel selection and assessment. *Missing Publication Name*, *30*(2), 67–74.

Allden, N., & Harris, L. (2013). Building a positive candidate experience: Towards a networked model of e-recruitment. *The Journal of Business Strategy*, *34*(5), 36–47. doi:10.1108/JBS-11-2012-0072

Apers, C., & Derous, E. (2017). Are they accurate? Recruiters' personality judgments in paper versus video resumes. *Computers in Human Behavior*, *73*, 9–19. doi:10.1016/j.chb.2017.02.063

Bangerter, A. A., Roulin, N., & König, C. J. (2012). Personnel selection as a signalling game. *The Journal of Applied Psychology*, *97*(4), 719–738. doi:10.1037/a0026078 PMID:22040263

Baruch, Y., & Altman, Y. (2016). The ecosystem of labor markets and careers. *People and Strategy*, *39*(3), 16.

Batagan, L., & Boja, C. (2015). Social networks a real solution for students' future jobs. *Amfiteatru Economic*, *17*(9), 1303–1313.

Berkelaar, B. L., & Buzzanell, P. M. (2014). Cybervetting, person–environment fit, and personnel selection: Employers' surveillance and sensemaking of job applicants' online information. *Journal of Applied Communication Research*, *42*(4), 456–476. doi:10.1080/00909882.2014.954595

Berkelaar, B. L., Scacco, J. M., & Birdsell, J. L. (2015). The worker as politician: How online information and electoral heuristics shape personnel selection and careers. *New Media & Society*, *17*(8), 1377–1396. doi:10.1177/1461444814525739

Black, S., Stone, D., & Johnson, A. (2015). Use of Social Networking Websites on Applicants' Privacy. *Employee Responsibilities and Rights Journal*, *27*(2), 115–159. doi:10.100710672-014-9245-2

Bohmova, L., & Pavlicek, A. (2015). The influence of social networking sites on recruiting human resources in the Czech Republic. *Organizacija*, *48*(1), 23–31. doi:10.1515/orga-2015-0002

Bonsón, E., & Bednárová, M. (2013). Corporate LinkedIn practices of Eurozone companies. *Online Information Review*, *37*(6), 969–984. doi:10.1108/OIR-09-2012-0159

Broughton, A., Foley, B., Ledermaier, S., & Cox, A. (2013). *The use of social media in the recruitment process*. Brighton, CA: Institute for Employment Studies.

Capiluppi, A., Serebrenik, A., & Singer, L. (2012). Assessing technical candidates on the social web. *IEEE Software*, *30*(1), 45–51. doi:10.1109/MS.2012.169

Castells, M. (1996). The rise of network society. Oxford, UK: Blackwell.

Castells, M., Fernandez-Ardevol, M., Qiu, J. L., & Sey, A. (2009). *Mobile communication and society: A global perspective*. MIT Press.

Chou, H.-T. G., Hammond, R. J., & Johnson, R. (2013). How Facebook might reveal users' attitudes toward work and relationships with coworkers. *Cyberpsychology, Behavior, and Social Networking*, *16*(2), 136–139. doi:10.1089/cyber.2012.0321 PMID:23276260

Daft, R. L., & Lengel, R. H. (1986). Organizational information requirements, media richness and structural design. *Management Science*, *32*(5), 554–571. doi:10.1287/mnsc.32.5.554

Davison, H. K., Bing, M. N., Kluemper, D. H., & Roth, P. L. (2016). Social Media as a Personnel Selection and Hiring Resource: Reservations and Recommendations. Academic Press.

Davison, H. K., Maraist, C., & Bing, M. N. (2011). Friend or foe? The promise and pitfalls of using social networking sites for HR decisions. *Journal of Business and Psychology*, *26*(2), 153–159. doi:10.100710869-011-9215-8

Drouin, M., O'Connor, K. W., Schmidt, G. B., & Miller, D. A. (2015). Facebook fired: Legal perspectives and young adults' opinions on the use of social media in hiring and firing decisions. *Computers in Human Behavior*, *46*, 123–128. doi:10.1016/j.chb.2015.01.011

El Ouirdi, M., Pais, I., Segers, J., & El Ouirdi, A. (2016). The relationship between recruiter characteristics and applicant assessment on social media. *Computers in Human Behavior*, *62*, 415–422. doi:10.1016/j.chb.2016.04.012

Gibbs, C., MacDonald, F., & MacKay, K. (2015). Social media usage in hotel human resources: Recruitment, hiring and communication. *International Journal of Contemporary Hospitality Management*, *27*(2), 170–184. doi:10.1108/IJCHM-05-2013-0194

Gough, D. A., Oliver, S., & Thomas, J. (2013). *Learning from research: Systematic reviews for informing policy decisions: A quick guide*. London: Nesta.

Harrison, J. A., & Budworth, M. H. (2015). Unintended consequences of a digital presence: Employment-related implications for job seekers. *Career Development International*, *20*(4), 294–314. doi:10.1108/CDI-06-2014-0080

Hoek, J., O'Kane, P., & McCracken, M. (2015). Publishing personal information online. *Personnel Review*, *45*(1), 67–83. doi:10.1108/PR-05-2014-0099

Hongladarom, S., Ess, C., & Nagenborg, M. (2007). Information technology ethics: Cultural perspectives. *Publizistik*, *52*(4), 565–567.

Hurrell, S. A., Scholarios, D., & Richards, J. (2017). 'The kids are alert': Generation Y responses to employer use and monitoring of social networking sites. *New Technology, Work and Employment*, *32*(1), 64–83. doi:10.1111/ntwe.12085

Jeske, D., & Shultz, K. S. (2016). Using social media content for screening in recruitment and selection: Pros and cons. *Work, Employment and Society*, *30*(3), 535–546. doi:10.1177/0950017015613746

Joachim, J., Martin, M., Lange, H., Schneiker, A., & Dau, M. (2018). Twittering for talent: Private military and security companies between business and military branding. *Contemporary Security Policy*, *39*(2), 298–316. doi:10.1080/13523260.2017.1420608

Karlen, J. M. (2014). Privacy: Expectations and employment. *Business Management Dynamics*, *3*(10), 28.

Kashi, K., Zheng, C., & Molineux, J. (2016). Exploring factors driving social recruiting: The case of Australian organizations. *Journal of Organizational Computing and Electronic Commerce*, *26*(3), 203–223. doi:10.1080/10919392.2016.1194055

Lam, H. (2016). Social media dilemmas in the employment context. *Employee Relations*, *38*(3), 420–437. doi:10.1108/ER-04-2015-0072

Lievens, F., Van Hoye, G., & Schreurs, B. (2005). Examining the relationship between employer knowledge dimensions and organizational attractiveness: An application in a military context. *Journal of Occupational and Organizational Psychology*, *78*(4), 553–572. doi:10.1348/09631790X26688

Luppicini, R. (2009). Technoethical inquiry: From technological systems to society. Global Media Journal, Canadian Edition, 2(1), 5–21.

Luppicini, R. (2010). *Technoethics and the Evolving Knowledge Society: Ethical Issues in Technological Design, Research, Development, and Innovation*. Information Science Reference.

Luppicini, R., & So, A. (2016). A technoethical review of commercial drone use in the context of governance, ethics, and privacy. *Technology in Society*, *46*, 109–119. doi:10.1016/j.techsoc.2016.03.003

Magnani, L. (2007). *Morality in a technological world: Knowledge as duty*. Cambridge, MA: Cambridge University Press; doi:10.4018/978-1-60566-952-6.

May, S. (2013). Introduction: Ethical perspectives and practices. In *Case Studies in Organizational Communication: Ethical Perspectives and Practices* (pp. 1–33). Thousand Oaks, CA: Sage Publications.

McKenna, C. (2014). *The use of social media in recruitment and its impact on diversity in services companies in Ireland* (Doctoral dissertation). Dublin Business School.

Micik, M., & Eger, L. (2015). Recruiting talents with social media. *Актуальні проблеми економіки*, (3), 266-274.

Nikolaou, I. (2014). Social networking web sites in job search and employee recruitment. *International Journal of Selection and Assessment*, *22*(2), 179–189. doi:10.1111/ijsa.12067

Priyadarshini, C., Kumar, Y. L. N., & Jha, R. R. (2017). Employer attractiveness through social media: A phenomenological study. *Qualitative Report*, *22*(4), 969–983.

Roberts, G. (2005). *Recruitment, selection and induction*. London: Chartered Institute of Personnel and Development.

Rosoiu, O. & Popescu, C. (2016). E-recruiting platforms: Features that Influence the efficiency of online recruitment systems. *Informatica Economica, 20*(2), 46–55.

Sameen, S., & Cornelius, S. (2015). Social networking sites and hiring: How social media profiles influence hiring decisions. *Journal of Business Studies Quarterly*, *7*(1), 27–35.

Schneider, T. J., Goffin, R. D., & Daljeet, K. N. (2015). Give us your social networking site passwords: Implications for personnel selection and personality. *Personality and Individual Differences*, *73*, 78–83. doi:10.1016/j.paid.2014.09.026

Schroeder, A. N., & Lile, C. R. (2016). Legal trends in organizational online social media use. *Industrial and Organizational Psychology: Perspectives on Science and Practice*, *9*(3), 682–689. doi:10.1017/iop.2016.71

Slovensky, R., & Ross, W. H. (2012). Should human resource managers use social media to screen job applicants? Managerial and legal issues in the USA. *Info*, *14*(1), 55–69. doi:10.1108/14636691211196941

Society for Human Resource Management. (2015). *Using social media for talent acquisition – recruitment and screening*. Retrieved from https://www.shrm.org/hr-today/trends-and-forecasting/research-and-surveys/Documents/SHRM-Social-Media-Recruiting-Screening-2015.pdf

Stoughton, J. W., Thompson, L. F., & Meade, A. W. (2015). Examining applicant reactions to the use of social networking websites in pre-employment screening. *Journal of Business and Psychology*, *30*(1), 73–88. doi:10.100710869-013-9333-6

Theurer, C. P., Tumasjan, A., Welpe, I. M., & Lievens, F. (2018). Employer branding: A brand equity-based literature review and research agenda. *International Journal of Management Reviews*, *20*(1), 155–179. doi:10.1111/ijmr.12121

Van Hoye, G., & Lievens, F. (2005). Recruitment-Related Information Sources and Organizational Attractiveness: Can Something Be Done About Negative Publicity? *International Journal of Selection and Assessment*, *13*(3), 179–187. doi:10.1111/j.1468-2389.2005.00313.x

Vasilescu, B., Capiluppi, A., & Serebrenik, A. (2013). Gender, representation and online participation: A quantitative study. *Interacting with Computers*, *26*(5), 488–511. doi:10.1093/iwc/iwt047

Walker, H. J., Feild, H. S., Giles, W. F., Armenakis, A. A., & Bernerth, J. B. (2009). Displaying employee testimonials on recruitment web sites: Effects of communication media, employee race, and job seeker race on organizational attraction and information credibility. *The Journal of Applied Psychology*, *94*(5), 1354–1364. doi:10.1037/a0014964 PMID:19702377

Zhao, X., Lampe, C., & Ellison, N. B. (2016). *The social media ecology: User perceptions, strategies and challenges*. New York: ACM Press. doi:10.1145/2858036.2858333

Zide, J., Elman, B., & Shahani-Denning, C. (2014). LinkedIn and recruitment: How profiles differ across occupations. *Employee Relations*, *36*(5), 583–604. doi:10.1108/ER-07-2013-0086

## APPENDIX

Final selection of articles to be considered for this review.


1. Aguado, D., Rico,R., Rubio, V. J., & Fernández, L. (2016). Applicant reactions to social network web use in personnel selection and assessment, *Missing Publication Name, 30*(2), 67–74.
2. Allden, N. & Harris, L. (2013). Building a positive candidate experience: Towards a networked model of e-recruitment. *Journal of Business Strategy, 34*(5), 36–47.
3. Batagan, L. & Boja, C. (2015). Social networks a real solution for students' future jobs. *Amfiteatru Economic, 17*(9), 1303–1313.
4. Berkelaar, B. L. & Buzzanell, P. M. (2014). Cybervetting, person–environment fit, and personnel selection: Employers' surveillance and sensemaking of job applicants' online information. *Journal of Applied Communication Research, Missing volume*(issue number), page numbers.
5. Berkelaar, B. L., Scacco, J. M., & Birdsell, J. L. (2015). The worker as politician: How online information and electoral heuristics shape personnel selection and careers. *New Media & Society, 17*(8), 1377–1396.
6. Bohmova, L., & Pavlicek, A. (2015). The influence of social networking sites on recruiting human resources in the Czech Republic. *Organizacija, 48*(1), 23–31.
7. Bonsón, E. & Bednárová, M. (2013). Corporate LinkedIn practices of Eurozone companies. *Online Information Review, 37*(6), 969–984.
8. Broughton, A., Foley, B., Ledermaier, S., & Cox, A. (2013). The use of social media in the recruitment process. Brighton, CA: Institute for Employment Studies.
9. Capiluppi, A., Serebrenik, A., & Singer, L. (2012). Assessing technical candidates on the social web. *IEEE software, 30*(1), 45–51.
10. Drouin, M., O'Connor, K. W., Schmidt, G. B., & Miller, D. A. (2015). Facebook fired: Legal perspectives and young adults' opinions on the use of social media in hiring and firing decisions. *Computers in Human Behavior, 46*, 123–128.
11. El Ouirdi, M., Pais, I., Segers, J., & El Ouirdi, A. (2016). The relationship between recruiter characteristics and applicant assessment on social media. *Computers in Human Behavior, 62*, 415–422.
12. El Ouirdi, M., Segers, J., El Ouirdi, A., & Pais, I. (2015). Predictors of job seekers' self-disclosure on social media. *Computers in Human Behavior, 53*, 1–12.

13. Gibbs, C., MacDonald, F., & MacKay, K. (2015). Social media usage in hotel human resources: recruitment, hiring and communication. *International Journal of Contemporary Hospitality Management*, *27*(2), 170–184.
14. Harrison, J. A. & Budworth, M.-H. (2015). Unintended consequences of a digital presence. *Career Development International*, *20*(4), 294–314.
15. Hoek, J., O'Kane, P., & McCracken, M. (2015). Publishing personal information online. *Personnel Review*, *45*(1), 67–83.
16. Hurrell, S. A., Scholarios, D., & Richards, J. (2017). The kids are alert: Generation Y responses to employer use and monitoring of social networking sites. *New Technology, Work and Employment*, *32*(1), 64–83.
17. Jeske, D. & Shultz, K. S. (2016). Using social media content for screening in recruitment and selection: Pros and cons. *Work, Employment and Society*, *30*(3), 535–546.
18. Joachim, J., Martin, M., Lange, H., Schneiker, A., & Dau, M. (2018). Twittering for talent: Private military and security companies between business and military branding. *Contemporary Security Policy, 39*(2), 298-316.
19. Kashi, K., Zheng, C., & Molineux, J. (2016). Exploring factors driving social recruiting: The case of Australian organizations. *Journal of Organizational Computing and Electronic Commerce*, *26*(3), 203–223.
20. Micik, M. & Eger, L. (2015). Recruiting talents with social media. *Actual Problems of Economics*, *165*(3), 266–274.
21. Nikolaou, I. (2014). Social networking web sites in job search and employee recruitment. *International Journal of Selection and Assessment*, *22*(2).
22. Priyadarshini, C., Kumar, Y. L. N., & Jha, R. R. (2017). Employer attractiveness through social media: A phenomenological study. *The Qualitative Report*, *22*(4), 969–983.
23. Rosoiu, O. & Popescu, C. (2016). E-recruiting platforms: Features that Influence the efficiency of online recruitment systems. *Informatica Economica*, *20*(2/2016), 46–55.
24. Roulin, N. (2014). The influence of employers' use of social networking websites in selection, online self-promotion, and personality on the likelihood of faux pas postings. *International Journal of Selection and Assessment*, *22*(1), 80–87.
25. Sameen, S. & Cornelius, S. (2015). Social networking sites and hiring: How social media profiles influence hiring decisions. *Journal of Business Studies Quarterly*, *7*(1), 27–35.
26. Schneider, T. J., Goffin, R. D., & Daljeet, K. N. (2015). Give us your social networking site passwords: Implications for personnel selection and personality. *Personality and Individual Differences*, *73*, 78–83.

27. Schroeder, A. N., & Lile, C. R. (2016). Legal trends in organizational online social media use. *Industrial and Organizational Psychology: Perspectives on Science and Practice*, *9*(3), 682–689.
28. Sivertzen, A.- M., Nilsen, E. R., & Olafsen, A. H. (2013). Employer branding: Employer attractiveness and the use of social media: A phenomenological study. *Journal of Product & Brand Management*, *22*(7), 473–483.
29. Slovensky, R., & Ross, W. H. (2012). Should human resource managers use social media to screen job applicants? Managerial and legal issues in the USA. *Info*, *14*(1), 55–69.
30. Stoughton, J. W., Thompson, L. F., & Meade, A. W. (2015). Examining applicant reactions to the use of social networking websites in pre-employment screening. *Journal of Business and Psychology*, *30*(1), 73–88.
31. Theurer, C. P., Tumasjan, A., Welpe, I. M., & Lievens, F. (2018). Employer branding: A brand equity-based literature review and research agenda. *International Journal of Management Reviews, 20*(1), 155-179.
32. Zide, J., Elman, B., & Shahani-Denning, C. (2014). LinkedIn and recruitment: How profiles differ across occupations. *Employee Relations*, *36*(5), 583–604.

# Chapter 11
# Bias and Discrimination in Artificial Intelligence:
## Emergence and Impact in E-Business

**Jan C. Weyerer**
*German University of Administrative Sciences, Speyer, Germany*

**Paul F. Langer**
https://orcid.org/0000-0001-8735-0035
*German University of Administrative Sciences, Speyer, Germany*

## ABSTRACT

*Artificial intelligence (AI) has become an integral part of e-business and our lives, promising significant benefits to e-business companies and society. However, at the same time, AI systems in e-business may produce biased outcomes, leading to discrimination of minorities and violating human rights. Against this background, this chapter first describes the foundations of bias and discrimination in AI, highlighting its scientific and practical relevance, as well as describing its meaning, emergence, functioning, and impact in the context of e-business. Based on these foundations, the chapter further provides implications for research and practice on how to deal with AI-related bias and discrimination in the future, opening up future research directions as well as outlining solutions and recommendations for eliminating and preventing AI-related bias and discrimination in e-business.*

DOI: 10.4018/978-1-7998-1879-3.ch011

## INTRODUCTION

In the course of the advancing digital transformation, artificial intelligence (AI) has recently become increasingly important for society, economy and government. The application of AI particularly opens up a variety of new opportunities for organizations in e-business, promising considerable gains in effectiveness and efficiency in the execution of business activities and processes. In e-business, AI is, for instance, already used to determine which and how online ads are presented to a person (Hillard, Schroedl, & Manavoglu, 2010), the order in which consumers are presented the news (Nalisnick, Mitra, Craswell, & Caruana, 2016), whether a person is hired for a job (Strohmeier & Piazza, 2015) or whether a person is eligible to get a loan (Khandani, Kim, & Lo, 2010). Accordingly, AI plays a significant role in e-business and our lives, increasingly impacting our opinions and behavior in everyday life (Leavy, 2018). To put it in a nutshell, AI is regarded as a key technology and economic growth engine that is essential for the future sustainable competitiveness of e-businesses.

At the same time, however, there are major risks and challenges associated with the application of AI in the context of e-business, which may be technological, legal, social and ethical in nature (Wirtz, Weyerer, & Geyer, 2019). An intensively discussed aspect refers to the ethics of AI and the question of whether the development and application of AI and its consequences are ethically and morally justifiable or how ethical basic values can be integrated into AI applications (Anderson & Anderson, 2011; Lin, Abney, & Bekey, 2012; Wirtz et al., 2019). A significant ethical risk refers to the issue of AI-related bias and discrimination, which is increasingly attracting public attention and has become subject to a great social debate. AI-related bias and discrimination refers to the fact that AI can adopt human prejudices or discriminatory values and behaviors and even reproduce them in an amplifying way (Basu, 2018; Thierer, O'Sullivan Castillo, & Russell, 2017).

In general, this means that AI applications may potentially not only violate national anti-discrimination laws implemented by a great number of democratic governments, such as the U.S. (e.g. The Age Discrimination in Employment Act of 1967.), UK (e.g. Equality Act 2010.) or Germany (e.g. General Act on Equal Treatment.), but also the prohibition of discrimination enshrined in Article 2 of the Universal Declaration of Human Rights.

The issue of AI-related bias and discrimination and its potentially adverse effects on people and society is particularly relevant and far-reaching in the e-business realm. Here, AI's scope of application is extremely high and constantly advancing, yielding new, more and more pervasive AI applications, which affect a particularly large number of individuals – not to say our whole society.

Against this background, awareness of the topic among consumers has increased, putting the beneficence of AI into question. As a consequence, the acceptance and usage of AI solutions in e-business may be at stake in the long term, if the concerns and uncertainties among consumers further disseminate and solidify, thus threatening the great potential of AI for organizations in e-business. As AI is a cross-sectional technology with an almost universal scope of application, a particularly far-reaching effect of AI-related bias and discrimination on society is to be expected, which will spread virally in society especially via digital media. Given that AI is also still in its infancy with regard to research, development and use, it is therefore crucial to scrutinize its emerging risks and challenges as well as to supervise its development and dissemination in society (Weyerer & Langer, 2019).

The above-mentioned deliberations indicate that the social implications and core aspects associated with AI-related bias and discrimination are particularly relevant and the latter is not only a technological but also a social matter. Thus, the technology-dominated point of view prevalent in research and practice to date demands an adjustment by social aspects. It is essential for researchers and practitioners in e-business to understand how AI-related bias and discrimination occurs, operates and affects their customers and society as a whole in order to develop and apply effective remedies to protect consumers and society from harm.

Against this background, the main objective of this chapter is to improve our understanding of how AI-related bias and discrimination emerges, takes place and manifests itself in the e-business environment. Therefore, the authors seek to provide a holistic conceptual representation of the complex and versatile mechanisms of action and core elements of AI-related bias and discrimination in a simplified way by presenting a theoretical explanatory model that illustrates the emergence, functioning and impact of AI-related bias and discrimination in the context of e-business. In doing so, a further objective of this chapter is to raise awareness of the particular importance and pervasive threat of AI-related bias and discrimination in e-business. In this connection, the authors not only seek to identify potentially fruitful research directions to further explore the role of AI-related bias and discrimination in the context of e-business, but also particularly highlight the vital necessity of AI-related safety and risk management approaches and measures. Here, a crucial objective of the chapter refers to providing implications and starting points for practitioners and researchers with regard to the development of respective solution approaches in order to prevent and eliminate biased or discriminatory AI outcomes in e-business in the future. Overall, the authors seek to make an initial contribution towards understanding the issue of AI-related bias and discrimination in the context of e-business and triggering further reflection, research and effective action on this vital challenge.

## BACKGROUND

Although discrimination has a far-reaching research history in economics and social sciences, it is a rather new but increasingly emerging branch of research within the field of AI that has attracted interest across many research disciplines in recent years. Despite its great relevance, research on this issue is scarce and heterogeneous, lacking consistent use of terminology and a uniform definition of terms. Just the fact that a lot of different terms are used relating to the same phenomenon or problem makes it difficult to fully grasp the field of research. For instance, previous literature in research and practice examines this phenomenon under different labels, such as AI discrimination (e.g. Daws, 2018; Hale, 2018), AI fairness (e.g. Bellamy et al., 2018; Trewin, 2018), algorithmic bias (e.g. Raub, 2018; Amini, Soleimany, Schwarting, Bhatia, & Rus, 2019), algorithmic discrimination (e.g. Goodman, 2016; Hacker, 2018), or algorithmic fairness (e.g. Woodruff, Fox, Rousso-Schindler, & Warshaw, 2018). Accordingly, it is essential to outline a common understanding of bias and discrimination in AI before addressing its emergence, functioning and impact in the e-business context.

Discrimination is rather a term of general use and refers to the unfair treatment of individuals based on their affiliation to a particular group. For instance, individuals may be subject to discrimination because of their gender, race, religion etc. (Hajian, Domingo-Ferrer, & Martínez-Ballesté, 2011). Bias also pertains to "undue prejudice" but here can be more understood in mathematical or statistical terms, according to which it describes "statistics that lead to a skew that inflicts an unjust outcome upon a population" (Lloyd, 2018, p. 1).

The few contributions available approaching AI-related bias and discrimination are scattered across different research disciplines and reflect the virtually universal scope of AI application. The roots of research on AI-related bias and discrimination lie in the field of information technology from which the vast majority of research contributions stem from (e.g. Hajian & Domingo-Ferrer, 2013; Kamiran, Calders, & Pechenizkiy, 2013; Berendt & Preibusch, 2014; Joseph, Kearns, Morgenstern, Neel, & Roth, 2016; Veale & Binns, 2017; Bellamy et al., 2018; Amini et al., 2019). According to this, there has been a one-sided focus on technological issues in connection with AI-related bias and discrimination so far, while social or business-related aspects have been rather neglected. However, as AI is more and more applied within social domains, it also increasingly involves social aspects and implications that cannot be adequately addressed by technological approaches, but rather demand research insights from the social sciences (Whittaker et al., 2018). With the increasing use and diffusion of AI in business and society, AI-related bias and discrimination have also been increasingly addressed in the areas of law (e.g. Flores, Bechtel, & Lowenkamp, 2016; Dalenberg, 2018; Hacker, 2018), government, politics and public management (e.g. Gasser & Almeida, 2017; Cath, Wachter, Mittelstadt, Taddeo, & Floridi, 2018 Agarwal, 2018; Williams, Brooks, & Shmargad, 2018; Wirtz et al., 2019), but also in other domains such as healthcare (e.g. Gianfrancesco, Tamang, Yazdany, & Schmajuk, 2018; Challen et al., 2019).

Apart from the above-mentioned technological approaches, AI-related bias and discrimination has received most research attention in the field of e-business, which is not surprising given that it represents the leading and largest application area of AI. Nonetheless, research on AI-related bias and discrimination in e-business is still in its infancy and an emerging field of research with overall little research insights. Looking at respective scientific literature in e-business reveals that nearly all studies focus on a specific case of application, particularly examining AI-related bias and discrimination in the context of online advertising (Sweeney, 2013; Datta, Tschantz, & Datta, 2015; Plane, Redmiles, Mazurek, & Tschantz, 2017; Eslami, Krishna Kumaran, Sandvig, & Karahalios, 2018; Speicher, Ali et al., 2018; Lambrecht & Tucker, 2019), hiring and employment decisions (Mann & O'Neil, 2016; Raub, 2018; Houser, 2019), Internet platforms (Sandvig, Hamilton, Karahalios, & Langbort, 2014), rating platforms (Eslami, Vaccaro, Karahalios, & Hamilton, 2017) or consumer lending (Bartlett, Morse, Stanton, & Wallace, 2017).

Although there are some studies that address AI-related bias and discrimination from a more general e-business perspective (Gillespie, 2014; Winter, 2015; Garcia, 2016 Diakopoulos, 2016; Gwagwa & Koene, 2018), e-business research as a whole provides rather anecdotal knowledge than empirical evidence in this context. In addition, previous research so far fails to provide an essential requirement for reaching the generally desired state of reliable empirical knowledge in science, namely a respective solid conceptual foundation that systematizes the concept of AI-related bias and discrimination and its key aspects in the context of e-business. Providing a comprehensive and thorough understanding of the concept in this way is not only essential for developing this research area and enable it to thrive, it is also vital for making it more clear and manageable to practitioners and researchers in e-business. Accordingly, this chapter presents a common framework in the following, which seeks to integrate and systematize respective previous research insights in particular with regard to the emergence, functioning and impact of AI-related bias and discrimination in e-business.

## CONCEPTUAL EXPLANATORY FRAMEWORK FOR AI-RELATED BIAS AND DISCRIMINATION IN E-BUSINESS

Grasping the emerging and complex field of AI-related bias and discrimination in e-business and understanding its emergence, functioning and impact is vital for being able to prevent and eliminate its occurrence. In research, a conceptual framework is a particularly suited tool for conceptualizing and simplifying complex issues, as well as enabling a more comprehensive research approach and contributing to a better understanding of the structure of the respective issue (Eisenhart, 1991; Shehabuddeen, Probert, Phaal, & Platts, 1999).

A conceptual framework compiles and systematically organizes central concepts, components and key factors as well as their interrelationships in narrative or graphical form to create a focus, rationale and instrument for the integration and interpretation of information in a certain field of research (McGregor, 2018; Miles, Huberman, & Saldaña, 2014). Against this background, this chapter presents an explanatory conceptual framework of AI-related bias and discrimination in the context of e-business, thus providing a comprehensive description of the key aspects and components in connection with its emergence, functioning and impact. In doing so, the framework is supposed to convey a common general understanding of this phenomenon and serve as a guidance for researchers and practitioners.

Unjustified discrimination as discrimination or degradation of individuals or groups through no fault of one's own by conscious, unconsidered, and sometimes unconscious attitudes, prejudices, or emotional associations is a social phenomenon that is politically and socially outlawed (Universal Declaration of Human Rights, 1948).

The emergence of AI technologies that independently draw conclusions and learn from data, is also linked to the hope of incorporating greater rationality into decision-making processes (Weiss, 1999). It should be noted, however, that both the technology and the database are of human origin and are thus subject to human biases in all layers of AI involvement.

Against this background, it is important to have a closer look at potential sources of and reasons for bias and discrimination through the use of AI technologies in the context of e-business, which should be subject to scrutiny in order to prevent AI-related bias and discrimination. The study of Hajian and Domingo-Ferrer (2013) suggests a differentiation of three different phases in the context of AI discrimination: pre-processing, in-processing and post-processing. Following this approach, those phases are extended to AI input (corresponding to some extent with pre-processing), AI learning (corresponding with in-processing) and AI output (corresponding to some extent with post-processing) in the explanatory framework at hand, representing the major stages through which AI-related bias and discrimination takes place. The following sections describe these stages in more detail and explain the process of AI-related bias and discrimination in e-business as well as its reproduction, which is illustrated by the explanatory framework presented in figure 1 (see for the following Weyerer & Langer, 2019).

## 1. AI Input

AI systems are initially developed and programmed by humans and fed by databases, which are also subject to human influences, for instance, in the context of data collection and processing. In analogy to the growing up and learning of small children, who are, to a certain extent, assumed to be born free of prejudices, it is the parents and the environment that shape prejudices and discrimination. Similarly, AI systems are a result of human design and human-influenced data and thus also shaped by certain human values and prejudices that are inherent to human AI engineering and programming as well as the data base. More specifically, artificial learning and

*Figure 1. Explanatory framework of AI-based bias and discrimination in e-business*

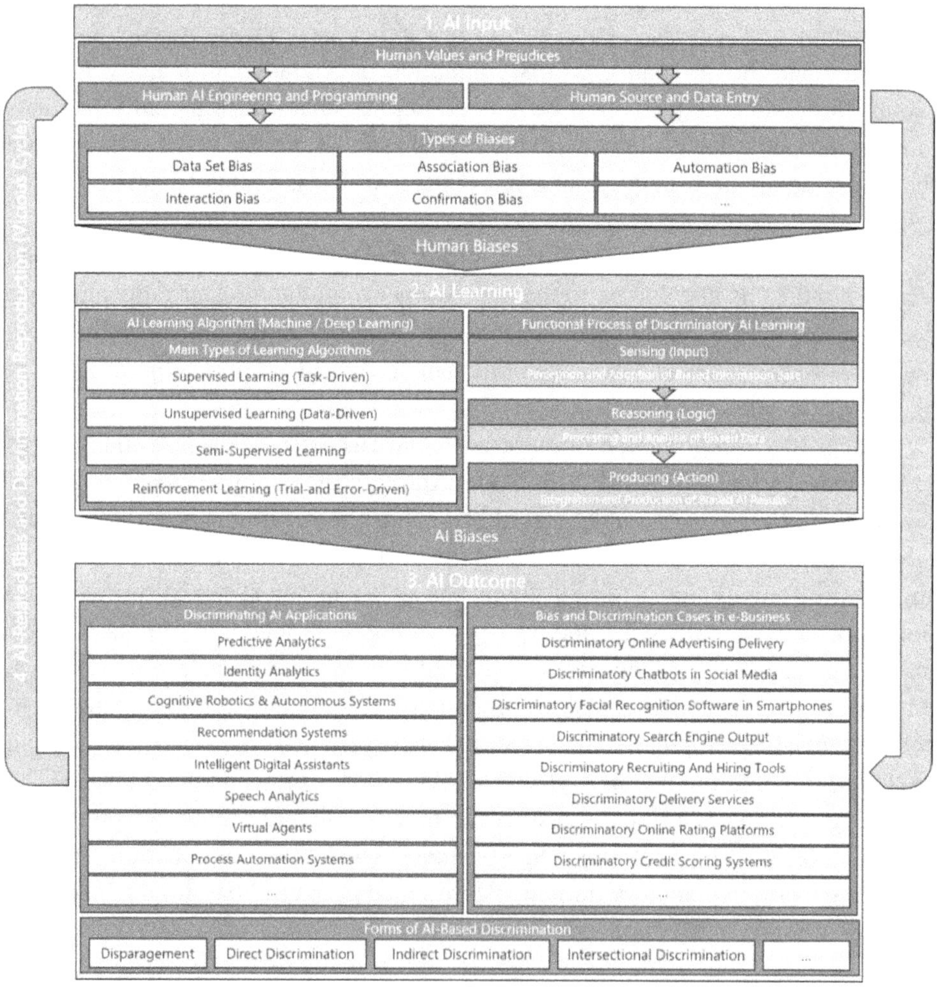

processing patterns are designed by humans with respective algorithms, software and other forms of artificial processing design.

Apart from the intelligence software that is a result of human AI engineering and programming, there is the database that is the fundament of all assessments and reasoning of AI. In this connection, humans themselves not only serve as a data source for AI systems but also influence relevant data processes such as data entry and develop the respective databases. To put it in the words of the renowned AI expert Kate Crawford: "Like all technologies before it, artificial intelligence will reflect the values of its creators." (Crawford, 2016). To sum up, the AI input is significantly determined by humans and thus also contains their prejudices, discriminatory attitudes and biases.

The above-mentioned concept of pre-processing refers to the use or selection of a given information source (Hajian & Domingo-Ferrer, 2013). The given source is a focal point of concern as most data is not free of biases and does not incorporate all necessary information. For instance, data on school performances could possibly show simple relations of poor school performance and migration background. However, there are good reasons to doubt the conclusion of a causal relation between migration background and poor school performance. Information on necessary language skills, educational background of the parents as well as the socio-economic status of the family could show that those aspects are key factors for school performance as a simple correlation does not imply a causal effect.

In general, there are three aspects that should be taken into consideration when looking at the information base of AI: (1) incompleteness of the given data source (e.g. historical data is missing, other general aspects are not reported etc.); (2) unrepresentative data sources (e.g. gender, race, age, education etc.); (3) data that contains intentional discrimination (racist reports, hate speech, discriminatory social media comments etc.). Regardless of the specific reason for biased data, the key reason of AI-related bias and discrimination is that most data sources do not fulfill basic requirements of reliability and validity and thus portray an incomplete and misleading image of the real world.

Against this background five different types of human biases can be distinguished, including dataset bias, associations bias, automation bias, interaction bias and confirmation bias (see for the following descriptions Chou, Murillo, & Ibars, 2017; Lloyd, 2018). Dataset bias occurs when the data basis of an AI system does not accurately reflect the diversity of a certain population, for instance, because a data set is too small or homogenous, perpetuating biased generalizations. An example in the context of e-business are AI systems for facial recognition that recognize women and darker-skinned individuals worse than male and lighter persons, since the AI algorithm was predominantly trained with faces characterized by the latter traits (Buolamwini & Gebru, 2018). Association bias exists when the training data for an AI system reinforce and perpetuate a cultural bias. Here, Google's online advertising algorithm may serve as an example in e-business, which has been found to display ads for high-income jobs less likely to women than men (Datta et al., 2015). Moreover, automation bias refers to a circumstance in which an AI system neglects social and cultural aspects, leading to decisions or actions that are contrary to human diversity and discriminate certain minorities. In this connection, Google's photo app provides an example of an automation bias in the context of e-business. Based on an AI algorithm, the app assigns automatic labels to images within digital photo albums and mistakenly classified several black people in pictures as gorillas (Crawford, 2016). In a similar way, an interaction bias may arise automatically when an AI system learns from biased communication with humans in the absence of safeguard measures against malicious or discriminatory beliefs. The most prominent example of an interaction bias in the context of e-business is the case of Microsoft's self-learning chatbot called Tay that was supposed to learn how young people communicate in social media. Within a few hours Tay turned into a racist, anti-Semitic, sexist chatbot because some human users fed it with respective content from which it learned, forcing Microsoft to shut it down (Scherer, 2016). Lastly, a confirmation bias is present when incaccurate assumptions or generalizations with regard to an individual or a group result from oversimplified, personalized information. Confirmation bias in the context of e-business often occurs in connection with AI-based recommender systems. For instance, shopping sites show users or customers recommendations for products that are the same or similar to those they have already purchased in the past (Chou et al., 2017).

All of the above-mentioned human biases may be part of the AI input that serves as a basis for the AI learning, which represents the next stage of AI-related bias and discrimination and is described in the following.

## 2. AI Learning

Classical algorithms based on programming codes can make traceable decisions. Therefore, it is relatively easy to trace back the reasons for any decision made. In times of deep learning and neural networks, it is by no means clear how computer-based results come about (Arel, Rose, & Karnowski, 2010; Benítez, Castro, & Requena, 1997; Lum & Isaac, 2016). The learning process of modern AI technologies is based on the evaluation of large data sets (big data) and the identification of corresponding patterns. As mentioned in the previous section, most databases do not correspond to representative samples, but rather to the totality of available data. Whether surveillance cameras, press content, social media posts or search queries, the given data can by no means be regarded as representative or undistorted. In addition, deep learning technologies are often based on human decisions and seek to reproduce them on the basis of historical data (Arel et al., 2010). Social bots, for instance, are set up to analyze, identify patterns and copy the argumentations and styles of user-generated content in order to mimic real people. Since the given user-generated content also contains hate speech, social bots also copy their argumentations (Ferrara, Varol, Davis, Menczer, & Flammini, 2016). This example shows that AI technologies can also be designed to accept if not even intend the production of discriminatory outcomes (Bessi & Ferrara, 2016; Shao, Ciampaglia, Varol, Flammini, & Menczer, 2017). As demonstrated in the previous section, there are two reasons for the discriminatory outcome: the programming of the algorithm as well as the given learning (data) base.

Against this background, the functional process of discriminatory AI learning consists of three main stages: sensing, reasoning and producing. The sensing stage refers to the perception and adoption of the input and biased information or data base. The following reasoning stage deals with the logic, processing and analysis of biased information or data. Finally, the producing stage pertains to the resulting action and describes the integration and production of biased AI results. The outlined process and an AI learning algorithm constitute in many cases a black box, since machine learning or AI includes approaches such as artificial neural networks and deep learning (independent data-driven self-learning technologies), which produce outcomes that are untraceable (Arel et al., 2010; Benítez et al., 1997; Castelvecchi, 2016). Against this background, four types of AI learning are distinguished in the literature,: supervised learning, unsupervised learning, semi-supervised learning and reinforcement learning (Nicolas, 2014; Ramasubramanian & Singh, 2017).

Supervised learning is commonly described as a task-driven learning approach as it is about finding a function to assign unknown observations in a data set to a certain category or value based on experiences with a training data set. The AI learning algorithm is confronted with a training example that contains clearly labeled input data from which the algorithm can learn and determine the output. When being confronted with unlabeled test data, the gained experience from the training is used to determine the missing information or label for the test data. Accordingly, the capability of a supervised AI learning algorithm strongly depends on the number of examples with which it is fed.

Unsupervised learning, in contrast, is characterized by a set of unlabeled input data which means that the corresponding output values of the input data are unknown and need to be discovered in the data. Accordingly, unsupervised learning is a data-driven learning approach that seeks to identify unknown patterns, deduce corresponding rules and recognize structure from the unlabeled input data. Therefore, the AI learning algorithm requires a large amount of data for meaningful clusterings and predictions.

Semi-supervised learning represents an intemediate form between supervised learning and unsupervised learning in which some data are labeled while a large bulk of data is unlabeled. In this approach, supervised and unsupervised learning methods are combined.

Reinforcement learning does not require any training data or human supervision in advance. Similar to humans learning a task, reinforcement learning is based on experience and interaction with the environment. The learning is characterized by a trial-and-error process and the AI-system learns optimal behavior or output by receiving rewards for correct results and penalties for wrong results.

## 3. AI Outcome

The previous sections outlined the weaknesses of artificial intelligence with regard to its human design, its potentially biased data sources and the resulting untraceable and flawed AI learning and abstraction process. The preceding AI input and AI learning stages determine the final AI outcome stage. Once the AI algorithm is programmed and learns human bias or discriminatory content, the latter become an integral part of the AI systems and its outcomes. The biased or discriminating algorithm may be part of any AI application used in e-business. Since AI represents a cross-sectional technology, there is a broad range of potential applications and areas that may be affected by AI bias or discrimination. Thus, discriminating AI applications may include predictive analytics, identity analytics, cognitive robotics and autonmous systems, recommendation systems, intelligent digital assistants, speech analytics, virtual agents or process automation systems (Wirtz et al., 2019). There is a number of well-documented supporting examples, illustrating the great scope of this problem in the context of e-business. Known discriminatory cases from the past, for instance, refer to online advertising delivery (Sweeney, 2013), chatbots in social media (Scherer, 2016), facial recognition software in smartphones (Buolamwini & Gebru, 2018), search engine output (Snow, 2018) recruiting and hiring tools (Dastin, 2018), delivery services (Ingold & Soper, 2016), online rating platforms (Eslami et al., 2017) or credit scoring systems (Citron & Pasquale, 2014).

In addition, discrimination itself is a complex field of research entailing many normative aspects that are difficult to integrate into technical solutions (Becker, 2010). Specific values such as equity of achievement, equity of needs or even equal opportunities are concepts that are set apart from pure rationality but rather by principals of morality or even humanity (ibid.). However, there are concrete concepts that are basic to discrimination research, in particular, the different forms of discrimination discussed in the literature. Although there are several forms of discrimination that each pose a different level of detectability and avoidability in the context of AI and machine learning, the discriminatory cases presented indicate that particularly four forms of discrimination are relevant in the context of e-business, including disparagement, direct discrimination, indirect discrimination and intersectional discrimination.

With the establishment of social media as important platforms for social interaction, various forms of disparagement in public and closed spaces are a common phenomenon, which are countered with automated, but also human-made deletion measures (Davidson, Warmsley, Macy, & Weber, 2017). In addition to the partly automated deletion, some of the social or chat bots, i.e. autonomous actors on the net, have already attracted attention through discriminatory contributions (Munger, 2017). A well-known example of disparagement by AI is Microsoft's "Tay" pilot project. Tay was a self-learning chat bot with its own Twitter profile that learned how young people communicate on social media platforms. Within a few hours on Twitter, Tay transformed himself into a racist, anti-Semitic, sexist chatbot, as his human interaction partners confronted him with appropriate content from which he learned and recognized which posts generated particularly successful interactions (Neff & Nagy, 2016; Scherer, 2016). Already after 16 hours, Microsoft was forced to take Tay offline to stop his discriminatory messages.

Furthermore, direct discrimination, refers to a less favorable treatment of an individual (e.g. in jobs, school, court etc.) based on an assessment-independent personal feature (e.g. gender, race etc.) (Ellis & Watson, 2012; UN Committee on Economic, Social and Cultural Rights [CESCR], 2009). This form of discrimination is relatively simple to be detected and avoided if the respective feature is reported. For instance, algorithms can be programmed in such a way that they do not consider data on race, age and gender to avoid unjust or prejudicial treatment of different groups of people (Dalenberg, 2018). This form of discrimination also occurs today through AI applications, such as an Amazon applicant management algorithm that did not suggest female candidates for highly paid jobs (Dastin, 2018)

More difficult is indirect discrimination that describes an unjust practice, policy or rule which applies to everyone, but has a negative effect on a particular group (Ellis & Watson, 2012; UN Committee on Economic, Social and Cultural Rights, 2009). Such indirect discrimination can be based on neutral selection criteria, such as the use of a specific register as database (e.g. a regional electoral register, a range of postcode, a list of current employees etc.) that has a unrepresentative population base (e.g. with regards to income levels, gender race etc.). Using neutral assessment criteria on an unrepresentative pool of individuals still delivers discriminatory results. Such discrimination often happens on grounds of registers with an overrepresentation of privileged people (male, white, employed, educated etc.). Such indirect discrimination can also be the result of AI technologies, for example, if algorithms decide upon customer groups based on their average income levels and therefore exclude certain minorities even if they do not consider race for their decision making. This is the case with Amazon Prime that is not available to all areas in the US. Those areas that are not eligible are mostly poor areas populated by people of color (Ingold & Soper, 2016).

Finally, intersectional discrimination describes a form of social differentiation based on individual characteristics such as gender, ethnicity, sexual orientation, which do not occur separately but are interrelated (Crenshaw, 1989). Specific combinations can lead to particular disadvantages. For example, women with a migration background may also be disadvantaged if, measured separately, there are no fundamental disadvantages against people with a migration background or against women. Inter-sectionality is therefore harder to be identified and can therefore also be the result of AI applications if these are designed not to discriminate against clearly defined minorities, for example by excluding differentiation according to gender or origin (Raji & Buolamwini, 2019). In addition to the forms mentioned, there are other forms, such as structural, linguistic or positive discrimination, which are less relevant in the context of AI applications.

## 4. AI-Related Bias and Discrimination Reproduction (Vicious Cycle)

AI-related bias and discrimination reproduction describes the reinforcing and endless repetition of bias and discriminatory content of an AI application by other AI applications and humans, constituting a vicious cycle. Accordingly, the reproduction may in general take place via two mechanisms: AI applications (AI-based reproduction) and humans (human-based reproduction). An AI outcome may serve as input for another AI application or a human. AI-based reproduction refers to a situation in which the biased or discriminatory content produced by an AI application is incorporated and reused by another AI application. In this connection, social bots are particularly relevant that already determine social media interaction and public sentiment (Ferrara et al., 2016). As already indicated, social bots are AI-based programs that mimic individuals on social media platforms, creating or sharing content independently. A careful analysis of the content produced or shared by social bots indicates that there have also been cases of bots sharing other bots' content (Appling & Briscoe, 2017). In the case of human-based reproduction people perceive, believe and share a biased or discriminatory AI outcome, which then acts directly or indirectly as a data source or input for other AI learning processes. A further reinforcement of the AI-related bias and discrimination effect and its reproduction occurs through humans' technological leap of faith, according to which people often perceive technology or in this context AI-based outcomes as more reliable and objective than those generated by humans. Discriminatory results of AI applications thus not only represent actual discrimination, but also become a feedback mechanism through which prejudices are reinforced by corresponding confirmation of supposedly neutral machines or AI applications (Weiss, 1999).

## Solution Approaches

As can be seen from the above-mentioned deliberations, AI-related bias and discrimination constitutes a major challenge to society, economy and government, threatening not only the observance of human rights and compliance with anti-discrimination legislation, but also the success of e-business companies that use AI systems. Against this background, research and industry increasingly put great effort into developing remedies against AI-related bias and discrimination. In particular, global players in the IT industry such as Google, Facebook, IBM or Accenture have proposed a variety of attempts towards preventing and eliminating AI-related bias and discrimination in the context of e-business.

Google, for instance, has recently introduced a new research approach called "Testing with Concept Activation Vectors" (TCAV) to combat AI bias and discrimination (Google, 2019). TCAV is an analysis tool that helps developers to check their algorithms and understand how an AI systems makes it decisions. The management consulting company Accenture has launched new AI testing services powered by a "Teach and Test" methodology to help companies provide AI systems free of biases and discriminatory content (Accenture, 2018). One of the most comprehensive solution approaches to AI-related bias and discrimination is the "AI Fairness 360 toolkit" introduced by IBM (Bellamy et al., 2018; IBM, 2018a). This approach bridges knowledge of scientific research and industry, as it was deliberately designed by IBM as an open source project to encourage input from the worldwide research community. The toolkit currently comprises a set of more than 70 fairness metrics to check for biases in AI systems and ten state-of-the-art algorithms to mitigate bias in AI systems. These algorithms include optimized pre-processing, reweighing, adversarial debiasing, reject option classification, disparate impact remover, learning fair representations prejudice remover, calibrated equalized odds post-processing, equalized odds post-processing and meta fair classifier. The fairness metrics contain, for instance, statistical parity difference, equal opportunity difference, average odds difference, disparate impact, Theil index, Euclidean distance, Mahalanobis distance or Manhattan distance (IBM, n.y.).

In scientific research, a number of contributions have addressed the question of how to prevent or eliminate AI-related bias and discrimination in order to achieve fair AI systems. Most approaches come from the field of information technology and propose solely technological remedies, for instance, certain algorithms or fairness metrics (e.g. Calmon, Wei, Vinzamuri, Natesan Ramamurthy, & Varshney, 2017; Zhao, Wang, Yatskar, Ordonez, & Chang, 2017; Goel, Yaghini, & Faltings, 2018;

Speicher, Heidari et al., 2018; Zhang, Wu, & Wu, 2018; Amini et al., 2019), some of which included in IBM's AI Fairness 360 toolkit (for an overview of included research contributions, see IBM, 2018b). In addition, some attempts also highlight social, legal, policy-related or management-related aspects in addressing AI-related bias and discrimination (e.g. Kirkpatrick, 2016; Gwagwa & Koene, 2018; Williams et al., 2018; World Economic Forum, 2018; West, Whittaker, & Crawford, 2019). Williams et al. (2018, p. 108), for instance, present several remedies including "external auditing, designing algorithmic judgments to be valid across social categories, considering how legal and other limitations shape algorithmic systems, and focusing research on a particular group in order to gain insight." Overall, the scientific contributions indicate that it is not only important to consider the AI system itself as a technological entity, but also to take account of its broader context when it comes to dealing with AI-related bias and discrimination in e-business.

## FUTURE RESEARCH DIRECTIONS

The explanatory framework of AI-related bias and discrimination in e-business presented in this chapter may be of particular importance to this emerging field of research. Given the lack of conceptual and empirical research examining AI-related bias and discrimination in e-business, future research is spoilt for choice with regard to potential research endeavours, but at the same possibly also without orientation. Against this background, this chapter and the framework presented may serve as a conceptual landmark for future research in this context, providing a general structure of the core elements of this field of research and opening up new avenues of research. The so far rather anecdotal and heterogenous state of knowledge in particular calls for empirical evidence concerning the relevant structures, components and processes in connection with AI-related bias and discrimination in e-business. Above all, this requires a solid conceptual foundation, for instance, in the form of conceptual frameworks that systematize relevant aspects of AI-related bias and discrimination in e-business. For example, future research could further elaborate the emergence, functioning and impact and describe the respective dimensions AI input, AI learning and AI outcome as well as their interconnections in more detail. In this context, research particularly demands a stronger focus on non-technological issues, such as social, legal, policy-related and management-related aspects and implications of AI-related bias and discrimination. In light of the multidisciplinary and cross-sectional character of the topic at hand, research would also benefit from more integrative research approaches that combine the different research perspectives within a common research framework to draw the big picture of AI-related bias and discrimination in e-business. Moreover, systematic literature reviews organizing the field of research and giving more specific research direction could also be of great use to the research community. Overall, such conceptual reference points are a crucial requirement for achieving the generally desired state of reliable empirical knowledge and may serve in particular as a foundation for hypotheses development and testing in empirical research.

A first step for future empirical research endeavours could be an empirical examination of the framework proposed, challenging it with respective data from the field of e-business. Other than that, in general, both qualitative and quantitative empirical research is necessary to provide scientifically sound descriptive and explanatory insights on this phenomenon in order to advance the field of research and our little understanding of AI-related bias and discrimination in e-business. Qualitative research approaches such as expert interviews or exploratory case studies could be used to obtain highly differentiated findings with regard to the emergence and impact of AI-related bias and discrimination in special settings or environments characterized by certain AI applications, individuals, groups, branches etc. In addition, quantitative research approaches applying methods such as regressions or structural equation modeling are especially necessary to learn more about the cause-effect relationships, interdepencies and impact of AI-related bias and discrimination in e-business. Finally, an important focus of future research approaches should lie on the identification, development and assessment of remedies against AI-related bias and discrimination in e-business.

## CONCLUSION

AI holds great potential benefits and can significantly contribute to value creation and solving long-standing problems in e-business practice. Against this background, AI is becoming more and more important for e-business companies and has recently moved high on the political agenda of many countries. AI applications are increasingly finding their way into e-business, thus affecting our whole society and many people's lives.

However, the various challenges and risks associated with AI have so far received little attention in research and practice. In particular, the pervasive issue of AI-related bias and discrimination has been widely neglected and thus is not transparent and comprehensively understood. Given the frequency and variety of recent incidents of AI-related bias and discrimination in the context of e-business as well as the resulting violation of human rights and threat to companies' AI success, the results of AI-related evaluations should be treated with great caution regarding potential discriminatory content. In addition, it is essential to raise awareness of this issue among researchers, practitioners and consumers as well as to convey a fundamental understanding of its core aspects and social implications in order to enable a broad discussion of the topic.

The explanatory framework presented takes account of this unsatisfactory and alarming state of affairs, providing a systematic and comprehensive overview of the core aspects in connection with the emergence, functioning and consequences of AI-related bias and discrimination in e-business. More specifically, the chapter explained: (1) the origin and causes of AI-related discrimination, (2) the AI learning process and different forms of AI learning underlying AI-related discrimination, (3) the AI outcome in terms of the broad spectrum of discriminating AI applications, the scope of affected areas in e-business and the different forms of AI-related discrimination, as well as (4) the reproduction, reinforcement and solidification or establishment of AI-related discrimination in e-business.

All these aspects and their interdependencies essentially form the complex phenomenon of AI-related discrimination and represent in their own right serious challenges to e-business research and practice, requiring special consideration and scrutiny. Besides the great complexity of the phenomenon and AI applications in general, particularly the black box character of many AI applications and the lack of transparency in terms of how they come to results and decisions is a critical issue in combination with their potential biases and discrimination. It not only makes it difficult for all stakeholders, in particular, users to trust their results and decisions, but also impedes scientists and developers in effectively handling and tackling AI-related bias and discrimination.

Against this background, it becomes apparent that a responsible and promising implementation and use of AI in e-business requires a critical examination of the AI-related challenges and risks, and in particular AI-related bias and discrimination. In order to effectively approach this issue, it is essential for companies that apply AI to establish a systematic AI-based risk management and implement an AI governance body that particularly entails ethical and data guidelines in order to provide a framework of action (Wirtz et al., 2019).

After all, the general question arises as to whether the advantages of AI outweigh its disadvantages and whether its application is acceptable. While there is no general and definite answer to this question, a fair assessement of an AI solution and its algorithm in terms of whether it is morally, legally, and economically justifiable should always be made against its alternatives (Thierer et al., 2017).

In this connection, it is also important to bear in mind that the potential challenges depend on the area in which AI technologies are employed: "depending on the use case or scenario, the cost of failure could be highly asymmetric to the application, which should be assessed as part of normal risk management activities" (Lloyd, 2018, p. 2). The more significant the information, decision or behavior resulting from

AI applications is for e-business companies and consumers, the more serious is the threat of discrimination against individuals or groups and companies' AI success.

In addition to the technical and conceptual solution approaches presented in the chapter, the public discourse of the topic may constitute an initial but important step towards coping with and overcoming this vital challenge. In this connection, the chapter and particularly the framework provided represent an initial supporting contribution that makes the issue of AI-related bias and discrimination clearer and more manageable to researchers, practitioners and civil society, potentially serving them as map and benchmark in their efforts to address this issue effectively and sustainably in the future.

## REFERENCES

Accenture. (2018). Accenture Launches New Artificial Intelligence Testing Services. Retrieved from https://newsroom.accenture.com/news/accenture-launches-new-artificial-intelligence-testing-services.htm

Equality Act 2010.

Agarwal, P. K. (2018). Public Administration Challenges in the World of AI and Bots. *Public Administration Review*, *78*(6), 917–921. doi:10.1111/puar.12979

Amini, A., Soleimany, A., Schwarting, W., Bhatia, S., & Rus, D. (2019). Uncovering and Mitigating Algorithmic Bias through Learned Latent Structure. In Association for the Advancement of Artificial Intelligence (Chair), AAAI/ACM Conference on Artificial Intelligence, Ethics, and Society (AIES). Symposium conducted at the meeting of ACM, Honolulu, HI. doi:10.1145/3306618.3314243

Anderson, M., & Anderson, S. L. (Eds.). (2011). *Machine Ethics*. Cambridge, UK: Cambridge University Press; doi:10.1017/CBO9780511978036.

Appling, D. S., & Briscoe, E. J. (2017). The Perception of Social Bots by Human and Machine. In *Proceedings of the Thirtieth International Florida Artificial Intelligence Research Society Conference* (pp. 20–25). Academic Press.

Arel, I., Rose, D. C., & Karnowski, T. P. (2010). Deep machine learning - a new frontier in artificial intelligence research. *IEEE Computational Intelligence Magazine*, *5*(4), 13–18. doi:10.1109/MCI.2010.938364

Bartlett, R. P., Morse, A., Stanton, R., & Wallace, N. (2017). Consumer Lending Discrimination in the FinTech Era. SSRN Electronic Journal. Advance online publication. doi:10.2139srn.3063448

Basu, A. (2018). Discrimination in the Age of Artificial Intelligence. Retrieved from http://ohrh.law.ox.ac.uk/discrimination-in-the-age-of-artificial-intelligence/

Becker, G. S. (2010). *The economics of discrimination* (2nd ed.). Chicago: University of Chicago press.

Bellamy, R. K. E., Dey, K., Hind, M., Hoffman, S. C., Houde, S., Kannan, K., . . . Zhang, Y. (2018). AI Fairness 360: An Extensible Toolkit for Detecting, Understanding, and Mitigating Unwanted Algorithmic Bias. Retrieved from https://arxiv.org/abs/1810.01943

Benítez, J. M., Castro, J. L., & Requena, I. (1997). Are artificial neural networks black boxes? *IEEE Transactions on Neural Networks*, *8*(5), 1156–1164. doi:10.1109/72.623216 PubMed

Berendt, B., & Preibusch, S. (2014). Better decision support through exploratory discrimination-aware data mining: Foundations and empirical evidence. *Artificial Intelligence and Law*, *22*(2), 175–209. doi:10.1007/s10506-013-9152-0

Bessi, A., & Ferrara, E. (2016). Social bots distort the 2016 US Presidential election online discussion. *First Monday*, *21*(11). doi:10.5210/fm.v21i11.7090

Buolamwini, J., & Gebru, T. (2018). Gender Shades: Intersectional Accuracy Disparities inCommercial Gender Classification. *Proceedings of Machine Learning Research*, *81*, 1–15.

Calmon, F., Wei, D., Vinzamuri, B., Natesan Ramamurthy, K., & Varshney, K. R. (2017). Optimized Pre-Processing for Discrimination Prevention. In I. Guyon, U. V. Luxburg, S. Bengio, H. Wallach, R. Fergus, S. Vishwanathan, & R. Garnett (Eds.), Vol. 30, pp. 3992–4001). Advances in Neural Information Processing Systems Curran Associates, Inc.; Retrieved from http://papers.nips.cc/paper/6988-optimized-pre-processing-for-discrimination-prevention.pdf

Castelvecchi, D. (2016). Can we open the black box of AI? *NATNews*, *538*(7623), 20–23. PubMed

Cath, C., Wachter, S., Mittelstadt, B., Taddeo, M., & Floridi, L. (2018). Artificial Intelligence and the 'Good Society': The US, EU, and UK approach. *Science and Engineering Ethics*, *24*(2), 505–528. doi: PubMed doi:10.100711948-017-9901-7

Challen, R., Denny, J., Pitt, M., Gompels, L., Edwards, T., & Tsaneva-Atanasova, K. (2019). Artificial intelligence, bias and clinical safety. *BMJ Quality & Safety*, *28*(3), 231–237. doi:10.1136/bmjqs-2018-008370 PubMed

Chou, J., Murillo, O., & Ibars, R. (2017). What the Kids' Game "Telephone" Taught Microsoft about Biased AI. Retrieved from https://www.fastcompany.com/90146078/what-the-kids-game-telephone-taught-microsoft-about-biased-ai

Citron, D. K., & Pasquale, F. A. (2014). The Scored Society: Due Process for Automated Predictions. *Washington Law Review (Seattle, Wash.)*, *89*(1), 1–33.

Crawford, K. (2016, June 28). Artificial Intelligence's White Guy Problem. The New York Times, p. 11. Retrieved from https://www.nytimes.com/2016/06/26/opinion/sunday/artificial-intelligences-white-guy-problem.html

Crenshaw, K. (1989). Demarginalizing the intersection of race and sex: A black feminist critique of antidiscrimination doctrine, feminist theory and antiracist politics. *University of Chicago Legal Forum*, *139*(1). Retrieved from https://chicagounbound.uchicago.edu/cgi/viewcontent.cgi?article=1052&context=uclf

Dalenberg, D. J. (2018). Preventing discrimination in the automated targeting of job advertisements. *Computer Law & Security Review*, *34*(3), 615–627. doi:10.1016/j.clsr.2017.11.009

Dastin, J. (2018). Amazon scraps secret ai recruiting tool that shoed bias against women. Retrieved from https://www.reuters.com/article/us-amazon-com-jobs-automation-insight/amazon-scraps-secret-ai-recruiting-tool-that-showed-bias-against-women-idUSKCN1MK08G

Datta, A., Tschantz, M. C., & Datta, A. (2015). Automated Experiments on Ad Privacy Settings: A Tale of Opacity, Choice, and Discrimination. Proceedings on Privacy Enhancing Technologies, 2015(1), 92–112.

Davidson, T., Warmsley, D., Macy, M., & Weber, I. (2017). Automated hate speech detection and the problem of offensive language. *Proceedings of the Eleventh International AAAI Conference on Web and Social Media (ICWSM 2017)*, 512–516.

Daws, R. (2018). Editorial: Stopping AI's discrimination will be difficult, but vital. Retrieved from https://www.artificialintelligence-news.com/2018/05/17/editorial-stopping-ai-discrimination/

Diakopoulos, N. (2016). Accountability in algorithmic decision making. *Communications of the ACM*, *59*(2), 56–62. doi:10.1145/2844110

Eisenhart, M. A. (1991). Conceptual frameworks for research circa 1991: ideas from a cultural anthropologist; implications for mathematics education researchers. In R. G. Underhill (Chair), Proceedings of the 13th annual meeting of the North American chapter of the international group for the psychology of mathematics education. Blacksburg, VA: Academic Press.

Ellis, E., & Watson, P. (2012). *EU Anti-Discrimination Law* (2nd ed.). Oxford, UK: Oxford University Press; doi:10.1093/acprof:oso/9780199698462.001.0001.

Eslami, M., Krishna Kumaran, S. R., Sandvig, C., & Karahalios, K. (2018). Communicating Algorithmic Process in Online Behavioral Advertising. In R. Mandryk, M. Hancock, M. Perry, & A. Cox (Eds.), *Proceedings of the 2018 CHI Conference on Human Factors in Computing Systems - CHI '18* (pp. 1–13). New York: ACM Press; doi:10.1145/3173574.3174006.

Eslami, M., Vaccaro, K., Karahalios, K., & Hamilton, K. (2017). "Be Careful; Things Can Be Worse Than They Appear"—Understanding Biased Algorithms and Users' Behavior Around Them in Rating Platforms. Proceedings of the Eleventh International AAAI Conference on Web and Social Media (ICWSM 2017), 62–71.

Ferrara, E., Varol, O., Davis, C., Menczer, F., & Flammini, A. (2016). The rise of social bots. *Communications of the ACM*, *59*(7), 96–104. doi:10.1145/2818717

Flores, A. W., Bechtel, K., & Lowenkamp, C. T. (2016). False Positives, False Negatives, and False Analyses: A Rejoinder to Machine Bias: There's Software Used across the Country to Predict Future Criminals. And It's Biased against Blacks. *Federal Probation*, *80*, 38.

Garcia, M. (2016). Racist in the Machine. *World Policy Journal*, *33*(4), 111–117. doi:10.1215/07402775-3813015

Gasser, U., & Almeida, V. A. F. (2017). A Layered Model for AI Governance. *IEEE Internet Computing*, *21*(6), 58–62. doi:10.1109/MIC.2017.4180835

Gianfrancesco, M. A., Tamang, S., Yazdany, J., & Schmajuk, G. (2018). Potential Biases in Machine Learning Algorithms Using Electronic Health Record Data. *JAMA Internal Medicine*, *178*(11), 1544–1547. doi:10.1001/jamainternmed.2018.3763 PubMed

Gillespie, T. (2014). The Relevance of Algorithms. In T. Gillespie, P. J. Boczkowski, & K. A. Foot (Eds.), *Media Technologies* (pp. 167–194). Cambridge, MA: The MIT Press; doi:10.7551/mitpress/9780262525374.003.0009

Goel, N., Yaghini, M., & Faltings, B. (2018). Non-Discriminatory Machine Learning through Convex Fairness Criteria. In J. Furman, G. Marchant, H. Price, & F. Rossi (Eds.), *Proceedings of the 2018 AAAI/ACM Conference on AI, Ethics, and Society - AIES '18* (p. 116). New York: ACM Press; doi:10.1145/3278721.3278722.

Goodman, B. W. (2016). A step towards accountable algorithms? algorithmic discrimination and the european union general data protection. 29th Conference on Neural Information Processing Systems (NIPS 2016), Barcelona. NIPS Foundation.

Google. (2019). At I/O '19: Building a more helpful Google for everyone. Retrieved from https://www.blog.google/technology/developers/io19-helpful-google-everyone

Gwagwa, A., & Koene, A. (2018). Minimizing algorithmic bias and discrimination in the digital economy. Working group for IEEE Standard on Algorithm Bias Considerations UNCTAD Africa e-Commerce week, Nairobi, Kenya.

Hacker, P. (2018). Teaching Fairness to Artificial Intelligence: Exisiting and Novel Strategies against Algorithmic Discrimination under EU Law. *Common Market Law Review*, *55*(4), 1143–1185.

Hajian, S., & Domingo-Ferrer, J. (2013). A methodology for direct and indirect discrimination prevention in data mining. *IEEE Transactions on Knowledge and Data Engineering*, *25*(7), 1445–1459. doi:10.1109/TKDE.2012.72

Hajian, S., Domingo-Ferrer, J., & Martínez-Ballesté, A. (Eds.). (2011). Discrimination prevention in data mining for intrusion and crime detection. 2011 IEEE Symposium on Computational Intelligence in Cyber Security (CICS). doi:10.1109/CICYBS.2011.5949405

Hale, K. (2018). IBM's Unbiased Approach To AI Discrimination. Retrieved from https://www.forbes.com/sites/korihale/2018/09/25/ibms-unbiased-approach-to-ai-discrimination/#60c28e1d7118

Hillard, D., Schroedl, S., & Manavoglu, E. (2010). Improving ad relevance in sponsored search. Proceedings of the third ACM international conference on web search and data mining. Advance online publication. doi:10.1145/1718487.1718532

Houser, K. (2019). Can AI Solve the Diversity Problem in the Tech Industry? Mitigating Noise and Bias in Employment Decision-Making. *Stanford Technology Law Review : STLR : An Online High-Technology Law Journal from Stanford Law School*.

IBM. (2018a). AI Fairness 360. Retrieved from https://developer.ibm.com/open/projects/ai-fairness-360/

IBM. (2018b). Introducing AI Fairness 360. Retrieved from https://www.ibm.com/blogs/research/2018/09/ai-fairness-360/

IBM. (n.d.). AI Fairness 360 Open Source Toolkit. Retrieved from http://aif360.mybluemix.net/

Ingold, D., & Soper, S. (2016). Amazon Doesn't Consider the Race of Its Customers. Should It? Retrieved from https://www.bloomberg.com/graphics/2016-amazon-same-day/

Joseph, M., Kearns, M., Morgenstern, J., Neel, S., & Roth, A. (2016). Rawlsian fairness for machine learning. arXiv preprint arXiv:1610.09559

Kamiran, F., Calders, T., & Pechenizkiy, M. (2013). Techniques for discrimination-free predictive models. In B. Custers, T. Calders, B. Schermer, & T. Zarsky (Eds.), *Discrimination and privacy in the information society* (pp. 223–239). Berlin: Springer; doi:10.1007/978-3-642-30487-3_12.

Khandani, A. E., Kim, A. J., & Lo, A. W. (2010). Consumer credit-risk models via machine-learning algorithms. *Journal of Banking & Finance*, *34*(11), 2767–2787. doi:10.1016/j.jbankfin.2010.06.001

Kirkpatrick, K. (2016). Battling algorithmic bias. *Communications of the ACM*, *59*(10), 16–17. doi:10.1145/2983270

Lambrecht, A., & Tucker, C. (2019). Algorithmic Bias? An Empirical Study of Apparent Gender-Based Discrimination in the Display of STEM Career Ads. *Management Science*, *65*(7), 2966–2981; Advance online publication. doi:10.1287/mnsc.2018.3093

Leavy, S. (2018). Gender bias in artificial intelligence. In E. Abraham, E. Di Nitto, & R. Mirandola (Eds.), *Proceedings of the 1st International Workshop on Gender Equality in Software Engineering - GE '18* (pp. 14–16). New York: ACM Press; doi:10.1145/3195570.3195580.

Lin, P., Abney, K., & Bekey, G. A. (Eds.). (2012). *Robot Ethics: The Ethical and Social Implications of Robotics*. Cambridge, MA: MIT Press.

Lloyd, K. (2018). Bias Amplification in Artificial Intelligence Systems. AAAI FSS-18: Artificial Intelligence in Government and Public Sector, Arlington, VA.

Lum, K., & Isaac, W. (2016). To predict and serve? *Significance*, *13*(5), 14–19. doi:10.1111/j.1740-9713.2016.00960.x

Mann, G., & O'Neil, C. (2016). Hiring Algorithms Are Not Neutral. Harvard Business Review. Retrieved from https://hbr.org/2016/12/hiring-algorithms-are-not-neutral

McGregor, S. L. T. (2018). *Understanding and evaluating research: A critical guide*. Thousand Oaks, CA: SAGE Publications.

Miles, M. B., Huberman, A. M., & Saldaña, J. (2014). *Qualitative data analysis: A methods sourcebook* (3rd ed.). Los Angeles, CA: Sage.

Munger, K. (2017). Tweetment effects on the tweeted: Experimentally reducing racist harassment. *Political Behavior*, *39*(3), 629–649. doi:10.1007/s11109-016-9373-5

Nalisnick, E., Mitra, B., Craswell, N., & Caruana, R. (2016). Improving Document Ranking with Dual Word Embeddings. In Proceedings of the 25th International Conference Companion on World Wide Web 2016 (pp. 83–84). doi:10.1145/2872518.2889361

Neff, G., & Nagy, P. (2016). Automation, algorithms, and politicsl talking to Bots: Symbiotic agency and the case of Tay. *International Journal of Communication*, *10*, 4915–4931.

Nicolas, P. R. (2014). *Scala for machine learning: Leverage scala and machine learning to construct and study systems that can learn from data. Community Experience Distilled*. Birmingham, UK: Packt Publishing.

Plane, A. C., Redmiles, E. M., Mazurek, M. L., & Tschantz, M. C. (2017). Exploring User Perceptions of Discrimination in Online Targeted Advertising. *Proceedings of the 26th USENIX Security Symposium*, 935–951.

Raji, I. D., & Buolamwini, J. (2019). Actionable auditing: Investigating the impact of publicly naming biased performance results of commercial ai products. In Association for the Advancement of Artificial Intelligence (Chair), AAAI/ACM Conference on Artificial Intelligence, Ethics, and Society (AIES). Symposium conducted at the meeting of ACM, Honolulu, HI.

Ramasubramanian, K., & Singh, A. (2017). *Machine Learning Using R*. Berkeley, CA: Apress.

Raub, M. (2018). Bots, Bias and Big Data: Artificial Intelligence, Algorithmic Bias and Disparate Impact Liability in Hiring Practices. *Arkansas Law Review*, *71*(2), 529–570.

Sandvig, C., Hamilton, K., Karahalios, K., & Langbort, C. (2014). Auditing Algorithms: Research Methods for Detecting Discrimination on Internet Platforms. In Data and Discrimination: Converting Critical Concerns into Productive Inquiry. Preconference at the 64th Annual Meeting of the International Communication Association. Retrieved from https://pdfs.semanticscholar.org/b722/7cbd34766655dea10d0437ab10df3a127396.pdf

Scherer, M. (2016). Tay the Racist Chatbot: Who is responsible when a machine learns to be evil? Retrieved from https://futureoflife.org/2016/03/27/tay-the-racist-chatbot-who-is-responsible-when-a-machine-learns-to-be-evil/?cn-reloaded=1

Shao, C., Ciampaglia, G. L., Varol, O., Flammini, A., & Menczer, F. (2017). The spread of fake news by social bots. arXiv preprint arXiv:1707.07592

Shehabuddeen, N., Probert, D., Phaal, R., & Platts, K. (1999). Representing and Approaching Complex Management Issues: Part 1 - Role and Definition. SSRN Electronic Journal. Advance online publication. doi:10.2139srn.1923155

Snow, J. (2018). Bias already exists in search engine results, and it's only going to get worse. MIT Technology Review. Retrieved from https://www.technologyreview.com/s/610275/meet-the-woman-who-searches-out-search-engines-bias-against-women-and-minorities/

Speicher, T., Ali, M., Venkatadri, G., Ribeiro, F. N., Arvanitakis, G., Benevenuto, F., ... Mislove, A. (2018). Potential for Discrimination in Online Targeted Advertising. *Proceedings of Machine Learning Research, 81*, 1–15.

Speicher, T., Heidari, H., Grgic-Hlaca, N., Gummadi, K. P., Singla, A., Weller, A., & Zafar, M. B. (2018). A Unified Approach to Quantifying Algorithmic Unfairness. In Y. Guo & F. Farooq (Eds.), *Proceedings of the 24th ACM SIGKDD International Conference on Knowledge Discovery & Data Mining - KDD '18* (pp. 2239–2248). New York: ACM Press; doi:10.1145/3219819.3220046.

Strohmeier, S., & Piazza, F. (2015). Artificial Intelligence Techniques in Human Resource Management—A Conceptual Exploration. In C. Kahraman & S. Çevik Onar (Eds.), *Intelligent Systems Reference Library. Intelligent Techniques in Engineering Management* (Vol. 87, pp. 149–172). Cham: Springer International Publishing; doi:10.1007/978-3-319-17906-3_7.

Sweeney, L. (2013). Discrimination in Online Ad Delivery. Queue, 11(3), 10. doi:10.1145/2460276.2460278

The Age Discrimination in Employment Act of 1967.

Thierer, A., O'Sullivan Castillo, A., & Russell, R. (2017). Artificial Intelligence and Public Policy. Mercatus Research. Retrieved from https://www.mercatus.org/system/files/thierer-artificial-intelligence-policy-mr-mercatus-v1.pdf

Trewin, S. (2018). AI Fairness for People with Disabilities: Point of View. CoRR. Retrieved from https://arxiv.org/abs/1811.10670v1

UN Committee on Economic. Social and Cultural Rights (2009). General comment No. 20: Non-discrimination in economic, social and cultural rights (art. 2, para. 2, of the International Covenant on Economic, Social and Cultural Rights). Retrieved from https://www.refworld.org/docid/4a60961f2.html

Universal Declaration of Human Rights, United Nations General Assembly 1948.

Veale, M., & Binns, R. (2017). Fairer machine learning in the real world: Mitigating discrimination without collecting sensitive data. Big Data & Society, 4(2), 1–17. doi:10.1177/2053951717743530

Weiss, G. (1999). *Multiagent systems: a modern approach to distributed artificial intelligence*. Cambridge, MA: MIT Press.

West, S. M., Whittaker, M., & Crawford, K. (2019). Discriminating Systems: Gender, Race and Power in AI. Retrieved from https://ainowinstitute.org/discriminatingsystems.pdf

Weyerer, J. C., & Langer, P. F. (2019). Garbage In, Garbage Out: The Vicious Cycle of AI-Based Discrimination in the Public Sector. In Y.-C. Chen, F. Salem, & A. Zuiderwijk (Eds.), Proceeding dg.o 2019 20th Annual International Conference on Digital Government Research (pp. 509–511). New York, NY: Academic Press.

Williams, B. Brooks., & Shmargad. (2018). How Algorithms Discriminate Based on Data They Lack: Challenges, Solutions, and Policy Implications. Journal of Information Policy, 8, 78–115. doi:10.5325/jinfopoli.8.2018.0078

Winter, J. (2015). Algorithmic Discrimination: Big Data Analytics and the Future of the Internet. In J. Winter & R. Ono (Eds.), *The Future Internet* (pp. 125–140). Cham: Springer International Publishing; doi:10.1007/978-3-319-22994-2_8.

Wirtz, B. W., Weyerer, J. C., & Geyer, C. (2019). Artificial Intelligence and the Public Sector—Applications and Challenges. *International Journal of Public Administration*, *42*(7), 596–615. doi:10.1080/01900692.2018.1498103

Woodruff, A., Fox, S. E., Rousso-Schindler, S., & Warshaw, J. (2018). A Qualitative Exploration of Perceptions of Algorithmic Fairness. In R. Mandryk, M. Hancock, M. Perry, & A. Cox (Eds.), *Proceedings of the 2018 CHI Conference on Human Factors in Computing Systems - CHI '18* (pp. 1–14). New York: ACM Press; doi:10.1145/3173574.3174230.

World Economic Forum. (2018). How to Prevent Discriminatory Outcomes in Machine Learning: White Paper. Retrieved from http://www3.weforum.org/docs/WEF_40065_White_Paper_How_to_Prevent_Discriminatory_Outcomes_in_Machine_Learning.pdf

Zhang, L., Wu, Y., & Wu, X. (2018). Achieving Non-Discrimination in Prediction. In J. S. Rosenschein & J. Lang (Chairs), Twenty-Seventh International Joint Conference on Artificial Intelligence {IJCAI-18}. Stockholm, Sweden: Academic Press.

Zhao, J., Wang, T., Yatskar, M., Ordonez, V., & Chang, K.-W. (2017). Men Also Like Shopping: Reducing Gender Bias Amplification using Corpus-level Constraints. In M. Palmer, R. Hwa, & S. Riedel (Eds.), *Proceedings of the 2017 Conference on Empirical Methods in Natural Language Processing* (pp. 2979–2989). Stroudsburg, PA: Association for Computational Linguistics; doi:10.18653/v1/D17-1323.

## ADDITIONAL READING

Bellamy, R. K. E., Dey, K., Hind, M., Hoffman, S. C., Houde, S., Kannan, K., . . . Zhang, Y. (2018). AI Fairness 360: An Extensible Toolkit for Detecting, Understanding, and Mitigating Unwanted Algorithmic Bias. CoRR, abs/1810.01943. Retrieved from https://arxiv.org/abs/1810.01943

Bostrom, N., & Yudkowsky, E. (2014): The ethics of artificial intelligence. In: The Cambridge handbook of artificial intelligence. doi:10.1017/CBO9781139046855.020

Hacker, P. (2018). Teaching Fairness to Artificial Intelligence: Exisiting and Novel Strategies against Algorithmic Discrimination under EU Law. *Common Market Law Review*, *55*(4), 1143–1185.

Raso, F., Hilligoss, H., Krishnamurthy, V., Bavitz, C., & Kim, L. Y. (2018): Artificial Intelligence & Human Rights: Opportunities & Risks. In: SSRN Journal. Doi:10.2139srn.3259344

Risse, M. (2018): Human Rights and Artificial Intelligence: An Urgently Needed Agenda. In: Publicum 4 (1). Doi:10.12957/publicum.2018.35098

West, S. M., Whittaker, M., & Crawford, K. (2019): Discriminating Systems: Gender, Race and Power in AI. Hg. v. AI Now Institute. Online verfügbar unter https://ainowinstitute.org/discriminatingsystems.pdf, zuletzt geprüft am 21.05.2019.

Weyerer, J. C., & Langer, P. F. (2019). Garbage In, Garbage Out: The Vicious Cycle of AI-Based Discrimination in the Public Sector. In Y.-C. Chen, F. Salem, & A. Zuiderwijk (Eds.), Proceeding dg.o 2019 20th Annual International Conference on Digital Government Research (pp. 509–511). New York, NY, USA.

Wirtz, B. W., Weyerer, J. C., & Geyer, C. (2018). Artificial Intelligence and the Public Sector—Applications and Challenges. *International Journal of Public Administration, 13*(7), 1–20. doi:10.1080/01900692.2016.1242620

World Economic Forum. (2018): How to Prevent Discriminatory Outcomes in Machine Learning. White Paper. Online verfügbar unter http://www3.weforum.org/docs/WEF_40065_White_Paper_How_to_Prevent_Discriminatory_Outcomes_in_Machine_Learning.pdf, zuletzt geprüft am 21.05.2019.

## KEY TERMS AND DEFINITIONS

**AI-Related Bias and Discrimination Reproduction:** The reinforcing and endless repetition of bias and discriminatory content of an AI application by other AI applications and humans.

**Algorithm:** An explicit rule of action for solving a mathematical problem.

**Artificial Intelligence:** An IT system's capability to show human-like intelligence by perceiving, understanding, learning, or acting like humans.

**Bias:** Statistics leading to a distortion that adds an unfair result to a particular group.

**Discrimination:** The unfair treatment of individuals based on their affiliation to a particular group.

**E-Business:** Economic activities and exchange processes by means of information and communication technology.

**Framework:** A conceptual systematic compilation of key concepts of a phenomenon explaining its core aspects and their interrelationships.

**Social Bots:** AI-based programs that mimic individuals on social media platforms, creating or sharing content independently.

# Chapter 12
# Delivering Effective Cybersecurity Awareness Training to Support the Organizational Information Security Function

**Regner Sabillon**
https://orcid.org/0000-0003-1807-2208
*Universitat Oberta de Catalunya, Spain*

## ABSTRACT

*Traditional security education, training, and awareness (SETA); cybersecurity awareness programs; and information security awareness programs are falling behind to deal with the current cyberthreat landscape in any organizational environment. Human behaviors are the weakest links in cybersecurity, especially in situations where cyberthreats are not isolated, blocked, or reported to the information security specialists for further action. Moreover, the study compares recent awareness frameworks, approaches, and methodologies. An extended research that includes an awareness training model to deal with existing challenges when delivering cybersecurity to different levels of positions in any organization. The cybersecurity awareness training model (CATRAM) has been designed to deliver training to different organizational audiences, each of these groups with specific content and separate objectives. The study concluded by addressing the need for future and innovative research to target new approaches to keep cybersecurity awareness focused on the everchanging cyberthreat landscape.*

DOI: 10.4018/978-1-7998-1879-3.ch012

## INTRODUCTION

A comprehensive Information Security Program must include Security Education, Training and Awareness (SETA). One of the main responsibilities of the Chief Information Security Officer (CISO) is to design and implement a successful and measurable SETA in any corporate environment. SETA aims to reduce the ocurrence of accidental cybersecurity breaches by members of any organization including Board of Directors, C-Suite Executives, Managers, Employees, Consultants, Vendors and Business Partners who are working with its cyber assets.

Cybersecurity awareness training programs may not be productive if people's behavior remain unchanged, and as result a positive corporate impact cannot be achieved. A cybersecurity awareness program is an organizational long-term investment, that will help to establish a cybersecurity culture if training is delivered on a continuous basis with relevant content that is aligned with the current cyber threat landscape. Cybersecurity directors could envision a more interactive experience that will allow staff to be more proactive beyond the point of dealing with cyber incidents and cyberthreats.

We consider that the Cybersecurity Awareness TRAining Model (CATRAM) can represent a substantial foundation for the implementation of any organizational cybersecurity awareness program. CATRAM can also assess any awareness training model that is persistent and relevant with the current cyberthreat landscape.

Whitman and Mattord (2019) highlight that a SETA program consists of security education, security training and security awareness. Organizations may or may not able to develop an in-house SETA program and would consider to outsource to educational institutions or Companies specializing in delivering security training. SETA programs are implemented to build in-depth knowledge for organizational security programs and protection of information assets, to develop skills for end users so they can perform their job responsibilities with security in mind and by improving security awareness for the protection of information assets.

According to Whitman et al. (2019), a SETA is defined as:

*A managerial program designed to improve the security of information assets by providing targeted knowledge, skills, and guidance for organizational employees.*

*Security Awareness, Security Education and Security Training* are the main components of any SETA program. *Security Education* consists of delivering the knowledge of information security issues and operations, *Security Training* provides participants with skills, knowledge and abilities to user their assigned resources in a wise manner and *Security Awareness* refers to create consciousness when dealing with InfoSec or cybersecurity matters.

Regardless of any cybersecurity control or security measure, users are recognized as the existing weakest link in securing systems and in the aftermath, they are not aware that the consequences of their actions may affect information security (NIST SP 800-12, 2017). Pendergast (2016) also suggests that employees are responsible for the financial losses caused by cybersecurity incidents and data breaches.

According to the NTT Group (2017), integrating cybersecurity processes can empower digital transformations of any organization based on the maturity level of their digital transformation projects. By doing so, Companies will identify and control cyber risks, reduce complexity of cybersecurity architecture for their business operations, add value to their digital business, define the prioritization of areas to deal with business-critical cyber risks and plan a cyber resilience plan for the digital transformation.

A recent study from Fujitsu (2017), that evaluated the digital transformations from 1,625 decision makers in different industries and sector across fourteen countries concluded that 86% of the participating organizations have a clear digital strategy that includes abilities to address cybersecurity issues. Moreover, 52% of these organizations are investing in cybersecurity or 51% in implementing Internet of Things (IoT) to strengthen their digital transformations.

A study from LeClair et al. (2013), highlights the interdisciplinary approach to cybersecurity education by targeting the three major pillars of any cybersecurity program (People, Technology and Process). The authors point out that cybersecurity approaches to all industries and requires an interdisciplinary focus to obtain the necessary skills needed in any organization. Cybersecurity awareness programs are designed to train all employees from all different departments and every employee across any organization will be contributing to the overall business cybersecurity program. With the evolving cyberthreat landscape, companies must find ways to protect their cyber assets from continuous cyberattacks by implementing different layers of security in all cybersecurity domains. Cybersecurity awareness education programs should be aligned with constant innovation in order to protect brick-and-mortar businesses along with e-commerce transactions for e-businesses.

The chapter is structured as follows: We start with the chapter Introduction and Background Section. In the Literature Review Section, we study literature related to our research topics. In the next two Sections, we introduce the architecture of our cybersecurity model; the Cybersecurity Awareness TRAining Model (CATRAM). The research methodolody is explained in the next Section. The dataset from the validation scenario is discussed in the following Section where we summarize the study's results. Finally, the conclusions are provided in the last Section.

## BACKGROUND

This chapter studies a novel model for creating, developing, delivering and maintaining a CyberSecurity Education, Training and Awareness (CSETA) methodology that was validated in a Canadian Higher Education organization. The implementation was part of a multi-case study along with the CyberSecurity Audit Model (CSAM); another novel model to conduct cybersecurity audits for 18 security domains.

Existing frameworks and methodologies in Table 1, have different approaches to deliver security awareness training but not all of them include clear objectives and topics to be aligned with the current cyberthreat landscaping. Furthermore, the SETA frameworks and methodologies lack specific objectives and content aligned with roles and responsibilities of employees and stakeholders of any organization.

The Cybersecurity Awareness TRAining Model (CATRAM) was designed distinctively to deliver cybersecurity awareness training to specific groups within any organization. CATRAM was conceived to deliver the awareness training for the members of the Board od Directors, Top Executives, Managers, IT (Information Technology) staff and of course, end-users.

In this particular research scenario, CATRAM was implemented as the foundational model of our target organization. This organization did not have any SETA policy in place and CATRAM was validated to introduce cybersecurity awareness for their employees. Nowadays, CATRAM is being used to develop the future cybersecurity awareness training program of this institution.

## LITERATURE REVIEW

According to the Gartner Magic Quadrant (2016) for Security Awareness Computer-Based Training (CBT) where leaders, visionaries, challengers and niche players are positioned. The Leaders are SANS Institute, Wombat Security Technologies, PhishMe, MediaPro, Security Innovation, Inspired eLearning, Terranova WW, PhishLine, Global Learning Systems, The Security Awareness Co.; Visionary vendors are Popcom Training and Security Mentor; Challenger vendors are BeOne Development, KnowBe4 and Optiv Security and last but not least are niche players like Junglemap, Digital Defense, Symantec (Blackfn Security) and Secure Mentem. Two years later (Gartner, 2018), we have seen significant changes for the Security Awareness Computer-Based Training market where positions are different in all quadrants. The Leaders are Proofpoint (Wombat Security), MediaPro, Cofense, KnowBe and Terranova.; Visionary vendors are Inspired eLearning and Barracuda (PhishLine); Challenger vendors are SANS Institute, InfoSec Institute and Global Learning Systems and finally are the niche players like Junglemap, Security

Innovation and Sophos. The new vendors positioned as leaders, clearly identified market needs and incorporate new features on their CBT products, by including security topics aligned with the everchanging cyberthreat landscaping. Vendors continue to differentiate security awareness products and services by introducing a variety of formats, lengths and styles, by providing gamification, multilanguage support, supplemental internal marketing content like newsletters, intranet postings and security alerts, and integration with partnerships to offer endpoint detection and response, endpoint protection and data security. Another research study from Gartner (2018), suggests that by 2023 organizations that have implemented security awareness programs will go through 75% fewer account takeover attacks in comparison with other organizations, that is because effective security awareness programs must have a commitment from upper management and be in alignment with any organization's needs, practices and culture. Companies face many challenges when deciding, delivering, implementing and maintaining a cybersecurity awareness training that is tailored to their specific business environment, strategy, needs and objectives. For instance, selecting which topics to include when delivering the training, how to deliver training to personnel, how to verify the effectiveness of the training, updating the training program, implementing control measures to test cyber behaviors in the workplace and defining the frequency to re-train stakeholders.

A recent study from Ponemon Institute (2018) surveyed 1,021 IT and IT security practitioners in the USA and Europe, the Middle East and Africa (EMEA) to study Domain Name System (DNS) architecture, implementation and to identify responsibilities that manage cybersecurity activities in organizations. Based on the results of the study, Ponemon Institute and Infoblox created the DNS Risk Index by categorizing five different areas: visibility, DNS attack protection, data protection and malware mitigation, threat intelligence and security operations. The most salient findings of this study show that most companies do not have dedicated staff to address DNS security, most companies are not tracking or identifying cyber assets, traffic analysis from firewalls is mostly used for malware mitigation and data assets protection, use of threat intelligence feeds is ineffective, measures to protect data assets include antivirus, endpoint security and data encryption and most cyberthreat investigations are conducted manually. The results show that the greatest concerns in terms of cyberattacks are advanced malware (63%), Advanced Persistent Threats (APTs – 59%), DNS-based data exfiltration (54%), unauthorized network access (51%), Ransomware (46%) and phishing/social engineering (45%).

The Global Security Awareness Report from SANS (2017), points out that time and communication were identified as the critical takeaways to a thriving awareness program. The findings highlighted poor communication to engage people, the problem of time and lack of resources being assigned to a corporate awareness program. The participants reported that they implemented awareness and behavior change (54.6%), had a compliance awareness program (27.1%), achieved long-term sustainment and culture change (9.8%), defined a program with robust metrics (0.9%) and did not have a cybersecurity awareness program at all (7.6%).

Symantec (2014) argues that poorly trained personnel increases the risks of disclosure and loss of sensitive data like Personal Identifiable Information (PII) and Intellectual Property (IP). Its Security Awareness Program reduces vulnerabilities by creating a corporate culture and train employees to protect any organization critical assets from cyberattacks, exploitation, fraud and unauthorized access. The main topics of Symantec's training program are information security, threats, vulnerabilities, countermeasures, securing the workplace, securing mobile users, protecting Internet information, social media mobile device security.

A research study from Enterprise Management Associates (EMA, 2014) reported that 56% of personnel, not including IT and security staff, have not received any security awareness training in their organizations and 84% of participants recognized that the awareness training from their workplaces was also used to decrease cyber risks at home. Furthermore, the study findings confirmed that the existing security awareness programs lack the appropriate delivery periodicity, content and quality. In addition, Company size, market and budgets have a significant impact on the existence and maturity of their corporate awareness training.

ESET (2017) provides free online cybersecurity awareness training to train employees and get a certification. The topics include an overview of threats like malware, phishing and social engineering; best practices for password management; best practices for email protection and preventive measures that cover best practices for cyber hygiene at the workplace and at home. PhishMe also provides access to a free of charge Computer Based Training (CBT) course called PhishMe CBFree which contains seventeen security awareness modules and four compliance training modules. The course is available in seven languages (English, Chinese, French, German, Portuguese, Spanish and Japanese). The Compliance modules are General Data Protection Regulation (GDPR), Payment Data, Personal Data and Health Care; The security awareness modules cover cybersecurity awareness, cloud computing, advanced spear phishing, business email compromise, ransomware, surfing the Web, data protection, insider threats, malicious links, malware, mobile devices, security outside of the office, passwords, physical security, social engineering, social networking and spear phishing (PhishMe, 2017). Table 1 presents an overview of most models and frameworks linked to best practices for the definition and consolidation of cybersecurity awareness programs.

Industrial and critical infrastructure organizations can also be targets of any cyberattack, as these organizations rely their businesses on Industrial Control Systems (ICS). Global malware attacks such as NotPetya, WannaCry and Emotet as well more targeted ICS cyberattacks such as Industroyer and TRITON, are just a few examples that can impact production outages, clean-ups, catastrophic safety and environmental incidents. The Global ICS & Industrial Internet of Things (IIOT) risk report (Cyberx, 2019) analyzed data from 850 production ICS networks using Network Traffic Analysis (NTA) in conjunction with deep packet inspections. The major findings included that 40% of industrial sites have at least one direct connection to the Internet, 53% of sites have obsolete Windows systems, 69% of sites have plain-text passwords traversing their networks, 57% of sites are not running anti-virus solutions that include automatic signature updates, 16% of sites have at least a misconfigured Wireless Access Point (WAP) and 84% of industrial sites have at least one remotely accessible device without multifactor authentication controls.

Our literature review approach used mixed methods (Qualitative and Quantitative studies), to select the material as initial references in our multi-case study. The lead researcher used computerized databases and the Internet searching for keywords like "security training"; "information security training"; "SETA"; "cybersecurity awareness training"; "cybersecurity awareness training program"; "cybersecurity training framework" and "security awareness training program."

Axelos (2015) suggests that cyber-resilience specific training should be delivered on a regular basis, training should be designed and tailored to specific organizational roles and responsibilities of employees, awareness campaigns should be created to raise awareness and to address specific cyber risks. Nevertheless, we have to keep finding innovative ways to deliver cybersecurity awareness training and above all, keep people engaged with cybersecurity awareness activities.

## THE CYBERSECURITY AWARENESS TRAINING MODEL (CATRAM)

We want to introduce the proposed Cybersecurity Awareness TRAining Model (CATRAM), as an innovative model that can be implemented at any organization to consolidate the awareness foundations of a corporate Cybersecurity Awareness Program or to initiate the implementation of an organizational Cybersecurity Awareness Training Program (See Figure 1). The model design answers our main research question:

*Why it is necessary to increase cyber awareness at the organizational and personal levels?*

The outcome of this research was to design a model for delivering cyber awareness training to support awareness education in any organizational environment. The Cybersecurity Awareness TRAining Model (CATRAM) has been designed to deliver

*Table 1. Cybersecurity Awareness approaches*

| Framework or Methodology that Focuses on Cybersecurity Awareness | Phases |
|---|---|
| ISO/IEC 27001:2005 (2005) | There aren't any specific phases or recommendations for the security awareness delivery. Clause 5.2.2 highlights the importance of necessary personnel competencies to support the Information Security Management System (ISMS), providing training to satisfy needs, maintaining training records and that the organization is responsible for the awareness training of relevant personnel |
| ISO/IEC 27032 (2012) | Section 2.4 covers the training and awareness program.<br>Defining training needs, designing and planning training, defining awareness program requirements and setting up training and awareness evaluation |
| Hewlett Packard Progressive Engagement Framework (Beyer et al., 2015) | 1. Awareness Profiling<br>• Company profiling<br>• Awareness assessment<br>• Gap analysis<br>• Awareness maturity level report<br>2. Awareness Planning<br>• Communication, education and training concept<br>• Awareness improvement plan<br>3. Transformation<br>• Creation, production and measures implementation<br>• Support of internal core team<br>4. Optimization<br>• Comparison between target and actual state<br>• Adjust and optimize accordingly |
| SANS Security Awareness Maturity Model (2017) | This Awareness Maturity Model is organized in five areas:<br>1. Non-Existent: An awareness program does not exist<br>2. Compliance Focused: The awareness program is either aligned with a compliance or audit requirements<br>3. Promoting awareness and behavior change: This program is focused on training topics that have greatest impacts to support the mission of the organization<br>4. Long-Term Sustainment and Culture Change: This program is aligned with a corporate cybersecurity program. It has processes, resources and leadership support<br>5. Robust Metrics Framework: This is a mature awareness program with a robust metrics framework in place |
| MediaPro Adaptive Awareness Framework (2017) | 1. Analyze: Use data to inform about the program<br>2. Plan: Draw a roadmap for planning the awareness program<br>3. Train: Build training to achieve real behavior changes<br>4. Reinforce: Battle the forgetting curve |
| Beyer-Brummel Comparative Cybersecurity Training Framework (2015) | Organized by levels:<br>1. Targeted: To produce non-IT cybersecurity skills to exact role specific performance<br>2. Education: To cultivate IT security insight and understanding<br>3. Advanced: To equip IT security professionals to address assurance, policy and training |
| NIST- Key steps leading to the implementation of the awareness and training program (2014) | 1. Design Awareness and Training Program<br>2. Develop Awareness and Training material<br>3. Implement Program<br>4. Post-Implementation |
| PCI Data Security Standard (PCI DSS)- Best Practices for Implementing a Security Awareness Program (2014) | 1. Assemble the Security Awareness Team<br>2. Determine Roles for Security Awareness<br>3. Target delivery of relevant material to the appropriate audience in an efficient and timely way<br>4. Define the Security Awareness training content<br>5. Define assessment metrics of the awareness training<br>6. Follow the Security Awareness Program checklist |
| Cano – Basic Model of the level of maturity of the Organizational Information Security Culture (2016) | To measure the maturity of certain elements of the InfoSec culture<br>Elements:<br>1. Culture foundations<br>2. Access foundations<br>3. Information understanding<br>4. Basic Instruments<br>5. Management compromise<br>Maturity indicators:<br>1. Reactive<br>2. Unstable<br>3. Proactive<br>4. Sustainable |

*continued on following page*

*Table 1. Continued*

| Framework or Methodology that Focuses on Cybersecurity Awareness | Phases |
|---|---|
| MITRE- Model to question the validity of any email (2010) | EARNEST uses a series of questions to challenge the validity of an electronic message:<br>Expected: It the email expected?<br>Ambiguous: Is it asking to open attachments?<br>Relationship: Any prior relationship with sender?<br>Normal: Are context, grammar, syntax and spelling consistent from your contact?<br>Exposed: Any malicious links in your email?<br>Sense: Does it make sense to receive a link or attachment?<br>Time: Is there any time factor for responding? |
| Cyber Safe Workforce – Security Awareness and Training Program (2016) | The awareness lifecycle includes the following phases:<br>1. Identify and Define: Define and create training plan<br>2. Baseline: Gather initial set of data for future trainining scope<br>3. Train: Deliver guidance and information training<br>4. Track & Measure: Audit participation and gather good metrics<br>5. Evaluate & Update: Keep updating your training program all the time |
| Whitman and Mattord - Framework of Security Education, Training, and Awareness (2019) | The Framework provides six components that are applicable to Awareness, Training and Education:<br>1. Attributes: It seeks teaching members the importance of security by focusing on what, how and why<br>2. Levels: Knowledge transfer from basic, detailed and in-depth levels<br>3. Objectives: Based on threat recognition, effective responses using learned skills and engagement of active defense<br>4. Teaching methods: Some examples include informal training, hands-on practice and seminars<br>5. Assessments: By using different evaluation techniques like problem solving and essays<br>6. Impact timeframe: Short-term, intermediate and long-term |
| Nguyen et al. – InCAT (Intelligence-based Cybersecurity Awareness Training) (2018) | A model for conducting cybersecurity training with a strong focus on drilling deep into the shared contexts among collected cyber awareness training results, cyberthreat intelligence reports, and other cybersecurity related data logs. InCAT feedback loop includes 8 steps:<br>1. Threat reports<br>2. Annotation model flow to knowledge discovery model<br>3. Derived knowledge from threat reports<br>4. Tests for users<br>5. Initial user reports<br>6. User report verification by the annotation model<br>7. User assessment reports<br>8. Final results in control dashboard system |

the initial cybersecurity awareness training at any organization or to re-introduce a better awareness training approach to an existing cybersecurity or information security awareness training program.

CATRAM has been designed to provide specific cybersecurity awareness training for personnel:

1. **Board of Directors and Executives**: Members of this group are trained based on the organizational cybersecurity strategy, governance and program.
2. **Managers**: Department managers are trained to support and lead cybersecurity initiatives in their corporate environment.
3. **End Users**: This group gets awareness training to improve cybersecurity practices in the workplace and their personal lives.
4. **IT Staff**: Information Technology specialists are trained in the use of advanced cybersecurity techniques, methods, procedures and best practices to support the corporate awareness program and the cybersecurity program.

Each awareness course has been designed with a specific outline, objectives, content and cybersecurity topics in alignment with the target audience, the organizational scope and aim, the cybersecurity awareness program and the corporate cybersecurity program. Cybersecurity topic choice are based on the group's main responsibilities and also on the cybersecurity domains that this group will be dealing with in the workplace on their daily tasks.

## A. Awareness Course for the Board of Directors and Executives

The course lasts 2 hours and can be organized in two different sessions. It is advisable that this course could be delivered in a classroom or board meeting environment.

*Figure 1. The Cybersecurity Awareness TRAining Model (CATRAM)*

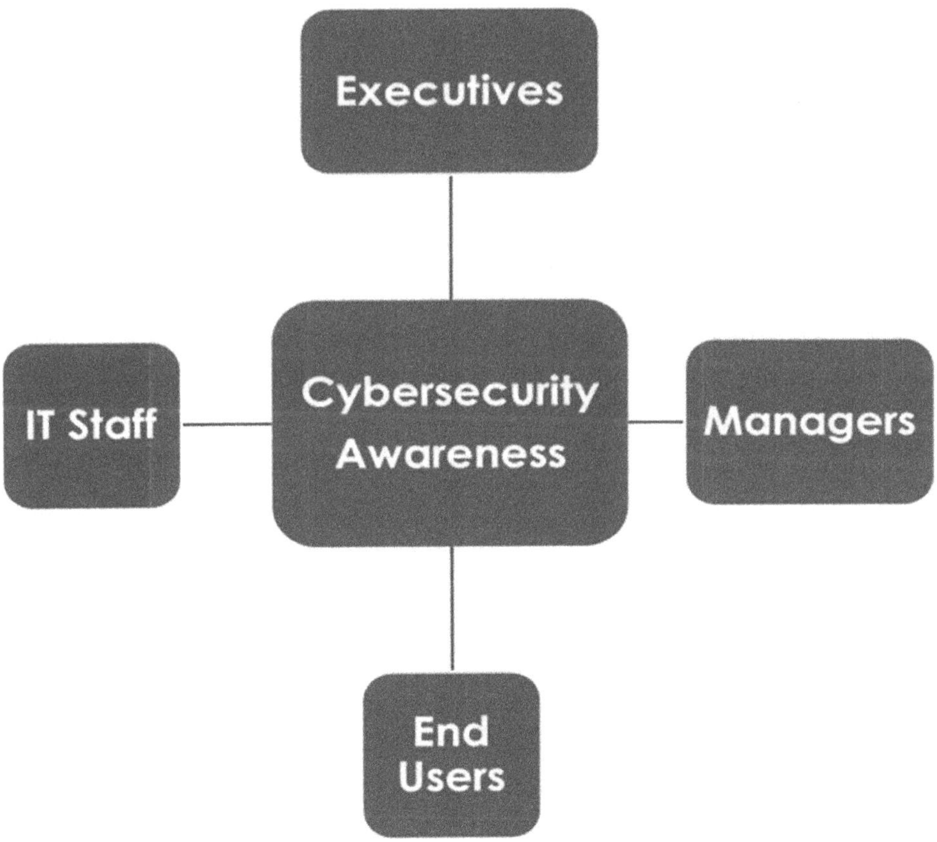

## Objectives

1. Provide a high-level overview of an effective cybersecurity awareness training for your organization.
2. Create cybersecurity awareness for the Board of Directors and C-Suite Executives.

Cybersecurity Awareness Topics

- Initial Survey
- Cybersecurity Introduction
- Cybersecurity and Cybercrime Statistics
- A Corporate Cybersecurity Program
- Cybersecurity Strategy
- Responsibilities of Stakeholders (Board of Directors and C-Suite Executives)
- Cyberthreat Landscape
- Cybersecurity Risk Management
- Cybersecurity Frameworks
- Cybersecurity Awareness and Training
- Cybersecurity Business Continuity
- Incident Response Management
- Conclusions
- Final Survey

## B. Awareness Course for Managers

The course lasts 2 hours and can be organized in two different sessions. The course can be delivered in a classroom setting, online or a blended environment.

## Objectives

1. Provide a high-level overview of an effective cybersecurity awareness training for your organization.
2. Create cybersecurity awareness for Managers.

Cybersecurity Awareness Topics

- Initial Survey
- Cybersecurity Introduction
- Cybersecurity and Cybercrime Statistics

- A Corporate Cybersecurity Program
- Cybersecurity Strategy
- Responsibilities of Stakeholders (Department Managers)
- Cyberthreat Landscape
- Cybersecurity Risk Management
- Cybersecurity Frameworks
- Cybersecurity Awareness and Training
- Cybersecurity Business Continuity
- Incident Response Management
- Conclusions
- Final Survey

## C. Awareness Course for End Users

The course lasts 4 hours and can be organized in two or four different sessions. The course can be delivered in a classroom setting, online or a blended environment. It is recommended to add a short video clip from YouTube as additional learning resource for your audiences.

### Objectives

1. Educate end users to help protecting the confidentiality, availability and integrity of your organization's information and cyber assets.
2. Create awareness of the importance of cybersecurity and cybersecurity controls.

Cybersecurity Awareness Topics

- Initial Survey
- Cybersecurity Introduction
- Cybersecurity and Cybercrime Statistics
- You are a target for cybercriminals
- Cybercrime
- Hackers
- Cyberthreats
- Social Engineering
- Phishing
- Internet Browsing
- Social Networks
- Mobile device security
- Passwords

- Encryption
- Data security
- Identity Theft
- Wi-Fi Security
- Working remotely
- Physical security
- Protecting your online profile
- Protecting your home network
- Protecting our children online
- Privacy
- Avoiding Scams
- Have you been hacked?
- Conclusions
- Final Survey

## D. Awareness Course for IT Professionals

The course lasts 20 hours and can be organized in ten or twenty different sessions. The course can be delivered in a classroom setting, online, self-paced e-doing or a blended environment.

### Objectives

1. Understand cybersecurity concepts.
2. Recognize key cybersecurity objectives for the protection of cyber assets.
3. Understand cybercrime operations.
4. Recognize cybersecurity threat agents that could impact your organization.
5. Understand any cyberattack architecture.
6. Identify most common cyberattacks.
7. Apply cybersecurity measures to defend against cyberattacks.
8. Understand a cybersecurity program architecture and operation.
9. Recognize the importance of developing, enforcing and maintaining cybersecurity policies.
10. Understand the fundamentals of ethical hacking.
11. Understand the architecture of penetration testing.
12. Get familiar with most cybersecurity frameworks.
13. Understand the basics of cyber threat intelligence.
14. Understand the importance of proper cybersecurity training.
15. Raise cybersecurity awareness in your organization.
16. Apply cybersecurity architecture principles.

17. Recognize the importance of hardening security in data, voice and video networks.
18. Recognize the importance of security hardening for information, systems and applications.
19. Identify cybersecurity vulnerabilities.
20. Remediate existing cybersecurity vulnerabilities.
21. Recognize the cybersecurity implications of new and evolving technologies.
22. Understand the principles of Cybersecurity Incident Response and Management.
23. Understand the fundamentals of Digital Forensics.
24. Recognize the importance of the continual evaluation of a corporate cybersecurity program.
25. Recognize the value of corporate cyber wargames to test cybersecurity.
26. Identify the opportunities for cybersecurity education and professional development.

Cybersecurity Awareness Topics

- Initial Survey
- Cybersecurity Fundamentals
- Cybercrime
- Cyberattacks
- Corporate Cybersecurity Program
- Cybersecurity Policies
- Ethical Hacking
- Penetration Testing
- Cyber Operations
- Cybersecurity Frameworks
- Cyber Threat Intelligence
- Cybersecurity Awareness and Training Program
- Architecture and Networks
- Information, Systems and Applications
- Vulnerability Management
- Evolving Technologies
- Incident Response Management
- Digital Forensics
- Enterprise Cybersecurity Assessment
- Cybersecurity Corporate Wargames
- Cybersecurity Education
- Final Survey

Alotaibi et al. (2016) highlight that one of the best ways to deal with cybercrime is by creating awareness and by adopting effective cybersecurity practices for people.

## MEASURING THE MODEL RESULTS

The results of CATRAM can be measured once all training courses have been delivered. Most of the assessment could be measured at the end user level by evaluating changes in security behaviors and alignment with corporate cybersecurity compliance. Ideally, end users must be advised that the effectiveness of the awareness training will be evaluated by performing announced assessments, and the delivery of non-announced assessment exercises as well.

Table 2 introduces suggested awareness areas and participating groups to assess the compliance and the impact of the cybersecurity awareness model.

Hayden (2016) introduces a model to measure the levels of security culture strength. The strength of the security culture could be a function of the organizational awareness and training program or it could the result of a highly regulated industry: A weak security culture (80% occurrence of a bad decision); a moderate security culture (50% occurrence of a bad decision) and a strong security culture (20% occurrence of a bad decision).

We evaluate the cybersecurity awareness by measuring compliance by addressing the following criteria:

- Does your organization have a cybersecurity awareness program?
- Do you provide some kind of cybersecurity training to your staff?
- Is training delivered on a regular recurring basis?
- Do employees are following security policies of the organization?
- Are you delivering training to recognize and deal with social engineering?
- Do your staff know how to recognize and report a security incident?
- Is your staff able to detect and respond to any cybersecurity emergency?
- Do you enforce privacy and confidentiality requirements in your organization?
- Are your employees following security procedures for data and information protection?
- Is your awareness training focused and delivered to specific audiences like end users, managers, IT, C-Suite executives and Board of Directors?
- Is your awareness training covering multidimensional topics?
- Does your training outline cover technical, social and user behavior areas?

*Table 2. CATRAM metric identifiers and objectives*

| Metric Identifier | Group | Metric Objectives |
|---|---|---|
| Cybersecurity Awareness and Training Effectiveness | Executives | Identify training gap needs and approve training courses |
| Cyber policy-making assessment | Executives | Review, update and approve cybersecurity policies |
| Cyber monitoring, metric definition and reporting | Executives | Approve required cybersecurity metrics |
| Awareness training completion | Managers | Verify that all staff completes training for every department |
| Communication flow | Managers | Enforce the distribution of awareness communication and proper training documentation |
| Cybersecurity incidents volume | IT | Evaluate Help Desk monthly report |
| Cybersecurity skills | IT | Evaluate new cybersecurity skills of technical staff that is consistent with the organization growth and operations |
| Infected digital devices | IT | Identify percentage on a monthly basis |
| Phishing awareness and detection | End Users | Identify phishing victims and users that are able to avoid phishing attacks |
| Social Media risks | End Users | Evaluate percentage of user's time |
| Password management | End Users | Assess user's behavior for password management |

## Evaluation Scorecard

We calculate the final cybersecurity maturity rating of the cybersecurity awareness training domain by using the criteria from Table 3. The score can be mapped to a specific maturity level.

Gartner (2018) indicates that security and risk management leaders must provision awareness training to employees in order to focus on protecting their online security and the personal aspects of cybersecurity, knowledge transfer of good practices to protect intellectual property and data in corporate environments. Gartner also suggests a series of best practices to develop and maintain a cybersecure worforce:

1. By nurturing a holistic cybersecure personal lifestyle that includes good hygiene for identity management and security awareness
2. By committing to training and awareness behavior that encircles corporate training, workshops and the use of the proper tools

3. By building trust that verifies employees' online behavior through timely tests of cybersecure hygiene

## METHODOLOGY

The Cybersecurity Awareness TRAining Model (CATRAM) has been tested, implemented and validated along with the CyberSecurity Audit Model (CSAM) in a Canadian higher education institution (Sabillon et al., 2019). The research project assessed the cybersecurity organizational strategy, implemented the CyberSecurity Audit Model (CSAM) and delivered cybersecurity awareness training to more than one hundred participants based on the Cybersecurity Awareness TRAining Model (CATRAM). The CyberSecurity Audit Model (CSAM) is a new exhaustive model that encloses the optimal assurance assessment of cybersecurity in any organization and it can verify specific guidelines for Nation States that are planning to implement a National Cybersecurity Strategy (NCS) or want to evaluate the effectiveness of its National Cybersecurity Strategy or Policy already in place. The CSAM has 18 domains; domain 1 is specific for Nation States and domains 2-18 can be implemented at any organization. The CyberSecurity Audit Model (CSAM) contains overview, resources, 18 domains, 26 sub-domains, 87 checklists, 169 controls, 429 sub-controls, 80 guideline assessment and an evaluation scorecard.

*Table 3. Cybersecurity awareness training maturity rating*

| Rating | Description |
|---|---|
| Immature (I): 0-30 | The organization does not have any plans to manage its cybersecurity. Controls for critical cybersecurity areas are inexistent or very weak. The organization has not implemented a comprehensive cybersecurity program nor an awareness training program. |
| Developing (D): 31-70 | The organization is starting to focus on cybersecurity matters. If technologies are in place, the organization needs to focus on key areas to protect cyber assets. Attention must be focused towards staff, processes, controls and regulations. The Awareness Education domain is developing. The organization has a foundation model for cybersecurity awareness and additional efforts are required to develop a complete cybersecurity awareness program. |
| Mature (M): 71-90 | While the organization has a mature cybersecurity awareness environment. Improvements are required to the key areas that have been identified with weaknesses. |
| Advanced (A): 91-100 | The organization has excelled in implementing cybersecurity awareness training best practices. There is always room for improvement. Keep documentation up-to-date and continually review cybersecurity processes through audits. |

The case study research included several phases like plan, design, preparation, collection, analysis, sharing and dissemination. We intended to perform qualitative research by utilizing interpretive material practices such as online and paper surveys, interviews, classroom and online training and analysis of documentation, processes and procedures of the target institution. We completed a multi-case study research following Yin's methodology (2018; 2014) to plan, design, prepare, collect, analyze and share phases by creating, implementing and validating two innovative cybersecurity models (CATRAM & CSAM). This initial validation of the CATRAM and CSAM took place in a Canadian Higher Education Institution. More recently, the CSAM has been validated for the second time in a larger Canadian Higher Education Institution.

The organization provided their staff time to support the case study research, resources to conduct the cybersecurity audit, the provision of classroom space and time, computer use, Internet access for the delivery of the cybersecurity awareness training courses, the access to their computer systems to conduct the research and to design the online courses in their Learning Management System (Moodle). The name of the target organization and participant's details cannot be disclosed to protect their cyber assets and due to an anonymity agreement.

We compare our cybersecurity awareness model with ten existing frameworks or methodologies that are based on cybersecurity awareness and training practices, these selected frameworks and methodologies are used by many organizations worldwide to design a corporate cybersecurity awareness training program. Table 4 shows a comparison of CATRAM against recent frameworks or methodologies.

On the other hand, we compare our model against not so recent frameworks and methodologies (Between 2012 and 2016) in Table 5. CATRAM has been designed to provide specific cybersecurity awareness training for personnel:

1. **Board of Directors and Executives**: Executives are trained based on the organizational cybersecurity strategy, governance and program.
2. **Managers**: Department managers are trained to support and lead cybersecurity initiatives in their corporate environment.
3. **End Users**: End users get awareness training to improve cybersecurity practices in the workplace and their personal lives.
4. **IT Staff**: Information Technology specialists are trained in the use of advanced cybersecurity techniques, methods, procedures and best practices to support the corporate awareness program and the cybersecurity program.

*Table 4. Frameworks or methodologies that focus on cybersecurity awareness training (Since 2017)*

| Frameworks Or Methodologies That Focus On Cybersecurity Awareness Training | | | | | |
|---|---|---|---|---|---|
| **Key Features** | **ESET (2017)** | **PhishMe (2017)** | **SANS Security Awareness Maturity Model (2017)** | **MediaPro (2017)** | **CATRAM (2017)** |
| | -Free cybersecurity awareness training based on Powepoint presentations, videos and webinars. It includes the following topics:<br>• Threats overview<br>• Password policies<br>• Web protection<br>• Email protection<br>• Preventive measures<br>-Participants obtain a training completion certificate | -Cofense provides a series of free cybersecurity awareness training resources including Cofense Phisme Free, Cofense CB Free, awareness presentations for encryption ransomware, business email compromise, awareness, spearphishing and awareness posters/ infographics | -This Awareness Maturity Model is organized in five areas:<br>1. Non-Existent: An awareness program does not exist<br>2. Compliance Focused: The awareness program is either aligned with a compliance or audit requirements<br>3. Promoting awareness and behavior change: This program is focused on training topics that have greatest impacts to support the mission of the organization<br>4. Long-Term Sustainment and Culture Change: This program is aligned with a corporate cybersecurity program. It has processes, resources and leadership support<br>5. Robust Metrics Framework: This is a mature awareness program with a robust metrics framework in place<br>-Time and communication are major takeaways that could affect the success of any cybersecurity awareness and training program | -MediaPro sells general security awareness packages based on the following topics:<br>• Threats<br>• Social engineering<br>• Malware<br>• Passwords<br>• Phishing<br>• Data classification<br>• Data storage and retention<br>• Cloud computing<br>• Physical security<br>• Mobile device protection<br>• Social media<br>-Courses are organized on essentials, advanced and adaptive packages | - CATRAM breaks traditional awareness training paradigms by targeting specific audiences and different sets of objectives in any organization<br>- CATRAM delivers cybersecurity awareness for members of the Board of Directors and Executives related to strategy and governance; for managers at the tactical level; a broad cybersecurity content for end users at the operational level and specialized training for IT staff<br>- CATRAM can be delivered and implemented to start a corporate cybersecurity awareness training program or to improve an existing one |

## RESULTS

Before conducting our case study research, our target organization did not have any cybersecurity awareness model nor any cybersecurity awareness education program whatsoever. The CATRAM delivery allowed the organization, to build a strong foundation for a future implementation of a comprehensive cybersecurity awareness training program. The cybersecurity audit of the awareness education

*Table 5. Comparison of CATRAM against older frameworks or methodologies that focus on cybersecurity awareness Training*

| Frameworks or Methodologies that Focus on Cybersecurity Awareness Training | | | | | | | |
|---|---|---|---|---|---|---|---|
| **Key Features** | **ISO/IEC 27032 (2012)** | **NIST- Key Steps Leading to the Implementation of the Awareness and Training Program (2014)** | **PCI DSS (2014)** | **Beyer-Brummel Comparative Cybersecurity Training Framework (2015)** | **Hewlett Packard Progressive Engagement Framework (Beyer et al., 2015)** | **Cyber Safe Workforce – Security Awareness and Training Program (2016)** | **CATRAM (2017)** |
| | The ISO/IEC 27032:2012 standard provides security techniques and guidelines for cybersecurity<br>- Clause 12.5.3.2 highlights the requirements for creating awareness and delivering cybersecurity training:<br>-Regular updating of knowledge and learning<br>-Any cybersecurity program should require minimum hours of training for employees and third-parties contractors (Related to their roles and responsibilities in cyberspace) | The NIST framework includes the folowing phases to design and implement the awareness program:<br>1. Design Awareness and Training Program<br>2. Develop Awareness and Training material<br>3. Implement Program<br>4. Post-Implementation | Payment Card Industry (PCI) Awareness training is specific for the PCI Data Security Standard (PCI DSS). Training complies with PCI DSS Requirement 12.6 for general security awareness education<br>- It is a four-hour course that can be delivered either online or for training groups | The Beyer-Brummel Comparative Cybersecurity Training Framework (2015) presents an approach based on cybersecurity levels and objectives for training and education:<br>-Targeted: To produce non-IT cybersecurity skills to exact role specific performance<br>-Education: To cultivate IT security insight and understanding<br>-Advanced: To equip IT security professionals to address assurance, policy and training | Personal commitment, budget and human resources are important to reach the awareness maturity level<br>- This framework proposes the following components in order to review the role of security for employees:<br>1. Awareness Profiling<br>• Company profiling<br>• Awareness assessment<br>• Gap analysis<br>• Awareness maturity level report<br>2. Awareness Planning<br>• Communication, education and training concept<br>• Awareness improvement plan<br>3. Transformation<br>• Creation, production and measures implementation<br>• Support of internal core team<br>4. Optimization<br>• Comparison between target and actual state<br>• Adjust and optimize accordingly | The awareness lifecycle includes the following phases:<br>Identify and Define: Define and create training plan<br>Baseline: Gather initial set of data for future trainining scope<br>Train: Deliver guidance and information training<br>Track & Measure: Audit participation and gather good metrics<br>Evaluate & Update: Keep updating your training program all the time | The results of CATRAM can be measured once all training courses have been delivered<br>-Most of the assessment could be measured at the end user level by evaluating changes in security behaviors and alignment with corporate cybersecurity compliance<br>- Ideally, end users must be advised that the effectiveness of the awareness training will be evaluated by performing announced assessments, and the delivery of non-announced assessment exercises as well<br>- CATRAM can be delivered in classroom, online and blended settings |

domain was conducted after the successful delivery and implementation of CATRAM. We conducted the audit of the awareness education based on the CSAM and the most relevant noncompliances are the lack of a corporate cybersecurity awareness training program and the confirmation that staff are aware that cyber training is not being delivered and it exists the necessity to eradicate this weakness (Table 6). The critical controls that need immediate attention are that the target organization does not have a valid cybersecurity awareness program, and employees were not aware how important is to keep training them in cyber topics to increase awareness and show a proactive participation in order for their potential awareness training program to be successful.

A series of tables are included to present the findings in this research scenario. Table 6 illustrates the assessment of the main cybersecurity awareness education controls. Table 7 contains the sub-controls findings based on the audit checklist. For further details you are referred to our dataset 3 in Mendeley. Major nonconformities need to addressed and corrected. Staff need to able to identify and report any cybersecurity incident, enforce privacy, confidentialy and protection for any Personally Identifiable Information (PII) for the internal and stakeholders of the institution.

Table 8 confirms that the cybersecurity awareness training is at a '*developing stage*' and consequently needs improvement in our target organization. The higher education institution needs to implement a full cybersecurity awareness and training program for all stakeholders. Partial awareness training is ineffective. The validation of the CATRAM helped the target organization to implement a foundation for their future cybersecurity awareness training program. While the CATRAM implementation was delivered for the Board of Directors, C-Suite Executives, Managers, IT staff and end users thus a critical recommendation was to train their students and external stakeholders as well.

In terms of recommendations, we suggested the creation and implementation of the corporate cybersecurity awareness training program, to define ownership to maintain the training program and the CATRAM update on an annual basis or as new cyberthreats emerge, to conduct "Train the Trainer" sessions for designated instructors or facilitators and, last but not least the constant evaluation of staff using

*Table 6. Control evaluation of the cybersecurity awareness education domain*

| Reference | Sub Area | Clause | Steps | Control Evaluation | | Checklist |
|---|---|---|---|---|---|---|
| | | | | Yes | No | |
| 13.1 | Awareness | 13.1.1 | Organization deploys a cybersecurity awareness program | ☐ | X | CSAM-Awareness |
| | | 13.1.2 | The awareness training program is delivered on an annual basis | X | ☐ | |
| | | 13.1.3 | Employees are aware of the need of this kind of training program | ☐ | X | |
| | | 13.1.4 | The training program is designed for different staffing levels | X | ☐ | |
| | | 13.1.5 | Training material is constantly updated as new cyber threats emerge | X | ☐ | |

*Table 7 Control evaluation of the cybersecurity awareness education*

| Cybersecurity Audit Checklist: CSAM – Awareness Education (Domain 13) | | | | | |
|---|---|---|---|---|---|
| | | | Findings | | |
| **Clause** | **No.** | **Checklist Questions** | **Compliant** | **Minor Nonconformity** | **Major Nonconformity** |
| 13.1.1 | 1 | Does your organization have a cybersecurity awareness program? | ☐ | ☐ | X |
| 13.1.1 | 2 | Do you provide some kind of cybersecurity training to your staff? | X | ☐ | ☐ |
| 13.1.2 | 3 | Is training delivered on a regular recurring basis? | X | ☐ | ☐ |
| 13.1.1 | 4 | Do employees are following security policies of the organization? | X | ☐ | ☐ |
| 13.1.1 | 5 | Are you delivering training to recognize and deal with social engineering? | X | ☐ | ☐ |
| 13.1.1 | 6 | Do your staff know how to recognize and report a security incident? | ☐ | X | ☐ |
| 13.1.1 | 7 | Are your personnel able to detect and respond to any cybersecurity emergency? | ☐ | ☐ | X |
| 13.1.1 | 8 | Do you enforce privacy and confidentiality requirements in your organization? | ☐ | ☐ | X |
| 13.1.1 | 9 | Are your employees following security procedures for data and information protection? | ☐ | ☐ | X |
| 13.1.4 | 10 | Is your awareness training focused and delivered to specific audiences like end users, managers, IT, C-Suite executives and Board of Directors? | X | ☐ | ☐ |
| 13.1.1 | 11 | Is your awareness training covering multidimensional topics? | X | ☐ | ☐ |
| 13.1.1 | 12 | Does your training outline cover technical, social and user behavior areas? | X | ☐ | ☐ |

scheduled and non-scheduled assessments to evaluate and understand cybersecurity awareness and behaviors.

## CONCLUSION

The main objective of this case study was to design and validate a cybersecurity model; the Cybersecurity Awareness TRAining Model (CATRAM) to address the challenges to deliver cybersecurity awareness training based on staff roles. The

*Table 8. Overall cybersecurity awareness rating*

| Cybersecurity Awareness TRAining Model (CATRAM) | | | |
|---|---|---|---|
| **Domain** | **13-Awareness Education** | | |
| **Control Evaluation** | **Ratings** | | **Score** |
| | Immature | ☐ | |
| | Developing | X | 60% |
| | Mature | ☐ | |
| | Advanced | ☐ | |
| Developing (D): 31-70<br>The organization is starting to focus on cybersecurity matters. If technologies are in place, the organization needs to focus on key areas to protect cyber assets. Attention must be focused towards staff, processes, controls and regulations.<br>The Awareness Education domain is developing. The organization has a foundation model for cybersecurity awareness and additional efforts are required to develop a complete cybersecurity awareness program. | | | |

cybersecurity model including all its components were successfully validated by a case study performed in a Canadian higher education institution.

CATRAM can support the implementation of a foundation or for consolidating a cybersecurity awareness training program at any organization. The results of this study show that the delivery of cybersecurity training based on organizational roles and responsibilities tend to motivate personnel to create and maintain awareness in their workplaces as well in their personal lives.

The limitation of our study is that CATRAM was validated in a single organization, time constraints, lack of interest for the topics and lack of engagement were some of the challenges that we have to overcome from some of the participants. Hence, future testing will enhance the model results by engaging more organizations. The case study findings have implications for our target organization but at the same time, implications for future research to review and expand our proposed cybersecurity model.

## REFERENCES

Alotaibi, F., Furnell, S., Stengel, I., & Papadaki, M. (2016). A review of Using Gaming Technologies for Cyber-Security Awareness. *International Journal of Information Security Research*, *6*(2), 660–666. doi:10.20533/ijisr.2042.4639.2016.0076

Axelos. (2015). *Cyber Resilience Best Practices*. Norwich: Resilia.

Beyer, M., Ahmed, S., Doerlemann, K., Arnell, S., Parkin, S., Sasse, A., & Passingham, N. (2015). *Awareness is only the first step: A framework for progressive engagement of staff in cyber security*. Hewlett Packard Enterpise.

Beyer, R., & Brummel, B. (2015). *Implementing Effective Cyber Security Training for End Users of Computer Networks*. Society for Human Resource Management and Society for Industrial and Organizational Psychology.

Cano, J. (2016). La educación en seguridad de la información. Reflexión pedagógicas desde el pensamiento de sistemas. *Memorias 3er Simposio Internacional en "Temas y problemas de Investigación en Educación: Complejidad y Escenarios para la Paz"*.

Cano, J. (2016). Modelo de madurez de cultura organizacional de seguridad de la información. Una visión desde el pensamiento sistémico-cibernético. Actas de la XIV Reunión Española sobre Criptología y Seguridad de la Información, 24-29.

Cyber, X. (2019). 2019 Global ICS & IIoT Risk Report. A data-driven analysis of vulnerabilities in our industrial and critical infrastructure. *CyberX Labs*. Retrieved from https://cyberx-labs.com/resources/risk-report-2019/

ESET. (2017). *ESET Cybersecurity Awareness Training*. ESET Canada. Retrieved from https://www.eset.com/ca/cybertraining/

Fujitsu. (2017). *The Digital Transformation PACT*. Retrieved from https://www.fujitsu.com

Gartner. (2016). *2016 Gartner Magic Quadrant for Security Awareness Computer-Based Training Vendors*. Gartner, Inc.

Gartner. (2018). *How to Build an Enterprise Security Awareness Program*. Gartner, Inc.

Gartner. (2018). *How to Secure the Human Link*. Gartner, Inc.

Gartner. (2018). *Magic Quadrant for Security Awareness Computer-Based Training*. Gartner, Inc.

Group, N. T. T. (2017). *Embedding cybersecurity into digital transformation - a journey towards business resilience*. NTT Security. Retrieved from https://www.nttsecurity.com

Hayden, L. (2016). *People-Centric Security: Transforming your Enterprise Security Culture*. New York: Mc Graw Hill.

International Organization for Standardization - ISO. (2005). *ISO/IEC 27001:2005 – Information Technology – Security Techniques – Information Security Management Systems – Requirements*. ISO.

International Organization for Standardization -ISO. (2012). *ISO/IEC 27032:2012 – Information Technology – Security Techniques – Guidelines for Cybersecurity*. ISO.

LeClair, J., Abraham, S., & Shih, L. (2013). An Interdisciplinary Approach to Educating an Effective Cyber Security Worforce. In *Proceedings of Information Security Curriculum Development Conference* (pp. 71-78). Kennesaw, GA: Academic Press.

MediaPro. (2017). *A Best Practices Guide for Comprehensive Employee Awareness Programs*. MediaPro.

MITRE. (2010). *The Importance of Using EARNEST.* The MITRE Corporation. Retrieved from https://www.mitre.org/sites/default/files/pdf/mitre_earnest.pdf

MITRE. (2017). *Cybersecurity Awareness & Training*. The MITRE Corporation.

Monahan, D. (2014). *Security Awareness Training: It's not just for Compliance-Research Report Summary. Enterprise Management Associates*. EMA.

National Institute of Standards and Technology – NIST. (2003). *Building an Information Technology Security Awareness and Training Program.* NIST Special Publication 800-50.

National Institute of Standards and Technology – NIST. (2017). *An Introduction to Information Security.* NIST Special Publication 800-12 Revision 1.

Nguyen, T.N., Sbityakov, L., & Scoggins, S. (2018). *Intelligence-based Cybersecurity Awareness Training- an Exploratory Project*. CoRR, abs/1812.04234.

PCI Security Standards Council - PCI DSS. (2014). *Best Practices for Implementing a Security Awareness Program*. PCI DSS.

Penderdast, T. (2016). How to Audit the Human Element and Assess Your Organization's Security Risk. *ISACA Journal*, *5*, 1–5.

PhishMe. (2017). PhishMe CBFree. *PhishMe Headquarters*. Retrieved from https://phishme.com/resources/cbfree-computer-based-training/

Ponemon Institute. (2018). *Assessing the DNS Security Risk. Research report sponsored by Infoblox*. Ponemon Institute LLC.

Sabillon, R. (2018). *Scenario III: Data for a single cybersecurity domain audit (Awareness Education)*. Mendeley Data. doi:10.17632/m4dk8n9sx7.2

Sabillon, R., Serra-Ruiz, J., Cavaller, V., & Cano, J. (2017). A Comprehensive Cybersecurity Audit Model to Improve Cybersecurity Assurance: The CyberSecurity Audit Model (CSAM). *Proceedings of Second International Conference on Information Systems and Computer Science (INCISCOS)*. 10.1109/INCISCOS.2017.20

Sabillon, R., Serra-Ruiz, J., Cavaller, V., Jeimy, J., & Cano, M. (2019). An Effective Cybersecurity Training Model to Support an Organizational Awareness Program: The Cybersecurity Awareness TRAining Model (CATRAM). A Case Study in Canada. *Journal of Cases on Information Technology*, *21*(3), 26–39. doi:10.4018/JCIT.2019070102

SANS Institute. (2017). *2017 Security Awareness Report: It's time to communicate. SANS Security Awareness*. Retrieved from https://securingthehuman.sans.org/media/resources/STH-SecurityAwarenessReport-2017.pdf

SANS Security Awareness, . (2017). *2017 Security Awareness Report*. SANS Institute.

Symantec. (2014). *Symantec Security Awareness Program: Mitigate information risk by educating your employees*. Symantec Corporation.

Ward, M. (2016). *Security Awareness and Training: Solving the unintentional insider threat*. Cyber Safe Worforce LLC.

Whitman, M. E., & Mattord, H. J. (2019). *Management of Information Security* (6th ed.). Boston, MA: Cengage Learning, Inc.

Yin, R. K. (2014). *Case Study Research: Design and Methods* (5th ed.). Thousand Oaks, CA: Sage Publications.

Yin, R. K. (2018). *Case Study Research and Applications* (6th ed.). Thousand Oaks, CA: Sage Publications.

# Compilation of References

Abduh, D' Souza, Quazi, & Burley. (2007). Investigating and Classifying Clients' Satisfaction with Business Incubator Services. Managing Service Quality: An International Journal, 17(1).

Abetti, P. A. (2004). Government-Supported Incubators in the Helsinki Region, Finland: Infrastructure, Results, and Best Practices. *The Journal of Technology Transfer*, *29*(1), 19–40. doi:10.1023/B:JOTT.0000011179.47666.55

Abu-Shaqra, B., & Luppicini, R. (2018). A technoethical study of ethical hacking communication and management within a Canadian university. In R. Luppicini (Ed.), *The changing scope of technoethics in contemporary society* (pp. 307–326). Hershey, PA: IGI Global. doi:10.4018/978-1-5225-5094-5.ch017

Accenture. (2018). Accenture Launches New Artificial Intelligence Testing Services. Retrieved from https://newsroom.accenture.com/news/accenture-launches-new-artificial-intelligence-testing-services.htm

Adams, D. A., Nelson, R. R., & Todd, P. A. (1992). Perceived Usefulness, Ease of Use, and Usage of Information Technology: A Replication. *Management Information Systems Quarterly*, *16*(2), 227–247. doi:10.2307/249577

Agarwal, A., & Saxena, A. (2019). An accountability of various clustering techniques for improvement of page searching process and analysis. *International Conference on Sustainable Computing in Science, Technology & Management (SUSCOM-2019)* 10.2139srn.3354444

Agarwal, P. K. (2018). Public Administration Challenges in the World of AI and Bots. *Public Administration Review*, *78*(6), 917–921. doi:10.1111/puar.12979

Agarwal, S. P. (2002). *Strengthening technology incubation system for creating high technology-based enterprises in Asia and the Pacific*. Economic and Social Commission for Asia and the Pacific.

Aghaei, S., Nematbakhsh, M. A., & Farsani, H. K. (2012). Evolution of the world wide web: From web 1.0 to web 4.0. *International Journal of Web & Semantic Technology*, *3*(1), 1–10. doi:10.5121/ijwest.2012.3101

Agosti, M., & Di Nunzio, G. (2007). Web log mining: A study of user sessions. *Proceedings of the 10th DELOS Thematic Workshop on Personalized Access, Profile Management, and Context Awareness in Digital Libraries (PersDL).*

Aguado, D., Rico, R., Rubio, V. J., & Fernández, L. (2016). Applicant reactions to social network web use in personnel selection and assessment. *Missing Publication Name*, *30*(2), 67–74.

Ajzen, I. (1991). The theory of planned behavior. Organizational Behavior and Human Decision Processes, 50(2), 179–211.

Ajzen, I. (1991). The Theory of Planned Behavior. *Organizational Behavior and Human Decision Processes*, *50*(2), 179–211. doi:10.1016/0749-5978(91)90020-T

Al-Ali, A. G., & Phaal, R. (2019). Design Sprints for Roadmapping an Agile Digital Transformation. Proceedings - 2019 IEEE International Conference on Engineering, Technology and Innovation, ICE/ITMC 2019. doi:10.1109/ICE.2019.8792597

Aldás-Manzano, J., Lassala-Navarré, C., Ruiz-Mafé, C., & Sanz-Blas, S. (2009). The role of consumer innovativeness and perceived risk in online banking usage. *International Journal of Bank Marketing*, *27*(1), 53–75. doi:10.1108/02652320910928245

Aldekhail, M. (2016). Application and significance of web usage mining in the 21st century: A literature review. *International Journal of Computer Theory and Engineering*, *8*(1), 41–47. doi:10.7763/IJCTE.2016.V8.1017

Alexander, N., & Colgate, M. (2000). Retail financial services: Transaction to relationship marketing. *European Journal of Marketing*, *34*(8), 938–953. doi:10.1108/03090560010331432

Allden, N., & Harris, L. (2013). Building a positive candidate experience: Towards a networked model of e-recruitment. *The Journal of Business Strategy*, *34*(5), 36–47. doi:10.1108/JBS-11-2012-0072

Alotaibi, F., Furnell, S., Stengel, I., & Papadaki, M. (2016). A review of Using Gaming Technologies for Cyber-Security Awareness. *International Journal of Information Security Research*, *6*(2), 660–666. doi:10.20533/ijisr.2042.4639.2016.0076

Amini, A., Soleimany, A., Schwarting, W., Bhatia, S., & Rus, D. (2019). Uncovering and Mitigating Algorithmic Bias through Learned Latent Structure. In Association for the Advancement of Artificial Intelligence (Chair), AAAI/ACM Conference on Artificial Intelligence, Ethics, and Society (AIES). Symposium conducted at the meeting of ACM, Honolulu, HI. doi:10.1145/3306618.3314243

Anderson, M., & Anderson, S. L. (Eds.). (2011). *Machine Ethics*. Cambridge, UK: Cambridge University Press; doi:10.1017/CBO9780511978036.

Andreadis, E., Garza-Reyes, J. A., & Kumar, V. (2017). Towards a conceptual framework for value stream mapping (VSM) implementation: An investigation of managerial factors. *International Journal of Production Research*, *55*(23), 7073–7095. doi:10.1080/00207543.2017.1347302

Angelopoulos, M., Kontakou, C., & Pollalis, Y. (2019). *Digital Transformation and Lean Management. Challenges in the Energy Industry of Utilities. A Review*. Academic Press.

Anitha, V., & Isakki, P. (2016). A survey on predicting user behavior based on web server log files in a web usage mining. *2016 International Conference on Computing Technologies and Intelligent Data Engineering (ICCTIDE'16)*, 1-4. 10.1109/ICCTIDE.2016.7725340

APCO. (2010). *APCO Project 25: Statement of requirements*. APCO.

Apers, C., & Derous, E. (2017). Are they accurate? Recruiters' personality judgments in paper versus video resumes. *Computers in Human Behavior*, *73*, 9–19. doi:10.1016/j.chb.2017.02.063

Appling, D. S., & Briscoe, E. J. (2017). The Perception of Social Bots by Human and Machine. In *Proceedings of the Thirtieth International Florida Artificial Intelligence Research Society Conference* (pp. 20–25). Academic Press.

Arel, I., Rose, D. C., & Karnowski, T. P. (2010). Deep machine learning - a new frontier in artificial intelligence research. *IEEE Computational Intelligence Magazine*, *5*(4), 13–18. doi:10.1109/MCI.2010.938364

Aronica, V., Abbiati, E., Bonfiglio, F., Idone, G., & Ruffino, C. (2018). Digital Transformation. Retrieved October 31, 2019, from https://www.eng.it/resources/whitepaper/doc/digital-transformation/engineering-digital-transformation-whitepaper-en.pdf

Ashby, R. W. (1954). *Design for a Brain*. New York: John Wiley & Sons Inc. doi:10.5962/bhl.title.6969

Asian Development Bank. (2017). *Promoting Logistics Development in Rural Areas*. ADB.

Aslanyürek, M. (2016). İnternet ve sosyal medya kullanıcılarının internet güvenliği ve çevrimiçi gizlilik ile ilgili kanaatleri ve farkındalıkları. *Maltepe Üniversitesi İletişim Fakültesi Dergisi*, *3*(1), 80–106.

Awad, N. F., & Ragowsky, A. (2008). Establishing trust in electronic commerce through online word of mouth: An examination across genders. *Journal of Management Information Systems*, *24*(4), 101–121. doi:10.2753/MIS0742-1222240404

Axelos. (2015). *Cyber Resilience Best Practices*. Norwich: Resilia.

Babbar, S., Behara, R. S., Koufteros, X. A., & Huo, B. (2017). Emergence of Asia and Australasia in operations management research and leadership. *International Journal of Production Economics*, *184*, 80–94. doi:10.1016/j.ijpe.2016.11.018

Baer, M. D., Van Der Werff, L., Colquitt, J. A., Rodell, J. B., Zipay, K. P., & Buckly, F. (2018). Trusting the 'look and feel': Situational normality, situational aesthetics, and the perceived trustworthiness of organisations. *Academy of Management Journal*, *61*(5), 1718–1740. doi:10.5465/amj.2016.0248

Bagozzi, R., & Yi, Y. (1988). On the evaluation of structural equation models. *Academy of Marking Science*, *16*, 76–94.

Baheti, R., & Gill, H. (2011). Cyber-Physical Systems. In T. Samad & A. M. Anaswamy (Eds.), The Impact of Control Technology. Academic Press.

Balaji, M. S., & Roy, S. K. (2017). Value co-creation with Internet of things technology in the retail industry. *Journal of Marketing Management*, *33*(1-2), 7–31. doi:10.1080/0267257X.2016.1217914

Baldi, P., & Long, A. D. (2001). A Bayesian framework for the analysis of microarray expression data: Regularized t-test and statistical inferences of gene changes. *Bioinformatics (Oxford, England)*, *17*(6), 509–519. doi:10.1093/bioinformatics/17.6.509 PMID:11395427

Bange, C., Handford, M., & Janoschek, N. (2014). *Information culture leveraging the power of collective ıntelligence for better decision making. BARC Research Study*. BARC Institute.

Bangerter, A. A., Roulin, N., & König, C. J. (2012). Personnel selection as a signalling game. *The Journal of Applied Psychology*, *97*(4), 719–738. doi:10.1037/a0026078 PMID:22040263

Bansal, H., Taylor, S., & James, Y. (2005). 'Migrating' to new service providers: Toward a unifying framework of consumers' switching behaviors. *Journal of the Academy of Marketing Science*, *33*(1), 96–115.

Bartlett, R. P., Morse, A., Stanton, R., & Wallace, N. (2017). Consumer Lending Discrimination in the FinTech Era. SSRN Electronic Journal. Advance online publication. doi:10.2139srn.3063448

Baruch, Y., & Altman, Y. (2016). The ecosystem of labor markets and careers. *People and Strategy*, *39*(3), 16.

Basu, A. (2018). Discrimination in the Age of Artificial Intelligence. Retrieved from http://ohrh.law.ox.ac.uk/discrimination-in-the-age-of-artificial-intelligence/

Batagan, L., & Boja, C. (2015). Social networks a real solution for students' future jobs. *Amfiteatru Economic*, *17*(9), 1303–1313.

Batista, P., & Silva, M. (2001). Mining web access logs of an on-line newspaper. *The Proceedings of 12th International Meeting of the Euro Working Group on Decision Support Systems.*

Beaulah, S. A., Correll, M. A., Munro, R. E., & Sheldon, J. G. (2008). Addressing informatics challenges in Translational Research with workflow technology. *Drug Discovery Today*, *13*(17-18), 771–777. doi:10.1016/j.drudis.2008.06.005 PMID:18617017

Becker, G. S. (2010). *The economics of discrimination* (2nd ed.). Chicago: University of Chicago press.

Becker, K. G., Hosack, D. A., Dennis, G., Lempicki, R. A., Bright, T. J., Cheadle, C., & Engel, J. (2003). PubMatrix: A tool for multiplex literature mining. *BMC Bioinformatics*, *4*(1), 61. doi:10.1186/1471-2105-4-61 PMID:14667255

Beck, J. M., Ma, W. J., Kiani, R., Hanks, T., Churchland, A. K., Roitman, J., ... Pouget, A. (2008). Probabilistic population codes for Bayesian decision making. *Neuron*, *60*(6), 1142–1152. doi:10.1016/j.neuron.2008.09.021 PMID:19109917

Bellamy, R. K. E., Dey, K., Hind, M., Hoffman, S. C., Houde, S., Kannan, K., . . . Zhang, Y. (2018). AI Fairness 360: An Extensible Toolkit for Detecting, Understanding, and Mitigating Unwanted Algorithmic Bias. Retrieved from https://arxiv.org/abs/1810.01943

Belsey, C. (1980). *Critical Practice*. London: Methuen; doi:10.4324/9780203328439.

Benítez, J. M., Castro, J. L., & Requena, I. (1997). Are artificial neural networks black boxes? *IEEE Transactions on Neural Networks*, *8*(5), 1156–1164. doi:10.1109/72.623216 PubMed

Benitez, J., Chen, Y., Teo, T. S., & Ajamieh, A. (2018). Evolution of the impact of e-business technology on operational competence and firm profitability: A panel data investigation. *Information & Management*, *55*(1), 120–130. doi:10.1016/j.im.2017.08.002

Berendt, B., & Preibusch, S. (2014). Better decision support through exploratory discrimination-aware data mining: Foundations and empirical evidence. *Artificial Intelligence and Law*, *22*(2), 175–209. doi:10.1007/s10506-013-9152-0

Berendt, B., & Spiliopoulou, M. (2000). Analyzing navigation behaviour in web sites integrating multiple information systems. *The International Journal on Very Large Data Bases*, *9*(1), 56–75. doi:10.1007007780050083

Bergek, A., & Norrman, C. (2008). Incubator Best Practice: A Framework. *Technovation*, *28*(1–2), 20–28. doi:10.1016/j.technovation.2007.07.008

Berkelaar, B. L., & Buzzanell, P. M. (2014). Cybervetting, person–environment fit, and personnel selection: Employers' surveillance and sensemaking of job applicants' online information. *Journal of Applied Communication Research*, *42*(4), 456–476. doi:10.1080/00909882.2014.954595

Berkelaar, B. L., Scacco, J. M., & Birdsell, J. L. (2015). The worker as politician: How online information and electoral heuristics shape personnel selection and careers. *New Media & Society*, *17*(8), 1377–1396. doi:10.1177/1461444814525739

Berman, S. J. (2012). Digital transformation: Opportunities to create new business models. *Strategy and Leadership*, *40*(2), 16–24. doi:10.1108/10878571211209314

Bessi, A., & Ferrara, E. (2016). Social bots distort the 2016 US Presidential election online discussion. *First Monday*, *21*(11). doi:10.5210/fm.v21i11.7090

Beyer, M., Ahmed, S., Doerlemann, K., Arnell, S., Parkin, S., Sasse, A., & Passingham, N. (2015). *Awareness is only the first step: A framework for progressive engagement of staff in cyber security*. Hewlett Packard Enterpise.

Beyer, R., & Brummel, B. (2015). *Implementing Effective Cyber Security Training for End Users of Computer Networks*. Society for Human Resource Management and Society for Industrial and Organizational Psychology.

Bhattacherjee, A. (2001). Understanding information systems continuance: An expectation-confirmation model. *Management Information Systems Quarterly*, *25*(3), 351–370.

Black, S., Stone, D., & Johnson, A. (2015). Use of Social Networking Websites on Applicants' Privacy. *Employee Responsibilities and Rights Journal*, *27*(2), 115–159. doi:10.100710672-014-9245-2

Bohmova, L., & Pavlicek, A. (2015). The influence of social networking sites on recruiting human resources in the Czech Republic. *Organizacija*, *48*(1), 23–31. doi:10.1515/orga-2015-0002

Böhringer, A. (2006, June). *Knowledge specialization and transfer: a knowledge based framework for business incubation*. Copenhagen: DRUID Summer Conference.

Bøllingtoft, A., & Ulhøi, J. P. (2005). The Networked Business Incubator—Leveraging Entrepreneurial Agency? *Journal of Business Venturing*, *20*(2), 265–290. doi:10.1016/j.jbusvent.2003.12.005

Bonsón, E., & Bednárová, M. (2013). Corporate LinkedIn practices of Eurozone companies. *Online Information Review*, *37*(6), 969–984. doi:10.1108/OIR-09-2012-0159

Brand Finance. (2019). *Banking 500 2019*. Retrieved from https://brandirectory.com/rankings/banking-500-2019

Briggs, E., & Douglas, G. (2010). Service Performance—Loyalty Intentions Link in a Business to-Business Context: The Role of Relational Exchange Outcomes and Customer Characteristics. *Journal of Service Research*, *13*(1), 37–51. doi:10.1177/1094670509345683

Broughton, A., Foley, B., Ledermaier, S., & Cox, A. (2013). *The use of social media in the recruitment process*. Brighton, CA: Institute for Employment Studies.

Brown, J. R., Crosno, J. L., & Tong, P. Y. (2019). Is the theory of trust and commitment in marketing relationships incomplete? *Industrial Marketing Management*, *77*, 155–169. doi:10.1016/j.indmarman.2018.10.005

Brun, I., Rajaobelina, L., & Ricard, L. (2014). Online relationship quality: Scale development and initial testing. *International Journal of Bank Marketing*, *32*(1), 5–27. doi:10.1108/IJBM-02-2013-0022

Brun, Y., di Marzo Serugendo, G., & Gacek, C. (2009). Engineering Self-Adaptive Systems. Through Feedback Loops. In *Self-Adaptive Systems* (pp. 48–70). Heidelberg, Germany: Springer-Verlag Berlin. doi:10.1007/978-3-642-02161-9_3

Buchner, A., & Mulvenna, M. D. (1998). Discovering internet marketing intelligence through on line analytical web usage mining. *SIGMOD Record*, *27*(4), 54–61. doi:10.1145/306101.306124

Bucklin, R. (2008). Marketing models for electronic commerce. In B. Wierenga (Ed.), *Handbook of Marketing Decision Models* (pp. 327–369). Springer. doi:10.1007/978-0-387-78213-3_10

Buolamwini, J., & Gebru, T. (2018). Gender Shades: Intersectional Accuracy Disparities inCommercial Gender Classification. *Proceedings of Machine Learning Research*, *81*, 1–15.

Burgoon, J. K., Stoner, G. A., Bonito, J. A., & Dunbar, N. E. (2003, January). Trust and deception in mediated communication. *36th Annual Hawaii International Conference on System Sciences*.

Cadez, I., Heckerman, D., Meek, C., Smyth, P., & White, S. (2003). Model-based clustering and visualization of navigation patterns on a web site. *Data Mining and Knowledge Discovery*, *7*(4), 399–424. doi:10.1023/A:1024992613384

Calmon, F., Wei, D., Vinzamuri, B., Natesan Ramamurthy, K., & Varshney, K. R. (2017). Optimized Pre-Processing for Discrimination Prevention. In I. Guyon, U. V. Luxburg, S. Bengio, H. Wallach, R. Fergus, S. Vishwanathan, & R. Garnett (Eds.), Vol. 30, pp. 3992–4001). Advances in Neural Information Processing Systems Curran Associates, Inc.; Retrieved from http://papers.nips.cc/paper/6988-optimized-pre-processing-for-discrimination-prevention.pdf

Calvo-Porral, C., & Lévy-Mangin, J.-P. (2015). Switching behavior and customer satisfaction in mobile services: Analyzing virtual and traditional operators. *Computers in Human Behavior*, *49*, 532–540.

Campbell, C. (1989). Change agents in the new economy: Business Indubators and Economic Development. *Economic Development Review (Schiller Park, Ill.)*, *7*(3), 56–57.

Campbell, C., Kendrick, R., & Samuelson, D. (1985). Stalking the Latent Entrepreneur. *Economic Development Review (Schiller Park, Ill.)*, *3*(2), 43–48.

Campo, K., & Breugelmans, E. (2015). Buying groceries in brick and click stores: Category allocation decisions and the moderating effect of online buying experience. *Journal of Interactive Marketing*, *31*, 63–78.

Canhoto, I. A., & Arp, S. (2017). Exploring the factors that support adoption and sustained use of health and fitness wearables. *Journal of Marketing Management*, *33*(1-2), 1–2, 32–60. doi:10.1080/0267257X.2016.1234505

Cano, J. (2016). La educación en seguridad de la información. Reflexión pedagógicas desde el pensamiento de sistemas. *Memorias 3er Simposio Internacional en "Temas y problemas de Investigación en Educación: Complejidad y Escenarios para la Paz"*.

Cano, J. (2016). Modelo de madurez de cultura organizacional de seguridad de la información. Una visión desde el pensamiento sistémico-cibernético. Actas de la XIV Reunión Española sobre Criptología y Seguridad de la Información, 24-29.

Capiluppi, A., Serebrenik, A., & Singer, L. (2012). Assessing technical candidates on the social web. *IEEE Software*, *30*(1), 45–51. doi:10.1109/MS.2012.169

Carayannis, E. G., & von Zedtwitz, M. (2005). Architecting GloCal (global–local), Real-Virtual Incubator Networks (G-RVINs) as Catalysts and Accelerators of Entrepreneurship in Transitioning and Developing Economies: Lessons Learned and Best Practices from Current Development and Business Incubation Practices. *Technovation*, *25*(2), 95–110. doi:10.1016/S0166-4972(03)00072-5

Carletto, G., Covarrubias, K., Davis, B., Krausova, M., Stamoulis, K., & Winters, P. (2007). Rural income generating activities in developing countries: Re-assessing the evidence. *Electronic Journal of Agricultural and Development Economics*, *4*(1), 146–193.

Casalo, L., Flavian, C., & Guinaliu, M. (2007). The role of security, privacy, usability and reputation in the development of online banking. Online Information Review, 31(5), 583-603.

Castaneda, C., Nalley, K., Mannion, C., Bhattacharyya, P., Blake, P., Pecora, A., ... Suh, K. S. (2015). Clinical decision support systems for improving diagnostic accuracy and achieving precision medicine. *Journal of Clinical Bioinformatics*, *5*(1), 4. doi:10.118613336-015-0019-3 PMID:25834725

Castells, M. (1996). The rise of network society. Oxford, UK: Blackwell.

Castells, M., Fernandez-Ardevol, M., Qiu, J. L., & Sey, A. (2009). *Mobile communication and society: A global perspective*. MIT Press.

Castelvecchi, D. (2016). Can we open the black box of AI? *NATNews*, *538*(7623), 20–23. PubMed

Cath, C., Wachter, S., Mittelstadt, B., Taddeo, M., & Floridi, L. (2018). Artificial Intelligence and the 'Good Society': The US, EU, and UK approach. *Science and Engineering Ethics*, *24*(2), 505–528. doi: PubMed doi:10.100711948-017-9901-7

Centeno, C. (2004). Adoption of internet services in the acceding and candidate countries,lessons from the internet banking case. Telematics and Informatics, 21(4), 411-417.

Çetin, H., & Irmak, S. (2014). Elektronik alışverişte akademisyenlerin güvenlik ve risk algılarının belirlenmesi. *Yönetim Bilimleri Dergisi*, *12*(24), 275–294.

Chaffey, D. (2011). E-Business and E-Commerce Management - Strategy, Implementation and Practice. Financial Times Prentice Hall.

Chakrabarti, S., Dom, B. E., Gibsony, D., Kleinberg, J., Kumar, R., Prabhakar, R., ... Tomkins, A. (1999). Mining the link structure of the world wide web. *IEEE Computer*, *32*(8), 60–67. doi:10.1109/2.781636

Chakravorti, B., & Chaturvedi, R. S. (2017). *Digital Planet 2017; How competitiveness and trust in digital economies vary across the world*. Retrieved from https://sites.tufts.edu/digitalplanet/files/2017/05/Digital_Planet_2017_FINAL.pdf

Chakravorti, B., Bhalla, A., & Chaturvedi, R. S. (2018). The 4 Dimensions of Digital Trust, Charted Across 42 Countries. *Harvard Business Review Digital Articles*, 2-8.

Challen, R., Denny, J., Pitt, M., Gompels, L., Edwards, T., & Tsaneva-Atanasova, K. (2019). Artificial intelligence, bias and clinical safety. *BMJ Quality & Safety*, *28*(3), 231–237. doi:10.1136/bmjqs-2018-008370 PubMed

Chandra, A., & Chao, C. A. (2011). Growth and Evolution of High-Technology Business Incubation in China. *Human Systems Management*, *30*(1-2), 55–69. doi:10.3233/HSM-2011-0739

Chang, H. H., & Chen, S. W. (2008). The impact of online store environment cues on purchase intention: Trust and perceived risk as a mediator. *Online Information Review*, *32*(6), 818–841. doi:10.1108/14684520810923953

Cheng, Y.-M. (2014). Extending the expectation-confirmation model with quality and flow to explore nurses' continued blended e-learning intention. Information Technology & People, 27(3), 230-258.

Cheng, S., & Schaeffer, P. V. (2011). Evaluation Without Bias: A Methodological Perspective on Performance Measures for Business Incubators. *Région et Dévelоppement*, *33*, 211–225.

Chen, J., Qian, F., Yan, W., & Shen, B. (2013). Translational biomedical informatics in the cloud: Present and future. *BioMed Research International*. PMID:23586054

Chien, S.-H., Chen, Y.-H., & Hsu, C.-Y. (2012). Exploring the impact of trust and relational embeddedness in e-marketplaces: An empirical study in Taiwan. *Industrial Marketing Management*, *41*(3), 460–468. doi:10.1016/j.indmarman.2011.05.001

Chomeya, R. (2010). Quality of Psychology Test Between Likert Scale 5 and 6 Points. *Journal of Social Sciences*, *6*(3), 399–403.

Chong, A. Y., Darmawan, N. O., & Lee, V. H. (2011). Determinants of 3G adoption in Malaysia: A structural analysis. *Journal of Computer Information Systems*, *51*, 71–80.

Chou, J., Murillo, O., & Ibars, R. (2017). What the Kids' Game "Telephone" Taught Microsoft about Biased AI. Retrieved from https://www.fastcompany.com/90146078/what-the-kids-game-telephone-taught-microsoft-about-biased-ai

Chou, H.-T. G., Hammond, R. J., & Johnson, R. (2013). How Facebook might reveal users' attitudes toward work and relationships with coworkers. *Cyberpsychology, Behavior, and Social Networking*, *16*(2), 136–139. doi:10.1089/cyber.2012.0321 PMID:23276260

Chrisman, J. J. (1989). Strategic, administrative, and operating assistance: The value of outside consulting to pre-venture entrepreneurs. *Journal of Business Venturing*, *4*(6), 401–418. doi:10.1016/0883-9026(89)90010-4

Chrisman, J., Bauerschmidt, A., & Hofer, C. W. (1998). The Determinants of New Venture Performance - An extended model. *Entrepreneurship Theory and Practice*, *23*(1), 5–29. doi:10.1177/104225879802300101

Chung, J. E., Stoel, L. X., & Ren, J. (2012). Predicting Chinese consumers' purchase intentions for imported soy-based dietary supplements. *British Food Journal*, *114*(1), 143–161.

Citron, D. K., & Pasquale, F. A. (2014). The Scored Society: Due Process for Automated Predictions. *Washington Law Review (Seattle, Wash.)*, *89*(1), 1–33.

Clark, B. H., Abela, A. V., & Ambler, T. (2005). Organizational motivation, opportunity and ability to measure marketing performance. *Journal of Strategic Marketing*, *13*(4), 241–259.

Combe, C. (2006). *E-business management and strategy*. Elsevier Ltd.

Connected Nation. (2018). *Rural Broadband: A Texas Tour*. Connected Nation.

Constantin, L. (2016). *Thousands of online shops compromised for credit card theft*. Retrieved from https://www.computerworld.com/article/3131085/security/thousands-of-online-shops-compromised-for-credit-card-theft.html

Consulting, C. (2011). Digital transformation: A roadmap for billion-dollar organizations. Retrieved October 31, 2019, from MIT Center for Digital Business and Capgemini Consulting website: https://www.capgemini.com/wp-content/uploads/2017/07/Digital_Transformation__A_Road-Map_for_Billion-Dollar_Organizations.pdf

Cooley, R., Mobasher, B., & Srivastava, J. (1997). Web mining: Information and pattern discovery on the world wide web. In *Proceedings of the 9th IEEE International Conference on Tools with Artificial Intelligence (ICTAI'97)* (pp. 558-567). Computer Society. 10.1109/TAI.1997.632303

Cooley, R., Mobasher, B., & Srivastava, J. (1999). Data preparation for mining world wide web browsing patterns. *Knowledge and Information Systems*, *1*(1), 5–32. doi:10.1007/BF03325089

Corey, K. E., & Wilson, M. I. (2009). e-Business and e-Commerce. International Encyclopedia of Human Geography. doi:10.1016/B978-008044910-4.00148-6

Cornelius, B., & Bhabra-Remedios, R. (2003). *Cracks in the Egg: Improving Performance Measures in Business Incubator Research*. Academic Press.

Costa-David, J., Malan, J., & Lalkaka, R. (2002). Improving Business Incubator Performance Through Benchmarking and Evaluation: Lessons Learned from Europe. *Materiały*, *16*, 28–04.

Coulouris, G. F., Dollimore, J., & Kindberg, T. (2005). *Distributed systems: Concepts and design*. Pearson Education.

Coulter, K. S., & Coulter, R. A. (2002). Determinants of trust in a service provider: The moderating role of length of relationship. *Journal of Services Marketing*, *16*(1), 35–50. doi:10.1108/08876040210419406

Crawford, K. (2016, June 28). Artificial Intelligence's White Guy Problem. The New York Times, p. 11. Retrieved from https://www.nytimes.com/2016/06/26/opinion/sunday/artificial-intelligences-white-guy-problem.html

Crenshaw, K. (1989). Demarginalizing the intersection of race and sex: A black feminist critique of antidiscrimination doctrine, feminist theory and antiracist politics. *University of Chicago Legal Forum*, *139*(1). Retrieved from https://chicagounbound.uchicago.edu/cgi/viewcontent.cgi?article=1052&context=uclf

Cyber, X. (2019). 2019 Global ICS & IIoT Risk Report. A data-driven analysis of vulnerabilities in our industrial and critical infrastructure. *CyberX Labs*. Retrieved from https://cyberx-labs.com/resources/risk-report-2019/

Daft, R. L., & Lengel, R. H. (1986). Organizational information requirements, media richness and structural design. *Management Science*, *32*(5), 554–571. doi:10.1287/mnsc.32.5.554

Dalenberg, D. J. (2018). Preventing discrimination in the automated targeting of job advertisements. *Computer Law & Security Review*, *34*(3), 615–627. doi:10.1016/j.clsr.2017.11.009

Daniel, D., Swartz, W., & Fermar, A. (2004). *Economics of a cashless society: An analysis of costs and benefits of payment instruments*. AEI-Brookings Joint Centre.

Darby, M. R., & Karni, E. (1973). Free competition and the optimal amount of fraud. *The Journal of Law & Economics*, *16*(1), 67–88. doi:10.1086/466756

Dastin, J. (2018). Amazon scraps secret ai recruiting tool that shoed bias against women. Retrieved from https://www.reuters.com/article/us-amazon-com-jobs-automation-insight/amazon-scraps-secret-ai-recruiting-tool-that-showed-bias-against-women-idUSKCN1MK08G

Datta, A., Tschantz, M. C., & Datta, A. (2015). Automated Experiments on Ad Privacy Settings: A Tale of Opacity, Choice, and Discrimination. Proceedings on Privacy Enhancing Technologies, 2015(1), 92–112.

Davidson, T., Warmsley, D., Macy, M., & Weber, I. (2017). Automated hate speech detection and the problem of offensive language. *Proceedings of the Eleventh International AAAI Conference on Web and Social Media (ICWSM 2017)*, 512–516.

Davis, F. D. (1989). Perceived usefulness, perceived ease of use, and user acceptance of information technology. *Management Information Systems Quarterly*, *13*(3), 319–340.

Davis, F. D., Bagozzi, R. P., & Warshaw, P. R. (1989). User acceptance of computer technology: A comparison of two theoretical models. *Management Science*, *35*(8), 982–1003. doi:10.1287/mnsc.35.8.982

Davison, H. K., Bing, M. N., Kluemper, D. H., & Roth, P. L. (2016). Social Media as a Personnel Selection and Hiring Resource: Reservations and Recommendations. Academic Press.

Davison, H. K., Maraist, C., & Bing, M. N. (2011). Friend or foe? The promise and pitfalls of using social networking sites for HR decisions. *Journal of Business and Psychology*, *26*(2), 153–159. doi:10.100710869-011-9215-8

Daws, R. (2018). Editorial: Stopping AI's discrimination will be difficult, but vital. Retrieved from https://www.artificialintelligence-news.com/2018/05/17/editorial-stopping-ai-discrimination/

De Cremer, D., Nguyen, B., & Simkin, L. (2017). The integrity challenge of the Internet-of-Things (IoT): On understanding its dark side. *Journal of Marketing Management*, *33*(1-2), 145–158. doi:10.1080/0267257X.2016.1247517

Defence Science Board. (2012). Task force report: The role of autonomy in dod systems. US Department of Defense, Office of the Under Secretary of Defence for Acquisition, Technology and Logistics.

Defence Science Board. (2016). Summer study on autonomy. US Department of Defense. Office of the Under Secretary of Defence for Acquisition, Technology and Logistics.

Deloitte. (2017). *Digital ecosystems in telecom, media and entertainment.* Author.

Demirkan, H., Spohrer, J. C., & Welser, J. J. (2016). Digital innovation and strategic transformation. *IT Professional, 18*(6), 14–18. doi:10.1109/MITP.2016.115

Denscombe, M. (2007). *The good research guide.* McGraw-Hill Education.

Desautels, T., Calvert, J., Hoffman, J., Jay, M., Kerem, Y., Shieh, L., ... Barton, C. (2016). Prediction of sepsis in the intensive care unit with minimal electronic health record data: A machine learning approach. *JMIR Medical Informatics, 4*(3), e28. doi:10.2196/medinform.5909 PMID:27694098

Dhandi, M., & Chakrawarti, R. J. (2016). A comprehensive study of web usage mining. *2016 Symposium on Colossal Data Analysis and Networking (CDAN).* 10.1109/CDAN.2016.7570889

Dharmarajan, K., & Dorairangaswamy, M. A. (2016). Web usage mining: Improve the user navigation pattern using FP-growth algorithm. *Elysium Journal of Engineering Research & Management, 3*(4), 1–4.

Dhar, V., & Stein, R. (2017, October). Fintech platforms and strategy. *Communications of the ACM, 60*(10), 32–35. doi:10.1145/3132726

Diakopoulos, N. (2016). Accountability in algorithmic decision making. *Communications of the ACM, 59*(2), 56–62. doi:10.1145/2844110

Dick, A., & Basu, K. (1994). Customer loyalty: Toward an integrated conceptual framework. *Journal of the Academy of Marketing Science, 22*, 99–113.

Dictionary, C. E. (2019). *E-business technologies.* Author.

Dietrich, F., Harley, B., & Langbein, J. (2010). *Development Guidelines for Technology Business Incubators.* Academic Press.

Dijkstra, E. W. (1974). Self-Stabilizing Systems in Spite of Distributed Control. *Commun. ACM, 17*(11), 643-644.

Diker, A. & Varol, A. (2013). E-ticaret ve güvenlik. *1th International Syposium on Digital Forensics and Security Proceedings*, 29-33.

Dinakaran, R. (2016). *Why e-wallet wins over debit or credit cards.* Retrieved from https://www.thehindubusinessline.com/news/variety/why-ewallet-wins-over-debit-or-credit-cards/article9334954.ece

Dogne, V., Jain, A., & Jain, S. (2015). Evolving trends and its application in web usage mining: A survey. *International Journal of Soft Computing and Engineering, 4*(6), 98–101.

Donaldson, B., & O'toole, T. (2007). *Strategic market relationships: from strategy to implementation.* John Wiley & Sons.

Doney, P. M., & Cannon, J. P. (1997). An examination of the nature of trust in buyer–seller relationships. *Journal of Marketing*, *61*(2), 35–51.

Downes, L., & Nunes, P. F. (2013). Big bang disruption. *Harvard Business Review*, *91*(03), 44–56.

Drouin, M., O'Connor, K. W., Schmidt, G. B., & Miller, D. A. (2015). Facebook fired: Legal perspectives and young adults' opinions on the use of social media in hiring and firing decisions. *Computers in Human Behavior*, *46*, 123–128. doi:10.1016/j.chb.2015.01.011

Dwyer, F. R., Schurr, P. H., & Oh, S. (1987). Developing buyer-seller relationships. *Journal of Marketing*, *51*(2), 11–27. doi:10.1177/002224298705100202

Ecommerce Foundation. (2017). *European ecommerce report light version*. Retrieved from https://www.eurocommerce.eu/media/142202/c_european_ecommerce_report_2017_v170623-published_28basic_29.pdf

Ecommerce Foundation. (2018). *European ecommerce report 2018 edition*. Retrieved from https://www.haendlerbund.de/de/downloads/ecommerce-europe/european-ecommerce-report-2018.pdf

Edelman. (2017). *Edelman Trust Barometer*. Retrieved from https://www.edelman.com/trust2017/

Edelman. (2019). *Edelman Trust Barometer*. Retrieved from https://www.edelman.com/sites/g/files/aatuss191/files/201904/2019_Edelman_Trust_Barometer_Technology_Report_0.pdf

Egan, J. (2008). *Relationship marketing: Exploring relational strategies in marketing*. Pearson education.

Eirinaki, M., & Vazirgannis, M. (2003). Web mining for web personalization. *ACM Transactions on Internet Technology*, *3*(1), 1–27. doi:10.1145/643477.643478

Eisenhart, M. A. (1991). Conceptual frameworks for research circa 1991: ideas from a cultural anthropologist; implications for mathematics education researchers. In R. G. Underhill (Chair), Proceedings of the 13th annual meeting of the North American chapter of the international group for the psychology of mathematics education. Blacksburg, VA: Academic Press.

EIU. (2013). *Fostering a data-driven culture*. A report from the Economist Intelligence Unit. Retrieved from https://eiuperspectives.economist.com/technology-innovation/fostering-data-driven-culture

El Ouirdi, M., Pais, I., Segers, J., & El Ouirdi, A. (2016). The relationship between recruiter characteristics and applicant assessment on social media. *Computers in Human Behavior*, *62*, 415–422. doi:10.1016/j.chb.2016.04.012

Ellahi, A., & Bokhari, R. H. (2013). Key quality factors affecting users'perception of social networking websites. *Journal of Retailing and Consumer Services*, *20*, 120–129.

Ellis, E., & Watson, P. (2012). *EU Anti-Discrimination Law* (2nd ed.). Oxford, UK: Oxford University Press; doi:10.1093/acprof:oso/9780199698462.001.0001.

Equality Act 2010.

Erdağ, N., & Batuman, E. (2006). *Elektronik ticaret el kitabı*. İstanbul: Arıkan Yayınları.

ESET. (2017). *ESET Cybersecurity Awareness Training*. ESET Canada. Retrieved from https://www.eset.com/ca/cybertraining/

Eslami, M., Vaccaro, K., Karahalios, K., & Hamilton, K. (2017). "Be Careful; Things Can Be Worse Than They Appear"—Understanding Biased Algorithms and Users' Behavior Around Them in Rating Platforms. Proceedings of the Eleventh International AAAI Conference on Web and Social Media (ICWSM 2017), 62–71.

Eslami, M., Krishna Kumaran, S. R., Sandvig, C., & Karahalios, K. (2018). Communicating Algorithmic Process in Online Behavioral Advertising. In R. Mandryk, M. Hancock, M. Perry, & A. Cox (Eds.), *Proceedings of the 2018 CHI Conference on Human Factors in Computing Systems - CHI '18* (pp. 1–13). New York: ACM Press; doi:10.1145/3173574.3174006.

ETSI. (2013). *Electromagnetic compatibility and Radio spectrum Matters (ERM); Digital Mobile Radio (DMR) Systems; Part 1: DMR Air Interface (AI) protocol*. ETSI.

ETSI. (2019). *Terrestrial Trunked Radio (TETRA); Voice plus Data (V+D); Part 2: Air Interface (AI)*. ETSI.

Etzioni, E. (1996). The world wide web: Quagmire or gold mine. *Communications of the ACM*, *39*(11), 65–68. doi:10.1145/240455.240473

Facca, F. M., & Lanzi, P. L. (2005). Mining interesting knowledge from weblogs: A survey. *Data & Knowledge Engineering*, *53*(3), 225–241. doi:10.1016/j.datak.2004.08.001

FactsofIndonesia. (2017, March 4). *Largest Cities in Indonesia – Development and Growth*. Retrieved from FactsofIndonesia.com: https://factsofindonesia.com/largest-cities-in-indonesia

Falkenreck, C., & Wagner, R. (2017). The Internet of Things–Chance and challenge in industrial business relationships. *Industrial Marketing Management*, *66*, 181–195. doi:10.1016/j.indmarman.2017.08.007

Farah, M. F. (2017). Application of the theory of planned behavior to customer switching intentions in the context of bank consolidations. *International Journal of Bank Marketing*, *5*(1), 147–172.

Featherman, M. S., & Pavlou, P. A. (2003). Predicting e-services adoption: A perceived risk facets perspective. *International Journal of Human-Computer Studies*, *59*(4), 451–474. doi:10.1016/S1071-5819(03)00111-3

Fenollar, P., & Cuestas, P. J. (2010). Cómo es el innovador? Rasgos de la personalidad como antecedentes. *XXII Congreso Nacional de Marketing*.

Ferrara, E., Varol, O., Davis, C., Menczer, F., & Flammini, A. (2016). The rise of social bots. *Communications of the ACM*, *59*(7), 96–104. doi:10.1145/2818717

Ferreira, J. J. M., Fernandes, C. I., & Ferreira, F. A. F. (2019). To be or not to be digital, that is the question: Firm innovation and performance. *Journal of Business Research, 101*, 583–590. doi:10.1016/j.jbusres.2018.11.013

Fielding, R. T., & Taylor, R. N. (2000). *Architectural styles and the design of network-based software architectures* (PhD dissertation). University of California.

Fishbein, M., & Ajzen, I. (1975). *Belief, attitude, intention and behavior: An introduction to theory and research*. Reading, MA: Addison–Wesley.

Flores, A. W., Bechtel, K., & Lowenkamp, C. T. (2016). False Positives, False Negatives, and False Analyses: A Rejoinder to Machine Bias: There's Software Used across the Country to Predict Future Criminals. And It's Biased against Blacks. *Federal Probation, 80*, 38.

Fontanarosa, P. B., & DeAngelis, C. D. (2002). Basic science and translational research in JAMA. *Journal of the American Medical Association, 287*(13), 1728–1728. doi:10.1001/jama.287.13.1728 PMID:11926900

Francisco, L.-C., Juan, S.-F., & Francisco, M.-L. (2014). Antecedents of the adoption of the new mobile payment systems. *Computers in Human Behavior*, 464–478.

Frodeman, R., Klein, J. T., & Mitcham, C. (Eds.). (2010). The Oxford Handbook of Interdisciplinarity. Retrieved from http://www.universitystudy.ca/search-programs/?k=Interdisciplinary%20Studies&description_language=ENGLISH&cipcode_description=Multidisciplinary/interdisciplinary%20studies,%20other

Fuglsang, L., & Jagd, S. (2015). Making sense of institutional trust in organizations: Bridging institutional context and trust. *Organization, 22*(1), 23–39. doi:10.1177/1350508413496577

Fujitsu. (2017). *The Digital Transformation PACT*. Retrieved from https://www.fujitsu.com

Fu, Y., Creado, M., & Ju, C. (2001). Reorganizing web sites based on user access patterns. In *Proceedings of the Tenth International Conference on Information and Knowledge Management* (pp. 583–585). ACM Press. 10.1145/502585.502698

Galindo-Martín, M.-Á., Castaño-Martínez, M.-S., & Méndez-Picazo, M.-T. (2019). Digital transformation, digital dividends and entrepreneurship: A quantitative analysis. *Journal of Business Research, 101*, 522–527. doi:10.1016/j.jbusres.2018.12.014

Garbarino, E., & Johnson, M. S. (1999). The different roles of satisfaction, trust, and commitment in customer relationships. *Journal of Marketing, 63*(2), 70–87. doi:10.1177/002224299906300205

García, J. J. L., Lizcano, D., Ramos, C. M. Q., & Matos, N. (2019). Digital marketing actions that achieve a better attraction and loyalty of users: An analytical study. *Future Internet, 11*(6). doi:10.3390/fi11060130

Garcia, M. (2016). Racist in the Machine. *World Policy Journal, 33*(4), 111–117. doi:10.1215/07402775-3813015

Gartner. (2016). *2016 Gartner Magic Quadrant for Security Awareness Computer-Based Training Vendors*. Gartner, Inc.

Gartner. (2018). *How to Build an Enterprise Security Awareness Program*. Gartner, Inc.

Gartner. (2018). *How to Secure the Human Link*. Gartner, Inc.

Gartner. (2018). *Magic Quadrant for Security Awareness Computer-Based Training*. Gartner, Inc.

Gasser, U., & Almeida, V. A. F. (2017). A Layered Model for AI Governance. *IEEE Internet Computing*, *21*(6), 58–62. doi:10.1109/MIC.2017.4180835

Geetharamani, R., Revathy, P., & Jacob, S. G. (2015). Prediction of users webpage access behaviour using association rule mining. *Sadhana*, *40*(8), 2353–2365. doi:10.100712046-015-0424-0

Ghasrodashti, E. K. (2017). *Explaining brand switching behavior using pull–push–mooring theory and the theory of reasoned action*. Academic Press.

Gholami, R., Watson, R. T., Molla, A., Hasan, H., & Bjørn-Andersen, N. (2016). Information systems solutions for environmental sustainability: How can we do more? *Journal of the Association for Information Systems*, *17*(8), 521–536. doi:10.17705/1jais.00435

Gianfrancesco, M. A., Tamang, S., Yazdany, J., & Schmajuk, G. (2018). Potential Biases in Machine Learning Algorithms Using Electronic Health Record Data. *JAMA Internal Medicine*, *178*(11), 1544–1547. doi:10.1001/jamainternmed.2018.3763 PubMed

Gibbs, C., MacDonald, F., & MacKay, K. (2015). Social media usage in hotel human resources: Recruitment, hiring and communication. *International Journal of Contemporary Hospitality Management*, *27*(2), 170–184. doi:10.1108/IJCHM-05-2013-0194

Gillespie, T. (2014). The Relevance of Algorithms. In T. Gillespie, P. J. Boczkowski, & K. A. Foot (Eds.), *Media Technologies* (pp. 167–194). Cambridge, MA: The MIT Press; doi:10.7551/mitpress/9780262525374.003.0009

Gimpel, D. H., & Röglinger, D. M. (2015). Digital transformation: Changes and chances. Retrieved October 31, 2019, from https://www.fim-rc.de/wp-content/uploads/Fraunhofer-Studie_Digitale-Transformation.pdf

Goel, N., Yaghini, M., & Faltings, B. (2018). Non-Discriminatory Machine Learning through Convex Fairness Criteria. In J. Furman, G. Marchant, H. Price, & F. Rossi (Eds.), *Proceedings of the 2018 AAAI/ACM Conference on AI, Ethics, and Society - AIES '18* (p. 116). New York: ACM Press; doi:10.1145/3278721.3278722.

Goffman, E. (1967). *Interaction ritual*. New York: Pantheon.

Goforth, C. (2015). *Research Data Services + Science*. University of Virginia Library.

Gomez-Herrera, E., Martens, B., & Turlea, G. (2014). The drivers and impediments for cross-border e-commerce in the EU. *Information Economics and Policy*, *28*, 83–96. doi:10.1016/j.infoecopol.2014.05.002

Goodman, B. W. (2016). A step towards accountable algorithms? algorithmic discrimination and the european union general data protection. 29th Conference on Neural Information Processing Systems (NIPS 2016), Barcelona. NIPS Foundation.

Google. (2019). At I/O '19: Building a more helpful Google for everyone. Retrieved from https://www.blog.google/technology/developers/io19-helpful-google-everyone

Gough, D. A., Oliver, S., & Thomas, J. (2013). *Learning from research: Systematic reviews for informing policy decisions: A quick guide*. London: Nesta.

Gounaris, K. M., & Prout, M. F. (2009). Repairing relationships and restoring trust: Behavioral finance and the economic crisis. *Journal of Financial Service Professionals*, *63*(4).

Gozali, L. (2018b). *Framework Towards a Successful Business Incubator for Indonesian Public Universities* (Ph.D thesis). Universiti Teknologi Malaysia.

Gozali, L. (2015). A Framework of Successful E-Business Incubator for Indonesian Public Universities. *The Asian Journal of Technology Management*, *8*(2), 118–132.

Gozali, L., Masrom, M., Zagloel, T. Y. M., & Haron, H. N. (2016). A Framework of Successful Business Incubators for Indonesian Public Universities. *International Journal of Technology*, *7*(6), 1086–1096. doi:10.14716/ijtech.v7i6.3351

Gozali, L., Masrom, M., Zagloel, T. Y. M., Haron, H. N., & Tjahjadi, E. (2018a). A Framework toward Successful Business Incubator for Indonesian Public Universities: A Pilot Review. *International Conference of Industrial Engineering Operation Management 2018.*

Graff, J. C. (2016). Mixed methods research. *Evidence-Based Practice*, 47.

Grant, M. J., & Booth, A. (2009). A typology of reviews: An analysis of 14 review types and associated methodologies. *Health Information and Libraries Journal*, *26*(2), 91–108. doi:10.1111/j.1471-1842.2009.00848.x PubMed

Graze, J. (2017). *World FinTech Report 2017.* Available online: https://www.capgemini.com/service/introducing-the-world-fintech-report-2017/

Group, N. T. T. (2017). *Embedding cybersecurity into digital transformation - a journey towards business resilience.* NTT Security. Retrieved from https://www.nttsecurity.com

Gündüz-Ögüdücü, S. (2011). *Webpage recommendation models, theory and algorithms. Synthesis Lectures on Data Management #10.* Morgan & Claypool Publishers.

Gwagwa, A., & Koene, A. (2018). Minimizing algorithmic bias and discrimination in the digital economy. Working group for IEEE Standard on Algorithm Bias Considerations UNCTAD Africa e-Commerce week, Nairobi, Kenya.

Hacker, P. (2018). Teaching Fairness to Artificial Intelligence: Exisiting and Novel Strategies against Algorithmic Discrimination under EU Law. *Common Market Law Review*, *55*(4), 1143–1185.

Hackett, A. M., & Dilts, D. M. (2008). Inside the Black Box of Business Incubation. *The Journal of Technology Transfer, 33*(5), 439–471. doi:10.100710961-007-9056-9

Hacketts, S. M., & Dilts, D. M. (2004a). A Real Options - Driven Theory of Business Incubation. *The Journal of Technology Transfer, 29*(1), 41–54. doi:10.1023/B:JOTT.0000011180.19370.36

Hacketts, S. M., & Dilts, D. M. (2004b). A Systematic Review of Business Incubation Research. *The Journal of Technology Transfer, 29*(1), 55–82. doi:10.1023/B:JOTT.0000011181.11952.0f

Haffke, I., Kalgovas, B. J., & Benlian, A. (2016). *The Role of the CIO and the CDO in an Organization's Digital Transformation*. Academic Press.

Hajian, S., Domingo-Ferrer, J., & Martínez-Ballesté, A. (Eds.). (2011). Discrimination prevention in data mining for intrusion and crime detection. 2011 IEEE Symposium on Computational Intelligence in Cyber Security (CICS). doi:10.1109/CICYBS.2011.5949405

Hajian, S., & Domingo-Ferrer, J. (2013). A methodology for direct and indirect discrimination prevention in data mining. *IEEE Transactions on Knowledge and Data Engineering, 25*(7), 1445–1459. doi:10.1109/TKDE.2012.72

Hale, K. (2018). IBM's Unbiased Approach To AI Discrimination. Retrieved from https://www.forbes.com/sites/korihale/2018/09/25/ibms-unbiased-approach-to-ai-discrimination/#60c28e1d7118

Hamdani, D. (2006). *Conceptualizing and measuring business incubation*. Statistics Canada, Science, Innovation and Electronic Information Division.

Han, H. (2012). Image Congruence and Relationship Quality in Predicting Switching Intention. *Journal of Hospitality & Tourism Research (Washington, D.C.), 37*(3), 303–329.

Han, H., Kim, W., & Hyun, S. S. (2011). Switching intention model development: Role of service performances, customer satisfaction, and switching barriers in the hotel industry. *International Journal of Hospitality Management, 30*(3), 619–629.

Han, J., Kamber, M., & Pei, J. (2012). *Data mining: Concepts and techniques* (3rd ed.). Elsevier.

Harris, C. L., & Goode, M. M. H. (2004). The four levels of loyalty and the pivotal role of trust: A study of online service dynamics. *Journal of Retailing, 80*(2), 139–158. doi:10.1016/j.jretai.2004.04.002

Harrison, J. A., & Budworth, M. H. (2015). Unintended consequences of a digital presence: Employment-related implications for job seekers. *Career Development International, 20*(4), 294–314. doi:10.1108/CDI-06-2014-0080

Harris, P. A., Taylor, R., Thielke, R., Payne, J., Gonzalez, N., & Conde, J. G. (2009). Research electronic data capture (REDCap) - A metadata-driven methodology and workflow process for providing translational research informatics support. *Journal of Biomedical Informatics, 42*(2), 377–381. doi:10.1016/j.jbi.2008.08.010 PMID:18929686

Hasibovic, A. C., & Tanovic, A. (2019). *PRINCE2 vs Scrum in digital business transformation. In 2019 42nd International Convention on Information and Communication Technology, Electronics and Microelectronics, MIPRO 2019 - Proceedings* (pp. 1514–1518). Academic Press; doi:10.23919/MIPRO.2019.8756716.

Hayden, L. (2016). *People-Centric Security: Transforming your Enterprise Security Culture.* New York: Mc Graw Hill.

Heidegger, M. (1999). Zijn en Tijd [Being and Time]. Nijmegen: Uitgeverij SUN.

Heidegger, M. (1977). *The Question Concerning Technology and Other Essays.* New York: Harper & Row Publishers.

Heidegger, M. (2004). *What is Called Thinking?* (J. G. Gray, Trans.). Harpers Perennial.

Henseler, J., & Dijkstra, T. K. S. M., Ringle, C. M., Diamantopoulos, A., & Straub, D. W. (2014). Common Beliefs and Reality about Partial Least Squares: Comments on Rönkkö & Evermann. *Organizational Research Methods, 17*(2), 182–209.

Hillard, D., Schroedl, S., & Manavoglu, E. (2010). Improving ad relevance in sponsored search. Proceedings of the third ACM international conference on web search and data mining. Advance online publication. doi:10.1145/1718487.1718532

Hinton, P. R., Brownlow, C., McMurray, I., & Cozens, B. (2004). *SPSS explained.* London: Routledge, Taylor and Francis Group.

Hisrich, R. D. (1988). New business formation through the enterprise development center: A model for new venture creation. *IEEE Transactions on Engineering Management, 35*(4), 221–231. doi:10.1109/17.7444

Hissam, S., Klein, M., & Moreno, G. A. (2013). *Socio-Adaptive Systems Challenge Problem.* Workshop report. Carnegie Mellon Software Engineering Institute Chicago. CMU/SEI-2013-SR-010.

Hoek, J., O'Kane, P., & McCracken, M. (2015). Publishing personal information online. *Personnel Review, 45*(1), 67–83. doi:10.1108/PR-05-2014-0099

Hoffmann, C. P., Lutz, C., & Meckel, M. (2015). Content creation on the Internet: A social cognitive perspective on the participation divide. *Information Communication and Society, 18*(6), 696–716. doi:10.1080/1369118X.2014.991343

Hofstede Insights. (2019). *Values of Turkey's cultural dimensions.* Retrieved from https://www.hofstede-insights.com/product/compare-countries/

Höltl, A., & Trommer, S. (2013). Driver Assistance Systems for Transport System Efficiency: Influencing Factors on User Acceptance. *Journal of Intelligent Transportation Systems: Technology, Planning, and Operations, 17*(3), 245–254.

Holzinger, A. (2016). Interactive machine learning for health informatics: When do we need the human-in-the-loop? *Brain Informatics*, *3*(2), 119–131. doi:10.100740708-016-0042-6 PMID:27747607

Hongladarom, S., Ess, C., & Nagenborg, M. (2007). Information technology ethics: Cultural perspectives. *Publizistik*, *52*(4), 565–567.

Hotchkiss, R. S., Moldawer, L. L., Opal, S. M., Reinhart, K., Turnbull, I. R., & Vincent, J.-L. (2016). Sepsis and septic shock. *Nature Reviews. Disease Primers*, *2*(1), 16045. doi:10.1038/nrdp.2016.45 PMID:28117397

Hou, A. C., Chern, C. C., Chen, H. G., & Chen, Y. C. (2011). Migrating to a new virtual world: Exploring MMORPG switching through human migration theory. *Computers in Human Behavior*, *27*(5), 1892–1903.

Houser, K. (2019). Can AI Solve the Diversity Problem in the Tech Industry? Mitigating Noise and Bias in Employment Decision-Making. *Stanford Technology Law Review : STLR : An Online High-Technology Law Journal from Stanford Law School.*

Hrustek, L., Tomicic Furjan, M., & Pihir, I. (2019). *Influence of digital transformation drivers on business model creation. In 2019 42nd International Convention on Information and Communication Technology, Electronics and Microelectronics, MIPRO 2019 - Proceedings* (pp. 1304–1308). Academic Press; doi:10.23919/MIPRO.2019.8756666.

Hsiao, K.-L. (2017). What drives smartwatch adoption intention? Comparing Apple and non-Apple watches. *Library Hi Tech*, *35*(1), 186–206.

Hsieh, J.-K., & Hsieh, Y.-C. C. H.-C., & Feng, Y.-C. (2012). Post-adoption switching behavior for online service substitutes: A perspective of the push–pull–mooring framework. *Computers in Human Behavior*, *28*, 1912–1920.

Hughes, A. (2005). *Strategic database marketing: The masterplan for starting and managing a profitable, customer-based marketing program.* McGraw-Hill.

Hui, Beng, D., Fernandez, E. A. V., & Sio, D. B. K. (2011). Privatization of Business Incubation: Initiatives to Achieve sustainability and success. *Policy Brief*, *3*(1), 1–4.

Humphrey, D. (2004). Replacement of Cash by Cards in U.S. Consumer payments. *Journal of Economics and Business*, *56*, 211–225.

Hurrell, S. A., Scholarios, D., & Richards, J. (2017). 'The kids are alert': Generation Y responses to employer use and monitoring of social networking sites. *New Technology, Work and Employment*, *32*(1), 64–83. doi:10.1111/ntwe.12085

Husserl, E. (2014). Ideas I. Ideas for a Pure Phenomenology and Phenomenological Philosophy (D. O. Dahlstrom, Trans.). Indianapolis, IN: Hackett Publishing Company.

IBM. (2018a). AI Fairness 360. Retrieved from https://developer.ibm.com/open/projects/ai-fairness-360/

IBM. (2018b). Introducing AI Fairness 360. Retrieved from https://www.ibm.com/blogs/research/2018/09/ai-fairness-360/

IBM. (n.d.). AI Fairness 360 Open Source Toolkit. Retrieved from http://aif360.mybluemix.net/

ICC. (2018). *Online card payment transactions*. Retrieved from: https://bkm.com.tr/internetten-yapilan-kartli-odeme-islemleri/

I-Cheng, C., Chuang-Chun, L., & Kuanchin, C. (2014). The push, pull and mooring effects in virtual migration for social networking sites. *Information Systems Journal*, *24*, 323–346.

Industrial Internet Reference Architecture (2015) tech-arch.001, 2015-06-04 Version 1.7

Infodev, I. T. U. (2009). *ICT regulation toolkit*. Academic Press.

Ingold, D., & Soper, S. (2016). Amazon Doesn't Consider the Race of Its Customers. Should It? Retrieved from https://www.bloomberg.com/graphics/2016-amazon-same-day/

International Organization for Standardization - ISO. (2005). *ISO/IEC 27001:2005 – Information Technology – Security Techniques – Information Security Management Systems – Requirements*. ISO.

International Organization for Standardization -ISO. (2012). *ISO/IEC 27032:2012 – Information Technology – Security Techniques – Guidelines for Cybersecurity*. ISO.

Internet Society. (2016). *Global Internet Report 2016*. Internet Society.

Internet World Stats. (2018). *World internet user statistics*. Retrieved from http://www.internetworldstats.com/stats.htm

I-Scoop. (n.d.). Digital transformation: online guide to digital business transformation. Retrieved October 31, 2019, from https://www.i-scoop.eu/digital-transformation/

Isoraite, M., & Miniotiene, N. (2018). Electronic commerce: Theory and practice. *Integrated Journal of Business and Economics*, *2*(2), 73–79. doi:10.33019/ijbe.v2i2.78

ITU-R. Radiocommunication Sector of ITU. (2017). *Digital land mobile systems for dispatch traffic*. Report ITU-R M.2014-3. Geneva: ITU.

Iváncsy, R., & Vajk, I. (2006). Frequent pattern mining in web log data. *Acta Polytechnica Hungarica*, *3*(1), 77–90.

Ivanov, D., Dolgui, A., Das, A., & Sokolov, B. (2019). Digital Supply Chain Twins: Managing the Ripple Effect, Resilience, and Disruption Risks by Data-Driven Optimization, Simulation, and Visibility. In *Handbook of Ripple Effects in the Supply Chain* (pp. 309–332). Cham: Springer; doi:10.1007/978-3-030-14302-2_15.

Jafari, M., Sabzchi, F. S., & Irani, A. J. (2014). Applying web usage mining techniques to design effective web recommendation systems: A case study. *ACSIJ Advances in Computer Science: an International Journal*, *3*(02), 78–90.

Jagan, S., & Rajagopalan, S. P. (2015). A survey on web personalization of web usage mining. *International Research Journal of Engineering and Technology*, *2*(01), 6–12.

Jeske, D., & Shultz, K. S. (2016). Using social media content for screening in recruitment and selection: Pros and cons. *Work, Employment and Society*, *30*(3), 535–546. doi:10.1177/0950017015613746

Joachim, J., Martin, M., Lange, H., Schneiker, A., & Dau, M. (2018). Twittering for talent: Private military and security companies between business and military branding. *Contemporary Security Policy*, *39*(2), 298–316. doi:10.1080/13523260.2017.1420608

Johnston, R., & Lawrence, P. R. (1988). Beyond vertical integration: The rise of the value-adding partnership. *Harvard Business Review*, *66*(4).

Joseph, M., Kearns, M., Morgenstern, J., Neel, S., & Roth, A. (2016). Rawlsian fairness for machine learning. arXiv preprint arXiv:1610.09559

Joshi, K., Joshi, A., & Yesha, Y. (2003). On using a warehouse to analyze web logs. *Distrubuted and Parallel Databases,* (13), 161-180.

Joshi, A., & Krishnapuram, R. (2000). On mining web access logs. *Proc. of the 2000 ACM SIGMOD Workshop on Research Issues in Data Mining and Knowledge Discovery*, 63–69.

Joshila Grace, L. K., Maheswari, V., & Nagamalai, D. (2011). Web log data analysis and mining. In *CCSIT 2011, Part III, CCIS 133* (pp. 459–469). Springer-Verlag Berlin Heidelberg.

Jung, J., & Han, H. (2015). Travelers' switching behavior in the airline industry from the perspective of the push-pull-mooring framework. *Journal of Tourism Management, 59*, 139-153.

Kamiran, F., Calders, T., & Pechenizkiy, M. (2013). Techniques for discrimination-free predictive models. In B. Custers, T. Calders, B. Schermer, & T. Zarsky (Eds.), *Discrimination and privacy in the information society* (pp. 223–239). Berlin: Springer; doi:10.1007/978-3-642-30487-3_12.

Kantsperger, R., & Kunz, W. H. (2010). Consumer trust in service companies: A multiple mediating analysis. *Managing Service Quality: An International Journal*, *20*(1), 4–25. doi:10.1108/09604521011011603

Karlen, J. M. (2014). Privacy: Expectations and employment. *Business Management Dynamics*, *3*(10), 28.

Kashi, K., Zheng, C., & Molineux, J. (2016). Exploring factors driving social recruiting: The case of Australian organizations. *Journal of Organizational Computing and Electronic Commerce*, *26*(3), 203–223. doi:10.1080/10919392.2016.1194055

Kaya, F. (2009). *Türkiye'de kredi kartı uygulaması*. İstanbul: Graphis Matbaa.

Khandani, A. E., Kim, A. J., & Lo, A. W. (2010). Consumer credit-risk models via machine-learning algorithms. *Journal of Banking & Finance*, *34*(11), 2767–2787. doi:10.1016/j.jbankfin.2010.06.001

Kharpal, A. (2019). *Alibaba unveils its first A.I. chip as China pushes for its own semiconductor technology.* Retrieved from https://www.cnbc.com/2019/09/25/alibaba-unveils-its-first-ai-chip-called-the-hanguang-800.html

Kim, B. (2012). The diffusion of mobile data services and applications: exploring the role of habit and its antecedents. Telecommunication Policy, 36(1), 69–81.

Kim, B., & Kang, M. (2016). How user loyalty and nonconscious inertia influence the continued use of mobile communications platforms. *International Journal of Mobile Communications, 14*(4), 387.

Kim, G. S. B., & Lee, H. (2006). A study of factors that affect user intentions toward email service switching. *Information & Management, 43*, 884–893.

Kirkpatrick, K. (2016). Battling algorithmic bias. *Communications of the ACM, 59*(10), 16–17. doi:10.1145/2983270

Koilada, D. K. (2019). Value-based digital transformation: Innovating customer experiences. *2019 IEEE Technology and Engineering Management Conference, TEMSCON 2019*. 10.1109/TEMSCON.2019.8813559

Kokolakis, S. (2017). Privacy attitudes and privacy behaviour: A review of current research on the privacy paradox phenomenon. *Computers & Security, 64*, 122–134. doi:10.1016/j.cose.2015.07.002

Konuk, F. A., & Yıldırım, E. (2012). Sanal kart bilinirliliği üzerine bir araştırma: Sakarya ilinde bir uygulama. *11. National Business Congress Proceedings*, 85-87.

Kosala, R., & Blockeel, H. (2000). Web mining research: A survey. *SIGKDD Explorations: Newsletter of the Special Interest Group on Knowledge Discovery & Data Mining, 2*(1), 1–15. doi:10.1145/360402.360406

Koski, L. B., Gray, M. W., Lang, B. F., & Burger, G. (2005). AutoFACT: An auto matic functional annotation and classification tool. *BMC Bioinformatics, 6*(1), 151. doi:10.1186/1471-2105-6-151 PMID:15960857

Kotler, P., & Andreasen, A. (1996). *Strategic marketing for nonprofit organizations* (5th ed.). Englewood Cliffs, NJ: Prentice-Hall.

Kotler, P., & Armstrong, G. (2000). *Marketing: An introduction*. Upper Saddle River, NJ: Pearson Prentice Hall.

Kotler, P., & Armstrong, G. (2012). *Marketing* (14th ed.). Pearson Education.

Kotler, P., Kartajaya, H., & Setiawan, I. (2012). *Marketing 3.0* (3rd ed.). LID.

Kuang, C. F., Hsiao, H., Luong, L., Lin, G. C. I., & Wu, N. (2003). Development of a New Self-sufficient Model for University Incubator. *International Journal of Innovation and Incubation, 1*(1), 33–50.

Kuderer, M., Gulati, S., & Burgard, W. (2015). *Learning driving styles for autonomous vehicles from demonstration.* Paper presented at the Robotics and Automation (ICRA), 2015 IEEE International Conference on. 10.1109/ICRA.2015.7139555

Kueh, K., & Voon, B. H. (2007). Culture and service quality expectations: Evidence from generation y cosumers in Malaysia. *Managing Service Quality*, *17*(6), 656–680. doi:10.1108/09604520710834993

Kumar, B. S., & Rukmani, K. V. (2010). Implementation of web usage mining using apriori and fp growth algorithms, *Int. J. of Advanced Networking and Applications*, *1*(06), 400–404.

Kumra, R., & Mittal, R. K. (2004). Trust and its Determinants in Internet Banking: A Study of Private Sector Banks in India. *Decision, 31*(1).

Kutnjak, A., Pihiri, I., & Furjan, M. T. (2019, May). Digital Transformation Case Studies Across Industries–Literature Review. In 2019 42nd International Convention on Information and Communication Technology, Electronics and Microelectronics (MIPRO) (pp. 1293-1298). IEEE. doi:10.23919/MIPRO.2019.8756911

Laddaga, R. (1999). Creating Robust Software Through Self-Adaptation. *IEEE Intelligent Systems & their Applications*, *14*(3), 26–29. doi:10.1109/MIS.1999.769879

Laddaga, R., & Robertson, P. (2004). Self-Adaptive Software: A Position Paper. *ACM Transactions on Autonomous and Adaptive Systems*, *4*(2), 14.

Lai, J.-Y., & Wang, J. (2015). Switching attitudes of Taiwanese middle-aged and elderly patients toward cloud healthcare services: An exploratory study. *Technological Forecasting and Social Change*, *92*, 155–167.

Laing, A., & Hogg, G. (2008). Re-conceptualising the professional service encounter: Information empowered consumers and service relationships. *Journal of Customer Behaviour*, *7*(4), 333–346. doi:10.1362/147539208X386860

Lalkaka, R. (2001). Best practices in business incubation: Lessons (yet to be) learned. In *International Conference on Business Centers: Actors for Economic and Social Development.* Academic Press.

Lalkaka, R., & Abetti, P. A. (1999). Business incubation and enterprise support systems in restructuring countries. *Creativity and Innovation Management*, *8*(3), 197–209. doi:10.1111/1467-8691.00137

Lalkaka, R., & Bishop, J. (1996). *Business Incubators in Economic Development – An Initial Assessment in Industrialising Countries*. New York, NY: United Nations Development Programme.

Lamberte, M. (n.d.). Efficiency and Implications for the Conduct of Monetary Policy. *The Phillipine Payment Systems*.

Lambrecht, A., & Tucker, C. (2019). Algorithmic Bias? An Empirical Study of Apparent Gender-Based Discrimination in the Display of STEM Career Ads. *Management Science*, *65*(7), 2966–2981; Advance online publication. doi:10.1287/mnsc.2018.3093

Lam, H. (2016). Social media dilemmas in the employment context. *Employee Relations*, *38*(3), 420–437. doi:10.1108/ER-04-2015-0072

Lamport, L. (1974). *On Self-Stabilizing Systems.* Massachusetts Computer Associates Inc.

Lamport, L. (1978). The Implementation of Reliable Distributed Multiprocess Systems. *Computer Networks*, *2*(2), 95–114. doi:10.1016/0376-5075(78)90045-4

Lamport, L. (1978b, July). Time, Clocks and the Ordering of Events in a Distributed System. *Communications of the ACM*, *21*(7), 558–565. doi:10.1145/359545.359563

Lamport, L. (1983, July). The weak byzantine generals problem. *Journal of the Association for Computing Machinery*, *30*(3), 668–767. doi:10.1145/2402.322398

Lamport, L. (1998, May). The Part-Time Parliament. *ACM Transactions on Computer Systems*, *16*(2), 133–169. doi:10.1145/279227.279229

Lamport, L. (2001). Paxos made simple. *ACM Sigact News*, *32*(4), 18–25.

Lamport, L., Shostak, R., & Pease, M. (1982, July). The Byzantine Generals' Problem. *ACM Transactions on Programming Languages and Systems*, *4*(3), 382–401. doi:10.1145/357172.357176

Lam, S. Y. (2004). Customer value, satisfaction, loyalty, and switching costs: An illustration from a business-to business service context. *Journal of the Academy of Marketing Science*, *32*(3), 293–311.

Laukkanen, T. (2007). Internet vs mobile banking: Comparing customer value perceptions. *Business Process Management Journal*, *13*(6), 788–797. doi:10.1108/14637150710834550

Leavy, S. (2018). Gender bias in artificial intelligence. In E. Abraham, E. Di Nitto, & R. Mirandola (Eds.), *Proceedings of the 1st International Workshop on Gender Equality in Software Engineering - GE '18* (pp. 14–16). New York: ACM Press; doi:10.1145/3195570.3195580.

LeClair, J., Abraham, S., & Shih, L. (2013). An Interdisciplinary Approach to Educating an Effective Cyber Security Worforce. In *Proceedings of Information Security Curriculum Development Conference* (pp. 71-78). Kennesaw, GA: Academic Press.

Lee, E. S. (1966). A theory of immigration. *Demography*, *3*(1), 47–57.

Lee, H. H., Kim, J., & Fiore, A. M. (2010). Affective and cognitive online shopping experience: Effects of image interactivity technology and experimenting with appearance. *Clothing & Textiles Research Journal*, *28*(2), 140–154. doi:10.1177/0887302X09341586

Lee, I., & Shin, Y. J. (2018). Fintech: Ecosystem, business models, investment decisions, and challenges. *Business Horizons*, *61*(1), 35–46. doi:10.1016/j.bushor.2017.09.003

Lee, J. J., Hong, W. K., Hittelman, W. N., Mao, L., Lotan, R., Shin, D. M., ... Papadimitrakopoulou, V. M. (2000). Predicting cancer development in oral leukoplakia: Ten years of translational research. *Clinical Cancer Research*, *6*(5), 1702–1710. PMID:10815888

Lee, M. (2009). Factors influencing the adoption of internet banking: An integration of TAM and TPB with perceived risk and perceived benefit. *Electronic Commerce Research and Applications*, *8*(3), 130–141.

Lee, M.-C. (2009). Factors influencing the adoption of internet banking: An integration of TAM and TPB with perceived risk and perceived benefit. *Electronic Commerce Research and Applications*, *8*(3), 130–141. doi:10.1016/j.elerap.2008.11.006

Legters, B. (2013). *The evolution of payments*. Retrieved from FIS Payments Leader: https://www.paymentsleader.com/the-evolution-of-payments/

Leonard, K. (2018). Retrieved November 15, from, https://smallbusiness.chron.com/benefits-e-business-technology-4048.html

Lewicki, R. J., & Bunker, B. B. (1995). Trust in relationships. *Administrative Science Quarterly*, *5*(1), 583–601.

Lewicki, R. J., & Bunker, B. B. (1996). Developing and maintaining trust in work relationships. *Trust in organizations. Frontiers of Theory and Research*, *114*, 139.

Lewicki, R. J., McAllister, D. J., & Bies, R. J. (1998). Trust and distrust: New relationships and realities. *Academy of Management Review*, *23*(3), 438–458. doi:10.5465/amr.1998.926620

Lewis, D. A. (2001). Does Technology Incubation Work? A Critical Review. Washington, DC: Economic Development Administration. US Department of Commerce.

Lewis, D. A., Harper-Anderson, E., & Molnar, L. A. (2011). *Incubating success. Incubation best practices that lead to successful new ventures*. Ann Arbor, MI: Institute for Research on Labor, Employment, and Development.

Li, C. Y. (2018). The Power of a Thumbs-Up: Will E-commerce Switch to Social Commerce? *Information & Management*, *55*(3), 340–357.

Lievens, F., Van Hoye, G., & Schreurs, B. (2005). Examining the relationship between employer knowledge dimensions and organizational attractiveness: An application in a military context. *Journal of Occupational and Organizational Psychology*, *78*(4), 553–572. doi:10.1348/09631790X26688

Li, L., Su, F., Zhang, W., & Mao, J. Y. (2017). Digital transformation by SME entrepreneurs: A capability perspective. *Information Systems Journal*, ▪▪▪, 1–29.

Liljander, V., & Tore, S. (1995). The Nature of Customer Satisfaction in Service In Advances in Services Marketing and Management (vol. 4). London: JAI.

Lin, C., Wu, S., & Tsai, R. (2005). Integrating perceived playfulness into expectation-confirmation model for web portal context. Information & Management, 42(5), 683-693.

Lindgren, H. (2008). Decision support system supporting clinical reasoning process-an evaluation study in dementia care. *Studies in Health Technology and Informatics*, *136*, 315. PMID:18487750

Lin, P., Abney, K., & Bekey, G. A. (Eds.). (2012). *Robot Ethics: The Ethical and Social Implications of Robotics*. Cambridge, MA: MIT Press.

Liu, B. (2007). *Web data mining exploring hyperlinks, contents, and usage data* (1st ed.). Springer.

Liu, C., & Forsythe, S. (2010). Sustaining online shopping: Moderating role of online shopping motives. *Journal of Internet Commerce*, *9*(2), 83–103. doi:10.1080/15332861.2010.503848

Liu, C.-H., & Technology, N. C.-Y. (2015). The impact on switching intention of E-trading systems for the securities industry. *Global Journal of Business Research*, *9*(1), 75–88.

Liu, X., & Wei, K. K. (2003). An empirical study of product differences in consumers' e-commerce adoption behavior. *Electronic Commerce Research and Applications*, *2*(3), 229–239. doi:10.1016/S1567-4223(03)00027-9

Li, Y., Thomas, M. A., & Osei-Bryson, K.-M. (2017). Ontology-based data mining model management for self-service knowledge discovery. *Information Systems Frontiers*, *19*(4), 925–943. doi:10.100710796-016-9637-y

Lloyd, K. (2018). Bias Amplification in Artificial Intelligence Systems. AAAI FSS-18: Artificial Intelligence in Government and Public Sector, Arlington, VA.

Lopperi, K., & Sengupta, S. (2004). Are we ready? The state of wireless e-business in the USA. *Information Systems and e-Business Management*, *2*(4), 293–307. doi:10.1007/s10257-004-0044-4

Lowry, P. B., Cherrington, J. O., & Watson, R. R. (Eds.). (2001). *The E-Business Handbook*. CRC Press.

Lu, H. (2019). *Xinhua Headlines: New trends driving China's AI industry* Retrieved from http://www.xinhuanet.com/english/2019-08/31/c_138353742.htm

Lum, K., & Isaac, W. (2016). To predict and serve? *Significance*, *13*(5), 14–19. doi:10.1111/j.1740-9713.2016.00960.x

Lundahl, N., Vegholm, F., & Silver, L. (2009). Technical and functional determinants of customer satisfaction in the bank-SME relationship. *Managing Service Quality: An International Journal*, *19*(5), 581–594. doi:10.1108/09604520910984382

Luppicini, R. (2009). Technoethical inquiry: From technological systems to society. Global Media Journal, Canadian Edition, 2(1), 5–21.

Luppicini, R. (2010). *Technoethics and the Evolving Knowledge Society: Ethical Issues in Technological Design, Research, Development, and Innovation*. Information Science Reference.

Luppicini, R., & So, A. (2016). A technoethical review of commercial drone use in the context of governance, ethics, and privacy. *Technology in Society*, *46*, 109–119. doi:10.1016/j.techsoc.2016.03.003

Magnani, L. (2007). *Morality in a technological world: Knowledge as duty*. Cambridge, MA: Cambridge University Press; doi:10.4018/978-1-60566-952-6.

Mahmoodi, A., & Khataee, S. (2016). Observing the role of packaging handicrafts from the aspect of mobile outdoor advertising on branding and cultural exports. *International Research Journal of Applied and Basic Sciences*, *10*(3), 190–199.

Maier, E. (2016). Supply and demand on crowdlending platforms: Connecting small and medium-sized enterprise borrowers and consumer investors. *Journal of Retailing and Consumer Services*, *33*, 143–15. doi:10.1016/j.jretconser.2016.08.004

Malarvizhi, S. P., & Sathiyabhama, B. (2016). Frequent pagesets from web log by enhanced weighted association rule mining. *Cluster Computing*, *19*(1), 269–277. doi:10.100710586-015-0507-z

Maldonado-Guzman, G., Garza-Reyes, J. A., Rocha-Lona, L., & Kumar, V. (2017, September). Service innovation and performance in Mexican service SMEs. In *IFIP International Conference on Advances in Production Management Systems* (pp. 230-239). Springer. 10.1007/978-3-319-66926-7_27

Mann, G., & O'Neil, C. (2016). Hiring Algorithms Are Not Neutral. Harvard Business Review. Retrieved from https://hbr.org/2016/12/hiring-algorithms-are-not-neutral

Manrai, L. A., & Manrai, A. K. (2007). A field study of customers' switching behavior for bank services. *Journal of Retailing and Consumer Services*, *14*(3), 208–215. doi:10.1016/j.jretconser.2006.09.005

Marketing Science Institute. (n.d.). *2018-2020 Research Priorities: Marketers' Strategic Imperatives*. Retrieved from https://www.msi.org/articles/marketers-top-challenges-2018-2020-research-priorities/

Markov, Z., & Larose, D. (2007). *Data mining the web uncovering patterns in web content, structure, and usage*. Wiley-Interscience.

Marshall, M. N. (1996). Sampling for qualitative research. *Family Practice*, *13*(6), 522–526. doi:10.1093/fampra/13.6.522 PMID:9023528

Martínez, D. (2016). Key factors in marketing focused services businesses. *Scientific E-Journal of Human Sciences*, *12*(34), 42-58.

Martin, F. (1997). Business incubators and enterprise development: Neither tried or tested? *Journal of Small Business and Enterprise Development*, *4*(1), 3–11. doi:10.1108/eb020975

Matt, C., Hess, T., & Benlian, A. (2015). Digital transformation strategies. *Business & Information Systems Engineering*, *57*(5), 339–343. doi:10.100712599-015-0401-5

May, S. (2013). Introduction: Ethical perspectives and practices. In *Case Studies in Organizational Communication: Ethical Perspectives and Practices* (pp. 1–33). Thousand Oaks, CA: Sage Publications.

McAdam, M., Galbraith, B., McAdam, R., & Humphreys, P. (2006). Business processes and networks in university incubators: A review and research agendas. *Technology Analysis and Strategic Management, 18*(5), 451–472. doi:10.1080/09537320601019578

McAdam, M., & Marlow, S. (2008). A Preliminary Investigation into Networking Activities within the University Incubator. *International Journal of Entrepreneurial Behaviour & Research, 14*(4), 219–241. doi:10.1108/13552550810887390

McAdam, M., & McAdam, R. (2008). High Tech Start-ups in University Science Park Incubators: The Relationship between the Start-up's Lifecycle Progression and Use of the Incubator's Resources. *Technovation, 28*(5), 277–290. doi:10.1016/j.technovation.2007.07.012

McAllister, D. J. (1995). Affect and cognition-based trust as foundations for interpersonal cooperation in organizations. *Academy of Management Journal, 38*, 24–59.

McCarty, J., & Hastak, M. (2007). Segmentation approaches in data-mining: A comparison of RFM, CHAID, and logistic regression. *Journal of Business Research, 60*(6), 656–662. doi:10.1016/j.jbusres.2006.06.015

McGinn, R. (1978). What is technology. *Research in Philosophy and Technology, 1*, 179–197.

McGrath, R. G. (1999). Falling forward: Real options reasoning and entrepreneurial failure. *Academy of Management Review, 24*(1), 13–30. doi:10.5465/amr.1999.1580438

McGregor, S. L. T. (2018). *Understanding and evaluating research: A critical guide*. Thousand Oaks, CA: SAGE Publications.

McKenna, C. (2014). *The use of social media in recruitment and its impact on diversity in services companies in Ireland* (Doctoral dissertation). Dublin Business School.

McKnight, D. H., Choudhury, V., & Kacmar, C. (2002). Developing and Validating Trust Measures for e-Commerce: An Integrative Typology. *Information Systems Research, 13*(3), 334–359. doi:10.1287/isre.13.3.334.81

Mcknight, D. H., Cummings, L. L., & Chervany, N. L. (1998). Initial trust formation in new organizational relationships. *Academy of Management Review, 23*(3), 473–490. doi:10.5465/amr.1998.926622

McKnight, D. H., Kacmar, C. J., & Choudhury, V. (2004). Shifting Factors and the Ineffectiveness of Third Party Assurance Seals: A Two-stage Model of Initial Trust in a Web Business. *Electronic Markets, 14*(3), 252–265. doi:10.1080/1019678042000245263

McKnight, H., & Chervany, N. (2001). What Trust Means in ECommerce Customer Relationships: An Interdisciplinary Conceptual Typology. *International Journal of Electronic Commerce, 6*(2), 35–59. doi:10.1080/10864415.2001.11044235

McMahan, B., & Ramage, D. (2017). *Federated Learning: Collaborative Machine Learning without Centralized Training Data*. Retrieved from https://research.googleblog.com/2017/04/federated-learning-collaborative.html?m=1

MediaPro. (2017). *A Best Practices Guide for Comprehensive Employee Awareness Programs*. MediaPro.

Mendonça, C. M. C. D., & Andrade, A. M. V. D. (2018). Dynamic Capabilities and Their Relations with Elements of Digital Transformation in Portugal. *Journal of Information Systems Engineering & Management*, *3*(3), 23.

Mian, S. A. (1997). Assessing and Managing the University Technology Business Incubator: An Integrative Framework. *Journal of Business Venturing*, *12*(4), 251–285. doi:10.1016/S0883-9026(96)00063-8

Michailidou, E., Harper, S., & Bechhofer, S. (2008). Visual complexity and aesthetic perception of web pages. In Proceedings of the 26th annual acm international conference on design of communication (pp. 215-224). doi:10.1145/1456536.1456581

Micik, M., & Eger, L. (2015). Recruiting talents with social media. *Актуальні проблеми економіки*, (3), 266-274.

Miles, M. B., Huberman, A. M., & Saldaña, J. (2014). *Qualitative data analysis: A methods sourcebook* (3rd ed.). Los Angeles, CA: Sage.

Mirzanti, I. R., Simatupang, T. M., & Larso, D. (2015). Mapping on Entrepreneurship Policy in Indonesia. *Procedia: Social and Behavioral Sciences*, *169*(0), 346–353. doi:10.1016/j.sbspro.2015.01.319

Misztal, B. A. (2001). Normality and Trust in Goffman's Theory of Interaction Order. *Sociological Theory*, *19*(3), 312–324. doi:10.1111/0735-2751.00143

Mitchell, M. (2005). Self-Awareness and Control in Decentralized Systems. In *Spring Symposium on Metacognition in Computation*. Menlo Park, CA: AAAI Press.

MITRE. (2010). *The Importance of Using EARNEST*. The MITRE Corporation. Retrieved from https://www.mitre.org/sites/default/files/pdf/mitre_earnest.pdf

MITRE. (2017). *Cybersecurity Awareness & Training*. The MITRE Corporation.

Moe, W. W. (2003). Buying, searching, or browsing: Differentiating between online shoppers using in-store navigational clickstream. *Journal of Consumer Psychology*, *13*(1&2), 29–39. doi:10.1207/153276603768344762

Mokhtarian, P. L. (2003). A conceptual analysis of the transportation impacts of B2C e-commerce. *Transportation*, *31*(3), 257–284. doi:10.1023/B:PORT.0000025428.64128.d3

Molnar, L., Adkins, D., Yolanda, B., Grimes, D., Sherman, H., & Tornatzky, L. (1997). *Business Incubation Works*. Athens, Ohio: NBIA Publications.

Monahan, D. (2014). *Security Awareness Training: It's not just for Compliance-Research Report Summary. Enterprise Management Associates*. EMA.

Monreal, M. (2012). *Sistemas de pago para comercio electrónico*. Tesis de Licenciatura. CIMAT.

Moorman, C., Zaltman, G., & Deshpande, R. (1992). Relationships between providers and users of market research: The dynamics of trust within and between organizations. *JMR, Journal of Marketing Research*, *29*(3), 314–328. doi:10.1177/002224379202900303

Morgan, R. M., & Hunt, S. D. (1994). The commitment-trust theory of relationship marketing. *Journal of Marketing*, *58*(3), 20–38. doi:10.1177/002224299405800302

Motorola. (2014). *Project 25 and interoperability in land mobile radio*. White paper. Motorola.

Motoyama, Y., & Wiens, J. (2015). *Guidelines for local and state governments to promote entrepreneurship*. Ewing Marion Kauffman Foundation. doi:10.2139srn.2580321

Munger, K. (2017). Tweetment effects on the tweeted: Experimentally reducing racist harassment. *Political Behavior*, *39*(3), 629–649. doi:10.1007/s11109-016-9373-5

Mysen, T., Svensson, G., & Payan, G. M. (2011). The key role of opportunism in business relationships. *Marketing Intelligence & Planning, 29*, 426-449.

Nah, F. F.-H. (2004). A study on tolerable waiting time: How long are web users willing to wait? *Behaviour & Information Technology*, *23*(3), 153–163. doi:10.1080/01449290410001669914

Nakajima, M. (2012, February 20). The Evolution of Payment Systems. *The European Financial Review*, *11*.

Nakamoto, S. (2008). *Bitcoin: A Peer-to-Peer Electronic Cash System*. Academic Press.

Nalisnick, E., Mitra, B., Craswell, N., & Caruana, R. (2016). Improving Document Ranking with Dual Word Embeddings. In Proceedings of the 25th International Conference Companion on World Wide Web 2016 (pp. 83–84). doi:10.1145/2872518.2889361

National Bank of Serbia. (n.d.). *Payment System*. Retrieved from NBS: https://www.nbs.rs/internet/english/35/index.html

National Institute of Standards and Technology – NIST. (2003). *Building an Information Technology Security Awareness and Training Program*. NIST Special Publication 800-50.

National Institute of Standards and Technology – NIST. (2017). *An Introduction to Information Security*. NIST Special Publication 800-12 Revision 1.

Neff, G., & Nagy, P. (2016). Automation, algorithms, and politicsl talking to Bots: Symbiotic agency and the case of Tay. *International Journal of Communication*, *10*, 4915–4931.

Nesta / Berkley and Cambridge. (2013). Retrieved at https://www.nesta.org.uk/news/almost-1bn-raised-by-alternative-finance-in-2013/

Newman, S. (2015). *Building microservices: designing fine-grained systems*. O'Reilly Media, Inc.

Nguyen, T.N., Sbityakov, L., & Scoggins, S. (2018). *Intelligence-based Cybersecurity Awareness Training- an Exploratory Project*. CoRR, abs/1812.04234.

Nicolas, P. R. (2014). *Scala for machine learning: Leverage scala and machine learning to construct and study systems that can learn from data. Community Experience Distilled*. Birmingham, UK: Packt Publishing.

Nigam, B., Tokekar, S., & Jain, S. (2015). Evaluation of models for predicting user's next request in web usage mining. *International Journal on Cybernetics & Informatics*, *4*(1), 1–13. doi:10.5121/ijci.2015.4101

Nikolaou, I. (2014). Social networking web sites in job search and employee recruitment. *International Journal of Selection and Assessment*, *22*(2), 179–189. doi:10.1111/ijsa.12067

NIST 2015 76. US National Institute of Standards and Technology (2016) *Framework for Cyber-Physical Systems*, Release 1.0 May 2016

NIST. (2015). *Preliminary Discussion Draft Framework for Cyber-Physical Systems*. NIST.

Nof, S. Y., Ceroni, J., Jeong, W., & Moghaddam, M. (2015). Revolutionizing Collaboration through e-Work, e-Business, and e-Service. Springer. doi:10.1007/978-3-662-45777-1

Nwankpa, J. K., & Roumani, Y. (2016). *IT capability and digital transformation: A firm performance perspective*. Academic Press.

O'Neal, T., & Kulonda, D. (2005). The Role of Technology Incubators in Engineering Education. In *ASEE Southeast Section Conference 2005*. University of Tennessee at Chattanooga.

O'neill, E. S., Dluhy, N. M., & Chin, E. (2005). Modelling novice clinical reasoning for a computerized decision support system. *Journal of Advanced Nursing*, *49*(1), 68–77. doi:10.1111/j.1365-2648.2004.03265.x PMID:15610383

Obal, M., & Lancioni, R. A. (2013). Maximizing buyer–supplier relationships in the Digital Era: Concept and research agenda. *Industrial Marketing Management*, *42*(6), 851–854. doi:10.1016/j.indmarman.2013.06.002

Odior, E., & Banuso, F. (2012). Cashless banking in Nigeria: Challenges, benefits and policy. *European Scientific Journal*, *8*(12), 289–315.

OECD. (1999). *Business incubation: international case studies*. OECD Publishing.

OECD. (2010). *High-Growth Enterprises –What Governments Can Do to Make a Difference*. OECD.

OECD. (2018). *Rural 3.0. A framework for rural development*. Organization for Economic Co-operation and Development.

Oei, S. W. (2016). *Pengaruh Kesesuaian Tugas-Teknologi, Kualitas Informasi, Dan Kualitas Sistem Terhadap Kinerja Individu Pengguna Sistem Erp*. Retrieved from http://repository.unhas.ac.id/handle/123456789/21393

Ohlhorst, F. (2013). *Big data analytics turning big data into big money*. John Wiley & Sons, Inc.

Oliveira, T., Thomas, M., Baptista, G., & Campos, F. (2016). Mobile payment: Understanding the determinants of customer adoption and intention to recommend the technology. *Computers in Human Behavior*, *61*, 404–414.

Oliver, R. (1997). *Satisfaction: A Behavioral Perspective on the Consumer*. McGrawhill.

Olsina Santos, L. (1999). Website quality evaluation method: A case study on museums, Los Angeles, California, USA. *ICSE99 2nd Workshop on Software Engineering over the Internet*.

Orbik, Z., & Zozul'aková, V. (2019). Corporate Social and Digital Responsibility. Management Systems in Production Engineering, 27(2), 79–83. doi:10.1515/mspe-2019-0013

Oz, E. (2002). *Foundations of e-commerce*. Prentice Hall.

Özgüven, N. (2011). Tüketicilerin online alışverişe karşı tutumları ile demografik özellikleri arasındaki ilişkinin analizi. *KMÜ Sosyal ve Ekonomik Araştırmalar Dergisi*, *13*(21), 47–54.

Pacchini, A. P. T., Lucato, W. C., Facchini, F., & Mummolo, G. (2019). The degree of readiness for the implementation of Industry 4.0. *Computers in Industry*, *113*, 103125. doi:10.1016/j.compind.2019.103125

Pal, A., Herath, T., De, R., & Rao, H. R. (2018). Factors facilitating Adoption of Mobile Payment. *PACIS 2018 Proceedings, 337*.

Pan, W. (2002). A comparative review of statistical methods for discovering differentially expressed genes in replicated microarray experiments. *Bioinformatics (Oxford, England)*, *18*(4), 546–554. doi:10.1093/bioinformatics/18.4.546 PMID:12016052

Patterson, A. (2012). Social-networks of the world, unite and take over: A meta-ıntrospective perspective on the Facebook brand. *Journal of Business Research*, *65*(4), 527–534. doi:10.1016/j.jbusres.2011.02.032

Patterson, P., & Smith, T. (2003). A cross-cultural study of switching barriers and propensity to stay with service providers. *Journal of Retailing*, *79*, 107–120.

Pavlou, P. A. (2003). Consumer Acceptance of Electronic Commerce: Integrating Trust and Risk with the Technology Acceptance Model. *International Journal of Electronic Commerce*, *7*(3), 101–134. doi:10.1080/10864415.2003.11044275

Pavlou, P. A., & Gefen, D. (2004). Building Effective Online Marketplaces with Institution-Based Trust. *Information Systems Research*, *15*(1), 37–59. doi:10.1287/isre.1040.0015

Payne, P. R., Embi, P. J., & Sen, C. K. (2009). Translational informatics: Enabling high-throughput research paradigms. *Physiological Genomics*, *39*(3), 131–140. doi:10.1152/physiolgenomics.00050.2009 PMID:19737991

PCI Security Standards Council - PCI DSS. (2014). *Best Practices for Implementing a Security Awareness Program*. PCI DSS.

Pease, M., Shostak, R., & Lamport, L. (1980). Reaching agreement in the presence of faults. *Journal of the Association for Computing Machinery*, *27*(2), 228-234.

Penderdast, T. (2016). How to Audit the Human Element and Assess Your Organization's Security Risk. *ISACA Journal*, *5*, 1–5.

Perroni, M. G., da Costa, S. E. G., de Lima, E. P., & da Silva, W. V. (2017). The relationship between enterprise efficiency in resource use and energy efficiency practices adoption. *International Journal of Production Economics*, *190*, 108–119. doi:10.1016/j.ijpe.2016.08.023

Peters, L., Rice, M., & Sundararajan, M. (2004). The role of incubators in the entrepreneurial process. *The Journal of Technology Transfer*, *29*(1), 83–91. doi:10.1023/B:JOTT.0000011182.82350.df

Petrova, A., & Valles, C. (2012). *Perceived risks as adoption barriers of Facebook commerce: Exploring Facebook users' perception* (Master's Thesis). Jönköping University.

PhishMe. (2017). PhishMe CBFree. *PhishMe Headquarters*. Retrieved from https://phishme.com/resources/cbfree-computer-based-training/

Pihir, I., Oreski, D., & Zestic, M. (2017). Cost and benefits of webshop implementation for sme's in first three years. In Economic and Social Development: Book of Proceedings (pp. 685-694). Retrieved December 1 from, https://search.proquest.com/docview/2070392460?accountid=14701

Ping, R. Jr. (1993). The effects of satisfaction and structural constraints on retailer exiting, voice, loyalty, opportunism, and neglect. *Journal of Retailing*, *69*(Fall), 320–352.

Plane, A. C., Redmiles, E. M., Mazurek, M. L., & Tschantz, M. C. (2017). Exploring User Perceptions of Discrimination in Online Targeted Advertising. *Proceedings of the 26th USENIX Security Symposium*, 935–951.

Plattner, H., Meinel, C., & Leifer, L. (2010). *Design thinking: Understand–improve–apply*. Springer Science & Business Media.

Ponemon Institute. (2018). *Assessing the DNS Security Risk. Research report sponsored by Infoblox*. Ponemon Institute LLC.

Poovendran, R. (2010). Cyber-Physical Systems: Close Encounters Between Two Parallel Worlds. *Proceedings of the IEEE*, *98*(8), 1363–1366. doi:10.1109/JPROC.2010.2050377

Porollo, A. A., Adamczak, R., & Meller, J. (2004). POLYVIEW: A flexible visualization tool for structural and functional annotations of proteins. *Bioinformatics (Oxford, England)*, *20*(15), 2460–2462. doi:10.1093/bioinformatics/bth248 PMID:15073023

Porter, M. (1998). *Competitive Strategy: Techniques for Analyzing Industries and Competitors*. Free Press.

Pousttchi, K., Gleiss, A., Buzzi, B., & Kohlhagen, M. (2019). Technology impact types for digital transformation. In *Proceedings - 21st IEEE Conference on Business Informatics* (Vol. 1, pp. 487–494). CBI; doi:10.1109/CBI.2019.00063.

Pramanik, H. S., Kirtania, M., & Pani, A. K. (2019). Essence of digital transformation—Manifestations at large financial institutions from North America. *Future Generation Computer Systems*, *95*, 323–343. doi:10.1016/j.future.2018.12.003

Prasad, G. S., Reddy, N., & Acharya, U. (2010). Knowledge discovery from web usage data: A survey of web usage pre-processing techniques. In *Information Processing and Management BAIP 2010, CCIS 70* (pp. 505–507). Springer-Verlag Berlin Heidelberg. doi:10.1007/978-3-642-12214-9_88

Pratini, N. (2018). *How Fintech is (and isn't) leading the way in digital trust.* Available online https://fin.plaid.com/fintech-leading-the-way-in-digital-trust/

Priyadarshini, C., Kumar, Y. L. N., & Jha, R. R. (2017). Employer attractiveness through social media: A phenomenological study. *Qualitative Report*, *22*(4), 969–983.

Raji, I. D., & Buolamwini, J. (2019). Actionable auditing: Investigating the impact of publicly naming biased performance results of commercial ai products. In Association for the Advancement of Artificial Intelligence (Chair), AAAI/ACM Conference on Artificial Intelligence, Ethics, and Society (AIES). Symposium conducted at the meeting of ACM, Honolulu, HI.

Ramachandran, G. M., & Neelakrishnan, S. (n.d.). An approach to improving customer on-time delivery against the original promise date. *South African Journal of Industrial Engineering*, *28*(4), 109–119.

Ramasubramanian, K., & Singh, A. (2017). *Machine Learning Using R*. Berkeley, CA: Apress.

Rao, A. R., & Bergen, M. E. (1992). Price premium variations as a consequence of buyers' lack of information. *The Journal of Consumer Research*, *19*(3), 412–423. doi:10.1086/209311

Raub, M. (2018). Bots, Bias and Big Data: Artificial Intelligence, Algorithmic Bias and Disparate Impact Liability in Hiring Practices. *Arkansas Law Review*, *71*(2), 529–570.

Ravand, H. (2016). Partial Least Squares Structural Equation Modeling with R. *Practical Assessment, Research & Evaluation*, *21*(11), 1531–7714.

Read, L., & Rowe, D. (2003). *The key building blocks: A guide to business incubation and incubator management part II*. UKBI.

Resnick, P., & Varian, H. R. (1997). Recommender systems. *Communications of the ACM*, *40*(3), 56–58. doi:10.1145/245108.245121

Rice, M. P., & Matthews, J. (1995). *Growing New Ventures, Creating New Jobs: Principles and Practices of Successful Business Incubation*. Westport, CT: Quorum Books.

Rice, S. C. (2012). Reputation and uncertainty in online markets: An experimental study. *Information Systems Research*, *23*(2), 436–452. doi:10.1287/isre.1110.0362

Rivera, M., Seebacher, A., & Diaz Puente, J. M. (2015). Rural economy: a geographical distribution of scientific discourse. In A. Raupelienė (Ed.), *Proceedings of the 7th International Scientific Conference Rural Development 2015.* (pp. 1-6). Aleksandras Stulginskis University. 10.15544/RD.2015.083

Rivest, R. L. (1997). Perspectives on Financial Cryptography. In *International Conference on Financial Cryptography* (pp. 145-149). Berlin: Springer.

Roberts, G. (2005). *Recruitment, selection and induction.* London: Chartered Institute of Personnel and Development.

Rocha-Lona, L., Alvarez-Reyes, S. E., Garza-Reyes, J. A., & Kumar, V. (2015). An integrated approach of Six Sigma and QSAM methodologies for a pharmaceutical company: A shipment improvement process. *International Journal of Lean Enterprise Research, 1*(3), 266–283. doi:10.1504/IJLER.2015.071743

Rodríguez, M. (2015). *Nuevos Métodos de Pago Online, Seguridad y Confiabilidad.* Tesis de Licenciatura. Universidad de Cantabria.

Rosoiu, O. & Popescu, C. (2016). E-recruiting platforms: Features that Influence the efficiency of online recruitment systems. *Informatica Economica, 20*(2), 46–55.

Rotter, J. B. (1967). A new scale for the measurement of interpersonal trust 1. *Journal of Personality, 35*(4), 651–665. doi:10.1111/j.1467-6494.1967.tb01454.x PMID:4865583

Rousseau, D. M., Sitkin, S. B., Burt, R. S., & Camerer, C. (1998). Not so different after all: A cross-discipline view of trust. *Academy of Management Review, 23*(3), 393–404. doi:10.5465/amr.1998.926617

Rubio, D. M., Schoenbaum, E. E., Lee, L. S., Schteingart, D. E., Marantz, P. R., Anderson, K. E., . . . Esposito, K. (2010). Defining translational research: implications for training. *Academic Medicine: Journal of the Association of American Medical Colleges, 85*(3), 470.

Rusbult, C. (1980). Commitment and satisfaction in romantic associations a test of the investment model. *Journal of Experimental Social Psychology, 16*, 172–186.

Ryu, H.-S. (2018). What makes users willing or hesitant to use Fintech?: the moderating effect of user type. Industrial Management & Data Systems, 118(3), 541-569.

Saarikko, T., Westergren, U. H., & Blomquist, T. (2017). The Internet of Things: Are you ready for what's coming? *Business Horizons, 60*(5), 667–676. doi:10.1016/j.bushor.2017.05.010

Sabillon, R. (2018). *Scenario III: Data for a single cybersecurity domain audit (Awareness Education).* Mendeley Data. doi:10.17632/m4dk8n9sx7.2

Sabillon, R., Serra-Ruiz, J., Cavaller, V., & Cano, J. (2017). A Comprehensive Cybersecurity Audit Model to Improve Cybersecurity Assurance: The CyberSecurity Audit Model (CSAM). *Proceedings of Second International Conference on Information Systems and Computer Science (INCISCOS).* 10.1109/INCISCOS.2017.20

Sabillon, R., Serra-Ruiz, J., Cavaller, V., Jeimy, J., & Cano, M. (2019). An Effective Cybersecurity Training Model to Support an Organizational Awareness Program: The Cybersecurity Awareness TRAining Model (CATRAM). A Case Study in Canada. *Journal of Cases on Information Technology*, *21*(3), 26–39. doi:10.4018/JCIT.2019070102

Salehie, M., & Tahvildari, L. (2009). Self-Adaptive Software: Landscape and Research Challenges. *ACM Transactions on Autonomous and Adaptive Systems*, *4*(2), 1–42. doi:10.1145/1516533.1516538

Sameen, S., & Cornelius, S. (2015). Social networking sites and hiring: How social media profiles influence hiring decisions. *Journal of Business Studies Quarterly*, *7*(1), 27–35.

Samerski, S., & Müller, H. (2019). Digital health literacy in Germany – requested, but not supported? Results of the empirical study TK-DiSK [Digitale Gesundheitskompetenz in Deutschland - gefordert, aber nicht gefördert? Ergebnisse der empirischen Studie TK-DiSK]. *Zeitschrift für Evidenz, Fortbildung und Qualität im Gesundheitswesen*, *144–145*, 42–51. doi:10.1016/j.zefq.2019.05.006 PubMed

Sandvig, C., Hamilton, K., Karahalios, K., & Langbort, C. (2014). Auditing Algorithms: Research Methods for Detecting Discrimination on Internet Platforms. In Data and Discrimination: Converting Critical Concerns into Productive Inquiry. Preconference at the 64th Annual Meeting of the International Communication Association. Retrieved from https://pdfs.semanticscholar.org/b722/7cbd34766655dea10d0437ab10df3a127396.pdf

SANS Institute. (2017). *2017 Security Awareness Report: It's time to communicate. SANS Security Awareness.* Retrieved from https://securingthehuman.sans.org/media/resources/STH-SecurityAwarenessReport-2017.pdf

SANS Security Awareness, . (2017). *2017 Security Awareness Report.* SANS Institute.

Sari, R. D. (2015). *Pengaruh Penggunaan Celebrity Endorser Terhadap Keputusan Pembelian Konsumen Secra Online Pada Media Sosial Instagram Di Bandar Lampung.* Retrieved from http://digilib.unila.ac.id/7165/1/ABSTRAK.pdf

Sathikh, P. (2017). *Aesthetics of experience: Industrial design in the era of design thinking and user experience.* Advances in Transdisciplinary Engineering; doi:10.3233/978-1-61499-779-5-338

Schafer, J. B., Konstan, J. A., & Riedl, J. (2001). E-commerce recommendation applications. *Data Mining and Knowledge Discovery*, *5*(5), 115–153. doi:10.1023/A:1009804230409

Scharre, P. D. (2015). The Opportunity and Challenge of Autonomous Systems. In Autonomous Systems. Issues for Defence Policymakers. NATO Headquarters Supreme Allied Command. Allied Command Transformation.

Scherer, M. (2016). Tay the Racist Chatbot: Who is responsible when a machine learns to be evil? Retrieved from https://futureoflife.org/2016/03/27/tay-the-racist-chatbot-who-is-responsible-when-a-machine-learns-to-be-evil/?cn-reloaded=1

Schneider, T. J., Goffin, R. D., & Daljeet, K. N. (2015). Give us your social networking site passwords: Implications for personnel selection and personality. *Personality and Individual Differences*, *73*, 78–83. doi:10.1016/j.paid.2014.09.026

Schroeder, A. N., & Lile, C. R. (2016). Legal trends in organizational online social media use. *Industrial and Organizational Psychology: Perspectives on Science and Practice*, *9*(3), 682–689. doi:10.1017/iop.2016.71

Schurr, P. H., & Ozanne, J. L. (1985). Influences on exchange processes: Buyers' preconceptions of a seller's trustworthiness and bargaining toughness. *The Journal of Consumer Research*, *11*(4), 939–953. doi:10.1086/209028

Scillitoe, J. L., & Chakrabarti, A. K. (2010). The Role of Incubator Interactions in Assisting New Ventures. *Technovation*, *30*(3), 155–167. doi:10.1016/j.technovation.2009.12.002

Selnes, F. (1998). Antecedents and consequences of trust and satisfaction in buyerseller relationships. *European Journal of Marketing*, *32*(3/4), 305–322. doi:10.1108/03090569810204580

Shabbir, R., Zhang, J., Pitsaphol, C., & Hassan, W. (2015). The dynamics of value creation in online brand communities: Strategy map approach. *Information Management and Business Review*, *7*(1), 48–58.

Shao, C., Ciampaglia, G. L., Varol, O., Flammini, A., & Menczer, F. (2017). The spread of fake news by social bots. arXiv preprint arXiv:1707.07592

Sharma, N., & Patterson, P. G. (2000). Switching costs, alternative attractiveness and experience as moderators of relationship commitment in professional, consumer services. *International Journal of Service Industry Management*, *11*(5), 470–490.

Sha, W. (2009). Types of structural assurance and their relationships with trusting intentions in business-to-consumer e-commerce. *Electronic Markets*, *19*(1), 43–54. doi:10.100712525-008-0001-z

Sheehan, K. (1999). An investigation of gender differences in on-line privacy concerns and resultant behavior. In Internet Marketing. McGraw Hill.

Shehabuddeen, N., Probert, D., Phaal, R., & Platts, K. (1999). Representing and Approaching Complex Management Issues: Part 1 - Role and Definition. SSRN Electronic Journal. Advance online publication. doi:10.2139srn.1923155

Shen, Q.-L., & Li, Y.-T. (2010). Explore antecedent factors of switching costs and intentions and their impact on customer loyalty. *The 13th Conference on Interdisciplinary and Multifunctional Business Management, 13*, 271-288.

Shen, J. (2012). Social comparison, social presence, and enjoyment in the acceptance of social shopping websites. *Journal of Electronic Commerce Research*, *13*(3), 198–212.

Shen, J., Wang, Q., & Mei, H. (2004). Self-Adaptive Software: Cybernetic Perspective and an Application Server Supported Framework. *Proceedings of the 28th Annual International Computer Software and Applications Conference*.

Sheppard, B. H., & Sherman, D. M. (1998). The grammars of trust: A model and general implications. *Academy of Management Review*, *23*(3), 422–437. doi:10.5465/amr.1998.926619

Shih, Y.-Y., & Fang, K. (2006). Effects of network quality attributes on customer adoption intentions of internet banking. Total Quality Management & Business Excellence, 17(1), 61-77.

Siegel, A. I., & Fischl, M. (1971). Dimensions of visual information displays. *The Journal of Applied Psychology*, *55*(5), 470–476. doi:10.1037/h0031777 PMID:5113200

Singh, A., & Hess, T. (2017). How chief digital officers promote the digital transformation of their companies. *MIS Quarterly Executive*, *16*(01), 1–17.

Singleton, A. (2016). The Economics of Microservices. *IEEE Cloud Computing*, *3*(5), 16–20. doi:10.1109/MCC.2016.109

Slovensky, R., & Ross, W. H. (2012). Should human resource managers use social media to screen job applicants? Managerial and legal issues in the USA. *Info*, *14*(1), 55–69. doi:10.1108/14636691211196941

Smilor, R. W., Gibson, D. V., & Dietrich, G. B. (1990). University Spin-Out Companies: Technology Startups from UT–Austin. *Journal of Business Venturing*, *5*(1), 63–76. doi:10.1016/0883-9026(90)90027-Q

Snow, J. (2018). Bias already exists in search engine results, and it's only going to get worse. MIT Technology Review. Retrieved from https://www.technologyreview.com/s/610275/meet-the-woman-who-searches-out-search-engines-bias-against-women-and-minorities/

Society for Human Resource Management. (2015). *Using social media for talent acquisition – recruitment and screening*. Retrieved from https://www.shrm.org/hr-today/trends-and-forecasting/research-and-surveys/Documents/SHRM-Social-Media-Recruiting-Screening-2015.pdf

Song, K. F. (2007). Telepresence and fantasy in online apparel shopping experience. *Journal of Fashion Marketing and Management: An International Journal*, *11*(4), 553–570.

Speicher, T., Ali, M., Venkatadri, G., Ribeiro, F. N., Arvanitakis, G., Benevenuto, F., ... Mislove, A. (2018). Potential for Discrimination in Online Targeted Advertising. *Proceedings of Machine Learning Research*, *81*, 1–15.

Speicher, T., Heidari, H., Grgic-Hlaca, N., Gummadi, K. P., Singla, A., Weller, A., & Zafar, M. B. (2018). A Unified Approach to Quantifying Algorithmic Unfairness. In Y. Guo & F. Farooq (Eds.), *Proceedings of the 24th ACM SIGKDD International Conference on Knowledge Discovery & Data Mining - KDD '18* (pp. 2239–2248). New York: ACM Press; doi:10.1145/3219819.3220046.

Spiliopoulou, M., Mobasher, B., Berendt, B., & Nakagawa, M. (2003). A framework for evaluation of session reconstruction heuristic in web usage analysis. *INFORMS Journal on Computing*, *15*(2), 171–190. doi:10.1287/ijoc.15.2.171.14445

Srivastava, M., Srivastava, A. K., & Garg, R. (2019, February). Data preprocessing techniques in web usage mining: A literature review. In *Proceedings of International Conference on Sustainable Computing in Science, Technology and Management (SUSCOM)*. Amity University Rajasthan. 10.2139srn.3352357

Srivastava, J., Cooley, R., Deshpande, M., & Tan, P. (2000). Web usage mining: Discovery and applications of usage patterns from web data. *SIGKDD Explorations*, *1*(2), 12–23. doi:10.1145/846183.846188

Stanton, W., Etzel, M., & Walker, B. (2007). *Fundamentals of Marketing* (14th ed.). McGraw-Hill. doi:10.4324/9780203030783

Statista. (2017a, February). Online or in-store shopping preference for selected product categories by consumers worldwide as of 2017. *Statista - The Statistics Portal*. Retrieved from https://www.statista.com/statistics/692871/global-online-shopping-preference-by-category/

Statista. (2017b, October). Global markets with the highest online shopping penetration rate as of 2nd quarter 2017. *Statista - The Statistics Portal*. Retrieved from https://www.statista.com/statistics/274251/retail-site-penetration-across-markets/

Statista. (2018, March). E-commerce share of total global retail sales from 2015 to 2021. *Statista - The Statistics Portal*. Retrieved from https://www.statista.com/statistics/534123/e-commerce-share-of-retail-sales-worldwide/

Steiner, C. (2013). *Automate this: how algorithms took over our markets, our jobs, and the world*. New York: Penguin Group.

Stewart, K. (1998). An exploration of customer exit in retail banking. *International Journal of Bank Marketing*, *16*, 6–14.

Storbacka, K. (2011). A solution business model: Capabilities and management practices for integrated solutions. *Industrial Marketing Management*, *40*(5), 699–711. doi:10.1016/j.indmarman.2011.05.003

Stoughton, J. W., Thompson, L. F., & Meade, A. W. (2015). Examining applicant reactions to the use of social networking websites in pre-employment screening. *Journal of Business and Psychology*, *30*(1), 73–88. doi:10.100710869-013-9333-6

Strebe, R. (2016). Aesthetics on the web: Effects on approach and avoidance behaviour. *Behaviour & Information Technology*, *35*(1), 4–20. doi:10.1080/0144929X.2015.1070202

Strohmeier, S., & Piazza, F. (2015). Artificial Intelligence Techniques in Human Resource Management—A Conceptual Exploration. In C. Kahraman & S. Çevik Onar (Eds.), *Intelligent Systems Reference Library. Intelligent Techniques in Engineering Management* (Vol. 87, pp. 149–172). Cham: Springer International Publishing; doi:10.1007/978-3-319-17906-3_7.

Studdard, N. L. (2006). The effectiveness of entrepreneurial firm's knowledge acquisition from a business incubator. *The International Entrepreneurship and Management Journal*, *2*(2), 211–225. doi:10.100711365-006-8685-z

Sugiyono. (2015). *Statistik Nonparametris untuk penelitian.* Bandung: Alfabeta.

Sung, N. S., Crowley, W. F. Jr, Genel, M., Salber, P., Sandy, L., Sherwood, L. M., ... Getz, K. (2003). Central challenges facing the national clinical research enterprise. *Journal of the American Medical Association*, *289*(10), 1278–1287. doi:10.1001/jama.289.10.1278 PMID:12633190

Susanto, A., Chang, Y., & Ha, Y. (2016). Determinants of continuance intention to use the smartphone banking services. Industrial Management & Data Systems, 116(3), 508 - 525.

Sussman, S., Valente, T. W., Rohrbach, L. A., Skara, S., & Ann Pentz, M. (2006). Translation in the health professions: Converting science into action. *Evaluation & the Health Professions*, *29*(1), 7–32. doi:10.1177/0163278705284441 PMID:16510878

Sweeney, L. (2013). Discrimination in Online Ad Delivery. Queue, 11(3), 10. doi:10.1145/2460276.2460278

Symantec. (2014). *Symantec Security Awareness Program: Mitigate information risk by educating your employees*. Symantec Corporation.

Taherdoost, H. (2016). *Validity and Reliability of the Research Instrument; How to Test the Validation of a Questionnaire/Survey in a Research.* SSRN Electronic Journal. doi:10.2139srn.3205040

Tamasy, C. (2007). Rethinking Technology-Oriented Business Incubators: Developing a Robust Policy Instrument for Entrepreneurship, Innovation, and Regional Development? *Growth and Change*, *38*(3), 460–473. doi:10.1111/j.1468-2257.2007.00379.x

Tan, M., & Teo, T. (2000). Factors influencing the adoption of internet banking. *Journal of the AIS, 1*(5), 1-40.

Tan, P., & Kumar, V. (2000). *Modeling of web robot navigational patterns*. Army High Performance Computing Research Center.

Tan, Y. H., & Thoen, W. (2010). A logical model of trust in e-commerce. *Electronic Markets*, *10*(4), 258–263.

Taylor, N. J., Dennis, A. R., & Cummings, J. W. (2013). Situation normality and the shape of search: The effects of time delays and information presentation on search behavior. *Journal of the American Society for Information Science and Technology*, *64*(5), 909–928. doi:10.1002/asi.22782

Thakare, S. B., & Gawali, S. (2010). A effective and complete preprocessing for web usage mining. *International Journal on Computer Science and Engineering*, *2*(3), 848–851.

The Age Discrimination in Employment Act of 1967.

Theurer, C. P., Tumasjan, A., Welpe, I. M., & Lievens, F. (2018). Employer branding: A brand equity-based literature review and research agenda. *International Journal of Management Reviews*, *20*(1), 155–179. doi:10.1111/ijmr.12121

Thierer, A., O'Sullivan Castillo, A., & Russell, R. (2017). Artificial Intelligence and Public Policy. Mercatus Research. Retrieved from https://www.mercatus.org/system/files/thierer-artificial-intelligence-policy-mr-mercatus-v1.pdf

Thönes, J. (2015). Microservices. *IEEE Software*, *32*(1), 116–116. doi:10.1109/MS.2015.11

Thorlacius, L. (2007). The role of aesthetics in web design. *Nordicom Review*, *28*(1), 63–76.

Thüring, M., & Mahlke, S. (2007). Usability, aesthetics and emotions in human–technology interaction. *International Journal of Psychology*, *42*(4), 253–264. doi:10.1080/00207590701396674

Toh, K. T. K., Nagel, P., & Oakden, R. (2009). A business and ICT architecture for a logistics city. *International Journal of Production Economics*, *122*(1), 216–228. doi:10.1016/j.ijpe.2009.05.021

Towards a European Roadmap on Research and Innovation in Engineering and Management of Cyber-Physical Systems of Systems. (2015). *Cyber-Physical Systems of Systems: Research and Innovation Priorities*. Initial Document for Public Consultation.

*Trans-Atlantic Research and Education Agenda in Systems of Systems (T-AREA-SoS)*. (2013). Loughborough University. Document no. TAREA-PU-WP2-D-LU-9.

Trewin, S. (2018). AI Fairness for People with Disabilities: Point of View. CoRR. Retrieved from https://arxiv.org/abs/1811.10670v1

Tschimmel, K. (2012). Design Thinking as an effective Toolkit for Innovation. In ISPIM Conference Proceedings. The International Society for Professional Innovation Management (ISPIM).

TSI. (2015). *Household information technology usage report*. Retrieved from http://www.tuik.gov.tr/PreHaberBultenleri.do?id=18660

TSI. (2018). *Household information technology usage report*. Retrieved from http://www.tuik.gov.tr/PreTablo.do?alt_id=1028

Turban, E., King, D., Lee, K. Y., Liang, T.-P., & Turban, D. C. (2015). *Electronic commerce: a managerial and social networks perspective* (8th ed.). Springer. doi:10.1007/978-3-319-10091-3

Turkey, I. A. B. (2012). *Türkiye internet ölçümleme araştırması*. Retrieved from http://www.iab-turkiye.org/arastirma-raporlari.php

Uçar, A. (n.d.). *Bankacılık sektöründe bilişim suçları*. Retrieved from https://books.google.com.tr/books?id=Ka-fCQAAQBAJ&printsec=frontcover&hl=tr&source=gbs_ge_summary_r&cad=0#v=onepage&q&f=false

Umar, A., & Baharun, R. (2017). The mediating relationship of customer satisfaction between brand trust, brand social responsibility image wiith moderating role of switching cost. *Universiti Teknologi Malaysia*, *23*, 9020–9025.

UN Committee on Economic. Social and Cultural Rights (2009). General comment No. 20: Non-discrimination in economic, social and cultural rights (art. 2, para. 2, of the International Covenant on Economic, Social and Cultural Rights). Retrieved from https://www.refworld.org/docid/4a60961f2.html

United Nations. (2015). *Transforming our world: the 2030 Agenda for Sustainable Development*. UN.

Universal Declaration of Human Rights, United Nations General Assembly 1948.

University of Ottawa. (2019). E-Business Technology (2019). Retrieved October 31, 2019, from https://catalogue.uottawa.ca/en/graduate/doctorate-philosophy-electronic-business/

Ustinova, L., Ivanova, E., & Aletdinova, A. (2019). Innovative processes of digital transformation in radio electronics. *IOP Conference Series. Materials Science and Engineering*, *497*(1). doi:10.1088/1757-899X/497/1/012098

Utami, R. M., & Lantu, D. C. (2014). Development competitiveness model for small-medium enterprises among the creative industry in bandung. *Procedia: Social and Behavioral Sciences*, *115*, 305–323. doi:10.1016/j.sbspro.2014.02.438

Valls, J. F., Andrade, M. J., & Arribas, R. (2011). Consumer attitudes towards brands in times of great price sensitivity: Four case studies. *Innovative Marketing*, *7*(2), 60–70.

vAl-Ruithe, M., Benkhelifa, E., & Hameed, K. (2018). Key Issues for Embracing the Cloud Computing to Adopt a Digital Transformation: A study of Saudi Public Sector. *Procedia Computer Science*, *130*, 1037–1043. doi:10.1016/j.procs.2018.04.145

Van Doorn, J., Mende, M., Noble, S. M., Hulland, J., Ostrom, A. L., Grewal, D., & Petersen, J. A. (2017). Domo arigato Mr. Roboto: Emergence of automated social presence in organizational frontlines and customers' service experiences. *Journal of Service Research*, *20*(1), 43–58. doi:10.1177/1094670516679272

Van Esterik-Plasmeijer, P. W., & van Raaij, W. F. (2017). Banking system trust, bank trust, and bank loyalty. *International Journal of Bank Marketing*, *35*(1), 97–111. doi:10.1108/IJBM-12-2015-0195

Van Hoye, G., & Lievens, F. (2005). Recruitment-Related Information Sources and Organizational Attractiveness: Can Something Be Done About Negative Publicity? *International Journal of Selection and Assessment*, *13*(3), 179–187. doi:10.1111/j.1468-2389.2005.00313.x

van Lier, B. (2018a). *Thinking about ecologies of autonomous cyber-physical systems and their ethics. Inaugural Lecture*. Rotterdam University of Applied Sciences.

van Lier, B. (2018b). Cyber-Physical Systems of Systems and Complexity Science: The Whole is More than the Sum of Individual and Autonomous Cyber-Physical Systems. *Cybernetics and Systems*, *49*(7-8), 7–8, 538–565. doi:10.1080/01969722.2018.1552907 van Lier, B. (2018c). Blockchain Technology: The autonomy and Self-Organisation of Cyber-Physical Systems. In H. Treiblmaier & R. Beck (Eds.), *Business Transformation Through Blockchain*. Cham, Switzerland: Palgrave McMillan.

Van Wynsberghe, A. (2013). Designing robots for care: Care centered value-sensitive design. *Science and Engineering Ethics*, *19*(2), 407–433. doi:10.1007/s11948-011-9343-6 PubMed

Vance, A., Elie-Dit-Cosaque, C., & Straub, D. W. (2008). Examining trust in information technology artifacts: The effects of system quality and culture. *Journal of Management Information Systems*, *24*(4), 73–100. doi:10.2753/MIS0742-1222240403

Vasilescu, B., Capiluppi, A., & Serebrenik, A. (2013). Gender, representation and online participation: A quantitative study. *Interacting with Computers*, *26*(5), 488–511. doi:10.1093/iwc/iwt047

Veale, M., & Binns, R. (2017). Fairer machine learning in the real world: Mitigating discrimination without collecting sensitive data. Big Data & Society, 4(2), 1–17. doi:10.1177/2053951717743530

Vedel, B., Stephany, E. & Gabarret, I. (2011). *Selection Strategy, Assistance and Performance of Incubation Structures: A French Study*. Paper presented to ICSB.

Vedovello, C., & Godinho, M. (2003). Business incubators as a technological infrastructure for supporting small innovative firms' activities. *International Journal of Entrepreneurship and Innovation Management*, *3*(1/2), 4–21. doi:10.1504/IJEIM.2003.002215

Velasquez, J. D., & Palade, V. (2008). *Adaptive web sites a knowledge extraction from web data approach*. Amsterdam: IOS Press.

Venkatesh, V. (2008). Technology acceptance model 3 and a research agenda on interventions. *Decision Sciences*, *39*(2), 273–315.

Venkatesh, V., Morris, M. G., & Davis, G. B. (2003). User acceptance of information technology: Toward a unified view. *Management Information Systems Quarterly*, *27*(2), 425–478.

Verma, S. (2004). *Successful factors for Business Incubators: An Empirical Study of Canadian Business Incubators*. Ottawa, Canada: Eric Sprott School of Business, Carleton University.

Vize, R., Coughlan, J., Kennedy, A., & Ellis-Chadwick, F. (2013). Technology readiness in a B2B online retail context: An examination of antecedents and outcomes. *Industrial Marketing Management*, *42*(6), 909–918. doi:10.1016/j.indmarman.2013.05.020

Vogelsang, K., Liere-Netheler, K., Packmohr, S., & Hoppe, U. (2019). Success factors for fostering a digital transformation in manufacturing companies. Journal of Enterprise Transformation, 1-22.

Wade, M. (2015). *Digital Business Transformation*. Global Center for Digital Business Transformation.

Walker, H. J., Feild, H. S., Giles, W. F., Armenakis, A. A., & Bernerth, J. B. (2009). Displaying employee testimonials on recruitment web sites: Effects of communication media, employee race, and job seeker race on organizational attraction and information credibility. *The Journal of Applied Psychology*, *94*(5), 1354–1364. doi:10.1037/a0014964 PMID:19702377

Wang, Y.-S., Wang, Y.-M., Lin, H.-H., & Tang, T.-I. (2003). Determinants of user acceptance of Internet banking: An empirical study. *International Journal of Service Industry Management*, *14*(5), 501–519. doi:10.1108/09564230310500192

Ward, M. (2016). *Security Awareness and Training: Solving the unintentional insider threat*. Cyber Safe Worforce LLC.

Warner, K. S. R., & Wager, M. (2018). *Building dynamic capabilities for digital transformation: An ongoing process of strategic renewal*. Academic Press.

Waskowsky, Z. (2017). The strategy of coopetition. Value-creating networks of partnership relations: the case of sport markets. In *Proceedings of the Management International Conference*. (pp. 13-21). Monastier di Treviso.

Weiss, G. (1999). *Multiagent systems: a modern approach to distributed artificial intelligence*. Cambridge, MA: MIT Press.

Wei, Z., Song, X., & Wang, D. (2017). Manufacturing flexibility, business model design, and firm performance. *International Journal of Production Economics*, *193*, 87–97. doi:10.1016/j.ijpe.2017.07.004

West, S. M., Whittaker, M., & Crawford, K. (2019). Discriminating Systems: Gender, Race and Power in AI. Retrieved from https://ainowinstitute.org/discriminatingsystems.pdf

Westerman, G., Bonnet, D., & McAfee, A. (2014). The nine elements of digital transformation. *MIT Sloan Management Review*, *55*(3), 1–6.

Westfall, J. M., Mold, J., & Fagnan, L. (2007). Practice-based research: "Blue Highways" on the NIH roadmap. *Journal of the American Medical Association*, *297*(4), 403–406. doi:10.1001/jama.297.4.403 PMID:17244837

Weyerer, J. C., & Langer, P. F. (2019). Garbage In, Garbage Out: The Vicious Cycle of AI-Based Discrimination in the Public Sector. In Y.-C. Chen, F. Salem, & A. Zuiderwijk (Eds.), Proceeding dg.o 2019 20th Annual International Conference on Digital Government Research (pp. 509–511). New York, NY: Academic Press.

Whitman, M. E., & Mattord, H. J. (2019). *Management of Information Security* (6th ed.). Boston, MA: Cengage Learning, Inc.

Williams, B. Brooks., & Shmargad. (2018). How Algorithms Discriminate Based on Data They Lack: Challenges, Solutions, and Policy Implications. Journal of Information Policy, 8, 78–115. doi:10.5325/jinfopoli.8.2018.0078

Winter, J. (2015). Algorithmic Discrimination: Big Data Analytics and the Future of the Internet. In J. Winter & R. Ono (Eds.), *The Future Internet* (pp. 125–140). Cham: Springer International Publishing; doi:10.1007/978-3-319-22994-2_8.

Wirtz, B. W., Weyerer, J. C., & Geyer, C. (2019). Artificial Intelligence and the Public Sector—Applications and Challenges. *International Journal of Public Administration, 42*(7), 596–615. doi:10.1080/01900692.2018.1498103

Wonglimpiyarat, J. (2016). Government policies towards Israel's high-tech powerhouse. *Technovation, 52*, 18–27. doi:10.1016/j.technovation.2016.02.001

Woodruff, A., Fox, S. E., Rousso-Schindler, S., & Warshaw, J. (2018). A Qualitative Exploration of Perceptions of Algorithmic Fairness. In R. Mandryk, M. Hancock, M. Perry, & A. Cox (Eds.), *Proceedings of the 2018 CHI Conference on Human Factors in Computing Systems - CHI '18* (pp. 1–14). New York: ACM Press; doi:10.1145/3173574.3174230.

Woolf, S. H. (2008). The meaning of translational research and why it matters. *Journal of the American Medical Association, 299*(2), 211–213. doi:10.1001/jama.2007.26 PMID:18182604

Wooten, D. B., & Reed, A. II. (2004). Playing it safe: Susceptibility to normative influence and protective self-presentation. *The Journal of Consumer Research, 31*(3), 551–556. doi:10.1086/425089

World Bank, World Development Indicators. (2017). Rural population (% of total population) [Data file]. Retrieved from https://data.worldbank.org/indicator/sp.rur.totl.zs

World Economic Forum. (2018). How to Prevent Discriminatory Outcomes in Machine Learning: White Paper. Retrieved from http://www3.weforum.org/docs/WEF_40065_White_Paper_How_to_Prevent_Discriminatory_Outcomes_in_Machine_Learning.pdf

Wright, R., Worlf, D., Heneghan, L., & Clamp, A. (2017). Destination (un) known. Retrieved October 31, 2019, from https://home.kpmg/content/dam/kpmg/uk/pdf/2017/09/digital_transformation_guide_2017.pdf

Wu, Y.-L., & Tao, Y.-H. L. C.-P., Wang, S.-Y., & Chiu, C.-Y. (2014). User-switching behavior in social network sites: A model perspective with drill-down analyses. *Computers in Human Behavior, 33*, 92–103.

Xu, G., Zhang, Y., & Li, L. (2011). *Web mining and social networking techniques and applications*. Springer Science+Business Media.

Yaga, D., Mell, P., Roby, N., & Scarfone, K. (2018). *Blockchain Technology Overview. Draft report NISTIR 8202, January 2018*. National Institute of Standards and Technology.

Yang, Q., Liu, Y., Chen, T., & Tong, Y. (2019). Federated machine learning: Concept and applications. *ACM Transactions on Intelligent Systems and Technology, 10*(2), 12. doi:10.1145/3298981

Yıldırım, E. (2013). The effects of user comments on e-trust: An application on consumer electronics. *Journal of Economics. Business and Management, 1*(4), 360–364.

Yıldırım, E., & Barutçu, M. T. (2016). How uncertainty avoidance, power distance and indulgence affect social commerce expenditure? An investigation based on Facebook. *International Journal of Science Culture and Sport*, *4*(4), 403–421. doi:10.14486/IntJSCS593

Yin, R. K. (2014). *Case Study Research: Design and Methods* (5th ed.). Thousand Oaks, CA: Sage Publications.

Yin, R. K. (2018). *Case Study Research and Applications* (6th ed.). Thousand Oaks, CA: Sage Publications.

Yoon, S.-J. (2012). A social network approach to the influences of shopping experiences on e-wom. *Journal of Electronic Commerce Research*, *13*(3), 213–223.

Young, L. (2006). Trust: Looking forward and back. *Journal of Business and Industrial Marketing*, *21*(7), 439–445. doi:10.1108/08858620610708920

Zeng, Y., Jia, F., Wan, L., & Guo, H. (2017). E-commerce in agri-food sector: A systematic literature review. *The International Food and Agribusiness Management Review*, *20*(4), 439–459. doi:10.22434/IFAMR2016.0156

Zhang, L., Wu, Y., & Wu, X. (2018). Achieving Non-Discrimination in Prediction. In J. S. Rosenschein & J. Lang (Chairs), Twenty-Seventh International Joint Conference on Artificial Intelligence {IJCAI-18}. Stockholm, Sweden: Academic Press.

Zhao, J., Wang, T., Yatskar, M., Ordonez, V., & Chang, K.-W. (2017). Men Also Like Shopping: Reducing Gender Bias Amplification using Corpus-level Constraints. In M. Palmer, R. Hwa, & S. Riedel (Eds.), *Proceedings of the 2017 Conference on Empirical Methods in Natural Language Processing* (pp. 2979–2989). Stroudsburg, PA: Association for Computational Linguistics; doi:10.18653/v1/D17-1323.

Zhao, X., Lampe, C., & Ellison, N. B. (2016). *The social media ecology: User perceptions, strategies and challenges*. New York: ACM Press. doi:10.1145/2858036.2858333

Zhou, M., & Tian, D. (2010). *An Integrated Model of Influential Antecedents of Online Shopping Initial Trust: Empirical Evidence in a Low-Trust Environment*. Academic Press.

Zhou, T. (2011a). An empirical examination of initial trust in mobile banking. *Internet Research*, *21*(5), 527–540. doi:10.1108/10662241111176353

Zhou, T. (2011b). The effect of initial trust on user adoption of mobile payment. *Information Development*, *27*(4), 290–300. doi:10.1177/0266666911424075

Zhou, T. (2016). Examining user switch between mobile stores: A push-pull-mooring. *perspective. Information Resources Management Journal*, *29*(2), 1–13.

Zhou, W., Tsiga, Z., Li, B., Zheng, S., & Jiang, S. (2018). What influence users'e-finance continuance intention? The moderating role of trust. *Industrial Management & Data Systems*, *118*(8), 1647–1670. doi:10.1108/IMDS-12-2017-0602

Zhou, X., & Huang, Y. (2014). An improved parallel association rules algorithm based on mapreduce framework for big data. In *11th International Conference on Fuzzy Systems and Knowledge Discovery*, (pp. 284–288). 10.1109/FSKD.2014.6980847

Zide, J., Elman, B., & Shahani-Denning, C. (2014). LinkedIn and recruitment: How profiles differ across occupations. *Employee Relations*, *36*(5), 583–604. doi:10.1108/ER-07-2013-0086

Zimmerman, P., Gilbert, T., & Salvatore, F. (2019). Digital engineering transformation across the Department of Defense. *Journal of Defense Modeling and Simulation*, *16*(4), 325–338. doi:10.1177/1548512917747050

# About the Contributors

**Rocci Luppicini** is the Director of E-Business Technologies in the Faculty of Arts (2014-2019) and an Associate Professor in the Department of Communication at the University of Ottawa, Canada. He is the Founder and Editor-in-Chief of the International Journal of Technoethics and a leading expert in technology studies (TS) and Technoethics. Main research areas include: Ethical Dilemmas with New Technology (media ethics, e-democracy, cybercrime, digital divide, hacking, information theft, cyber espionage, cyberbullying), E-Business and E-Society (virtual organizations, online communities, Social Networking, e-trust, social responsibility, e-health, e-government, socio-technical change), Identity and Technology (human-computer interaction, e-identity management, human enhancement, post-human, social robotics, cyberculture, technotherapy), Educational Technology (program planning and development, distance education, blended education, instructional design, technology integration, organizational studies (systems theory, organizational communications, organizational change, social movements).

* * *

**Diya Abraham** is a Data Scientist with experience building data science products in healthcare retail. She is also a Research Affiliate at the Virginia Commonwealth University where she investigates different research projects involving data and publishes research papers. The combination of such a background helps focus on how to apply data science to build better business solutions.

**Ghadah Althawwad** is a PhD candidate in the Electronic Business program at the University of Ottawa. Her research interests focus on the influence of technology on communication. She has a special interest in using technoethical inquiry theory as a broad systems framework by organizations who used social media for recruitment purposes. She is interested in issues related to gender, online identity and discrimination created by virtual communication.

**Israel Alvarez-Velasquez** is a Master in Administrative Engineering from the Technological Institute of Orizaba (ITO) under the lines of Quality research, Continuous Improvement and Strategic Planning (2012). Graduate in Administration from the Universidad Veracruzana Campus Nogales cultivating lines of research on Marketing in the Children's Public, Quality and Leadership in the Human Resource (2010). Professor-Researcher of the career in Agroindustrial Engineering at the University of La Cañada. Professor-Researcher of the career in Administration at the University NovaUniversitas. Professor-Researcher of the career in Industrial Engineering at the Universidad del Istmo. Professor-Researcher of the Administrative Area at Bachelor's and Master's level in the Technological University of the Mixteca.

**Serra Çelik** is a research assistant at the Informatics Department at Istanbul University. She is also works as a researcher at the Computer Science Application and Research Center in Istanbul University. She received her MSc and Ph.D. degrees from the Quantitative Methods Department at the School of Business at Istanbul University. Serra's research interests are multivariate data analysis, statistical learning, data science, and digital transformation. She also gives courses in Quantitative Methods for Business, and Data Analysis.

**Jose Arturo Garza-Reyes** is a Professor of Operations Management and Head of the Centre for Supply Chain Improvement at the University of Derby, UK. He is actively involved in industrial projects where he combines his knowledge, expertise and industrial experience in operations management to help organisations achieve excellence in their internal functions and supply chains. As a leading academic, he has led and managed international research projects funded by the European Commission, British Academy, British Council and Mexico's National Council of Science and Technology (CONACYT). He has published extensively in leading scientific journals and four books in the areas of operations management and innovation, manufacturing performance measurement and quality management systems. He is Co-founder and current Editor of the Int. Journal of Supply Chain and Operations Resilience (Inderscience), Associate Editor of the Int. Journal of Production and Operations Management, Associate Editor of the Journal of Manufacturing Technology Management and Editor-in-Chief of the Int. Journal of Industrial Engineering and Operations Management. Professor Garza-Reyes has also led and guest edited special issues for Supply Chain Management: An International Journal, Int. Journal of Lean Six Sigma, Int. Journal of Lean Enterprise Research, Int. Journal of Engineering Management and Economics, and Int. Journal of Engineering and Technology Innovation. Areas of expertise and interest for Professor Garza-Reyes include general aspects of operations and manufacturing management, business excellence, quality improvement, and performance measurement. He is a Chartered Engineer (CEng), a certified Six Sigma-Green Belt, and has over eight years of industrial experience working as Production Manager, Production Engineer and Operations Manager for several international and local companies in both the UK and Mexico. He is also a fellow member of the Higher Education Academy (FHEA) and a member of the Institution of Engineering Technology (MIET).

**Lina Gozali** is a lecturer of Industrial Engineering Department at Universitas Tarumangara since 2006 and be a free-lance lecturer at Universitas Trisakti since 1993. She got Bachelor degree at Trisakti University, Jakarta - Indonesia, then she graduated Master Degree at STIE IBII, Jakarta – Indonesia, and graduated her Ph.D at Universiti Teknologi Malaysia, Kuala Lumpur – Malaysia in year 2018. She taught Production System and Supply Chain Management Subjects and her Ph.D research about Indonesian Business Incubator. She actively writing for almost 40 publication since 2008 in Industrial Engineering research sector such as: Production Scheduling, Plant Lay Out, Maintenance, Line Balancing, Supply Chain Management, Production and Inventory Control. She has been worked at PT. Astra Otoparts, Tbk as International Business Development Department for 4 years, Citibank, N.A as customer service for 1 year, PT. Pandrol as assistant marketing manager for 1 year. PT. Texmaco as merchandiser for 3 years.

**Habibah Norehan Haron** (Universiti Teknologi Malaysia, 401-500 World University Ranking, 1st Malayisa University Ranking, Kuala Lumpur, Malaysia) has achieved Presidential Award of Excellence 2015 from The Institution of Engineers Malaysia (IEM), Best student Award PhD (Engineering Education) 2012, Silver Award for Innovative Practices in Education and Industry (I-PEINX 2016), Bronze Award for Innovative Practices in Higher Education Expo (IPhex) 2014, Silver Award for New Academia Learning Innovation (NALI) 2018, Raja Tan Sri Ir. Zainal Prize for outstanding contribution to The Institution of Engineers Malaysia (IEM) article, Award for excellent services 2003 and 2014, Award of Excellence for Active Blended Learning from 2015 to 2018 and University Community Service Award 2018. She is a founder of STREAM Fun Learning Toy Library, founding member of an integrated private school and a humanitarian society. Her research project and publication are in Engineering Education, Supply Chain Management, Occupational Safety and Engineering Business Management.

**Paul F. Langer** is a researcher at the Chair for Information and Communication Management at the German University of Administrative Sciences Speyer, Germany. He frequently contributes with studies and reviews to international peer-reviewed journals and academic book projects. His main research interests are platform economics, artificial intelligence, social credits, switching costs, ubiquitous computing and digital government.

**Ben van Lier** is Director Strategy & Innovation at Centric, a Dutch IT company with offices in Belgium, Norway, Sweden, Germany, Romania and Lithuania. In this capacity, he focuses on research and analyses of developments in the interface between organizations and technology. Alongside his work at Centric, in 2009 he obtained his PhD from the Rotterdam School of Management (Erasmus University). In 2013, he was appointed Professor at Steinbeis University Berlin, a role he performs alongside his work at Centric and his ongoing research into areas such as systems and complexity theory, interoperability of information, the network centric approach and information and ethics. In 2015 he was also appointed Professor at Rotterdam University of Applied Sciences in which role he does research into the emerging (Industrial) Internet of Things, blockchain technology and cybersecurity.

**Dapeng Liu** holds a Ph.D. degree from the Department of Information Systems at Virginia Commonwealth University. His research appears in Decision Support Systems, Hawaii International Conference on System Sciences, and Americas Conference on Information Systems. His research interests include business intelligence, information security and privacy, and natural language processing.

**Michael Lo** is a President University Graduate. He currently work as the supervisor for logistic startup company covering Greater Jakarta area. His current research interests cover Logistic, Social Media, E-commerce, and etc.

**José Márquez-Domínguez** is a Professor-Researcher at the University of La Cañada since 2008, he obtained the degree of Master of Computer Science at the Autonomous University of Puebla. He is currently participating in different research projects as well as in congresses, has written and collaborated in the publication of refereed and indexed articles, book chapters and two books. The areas of interest are Pattern Recognition, Information Recovery, Augmented Reality, Technology in Education, Medical Informatics, Computer Graphics and Computer Systems Application.

**Maslin Masrom** (Universiti Teknologi Malaysia, 401-500 World University Ranking, 1st Malaysia University Ranking, Kuala Lumpur, Malaysia) has achieved Distinction Service Award, Excellence Service Award, McGraw-Hill Publishing Best Paper Award in Operations Management / Management Information Systems, Best Publishing Award (Category – Book Translation (English version to Malay version) – Multimedia and Communication Technology, 2006. Her teaching interests are Management Information Systems, Website Development, Qualitative and Quantitative Survey Data Analysis and Human Capital.

**Ma. del Rocio Morales-Salgado** obtained the degree of Doctor of Information Technology and Decision Analysis, by the Popular Autonomous University of the State of Puebla (UPAEP). He holds a Master's Degree in University Teaching, from the Universidad Iberoamericana Plantel Golfo Centro, and a Bachelor's degree from the Faculty of Mathematical Physics at the Computing College of the Benemérita Universidad Autónoma de Puebla (BUAP). His research focuses on Software Engineering and Data Science. She is currently academic director of the postgraduate courses in Information Technology and Electronic Business, as well as the Master's Degree in Data Science and Business Intelligence of the UPAEP. He was the leader of the project "System of Administration of the Resources of the International Airport of Puebla", as well as leader of the project of the "System of Tracking of Catalysts" for Volkswagen of Mexico. He served as leader of Cédula and Product Owner in the conformation of the best Testing Factory of the RELAIS Cluster of the Federal District, Michoacán and Puebla, and the implementation and transfer of the Brazilian model of Quality Evaluation MPS.Br and the SCRUM Methodology.

**Lesley Murphy** is a lecturer in Sales and Marketing in the School of Marketing, TU Dublin. She also regularly supervises undergraduate and post graduate projects. Her research interests include the role of trust – investigating relationships within organisations as well as understanding trust's role in business relationships.

**Tara Rooney** is a Lecturer in Strategic Marketing in Technological University Dublin. Research interests include consumer relationship dynamics, the process of relationship dissolution, digital and technology enabled relationships, mediated devices and customer experience, narrative and storytelling research.

**Regner Sabillon** is a Ph.D. Candidate in Network and Information Technologies (NIT) and researcher at the Internet Interdisciplinary Institute (IN3), Universitat Oberta de Catalunya, Spain. Canadian researcher in Cybersecurity, Cyber law, Cyberforensics and Cybercrime areas. Faculty member at the School of Computing and Information Systems (SCIS), Athabasca University and at the School of Information and Communication Technologies (ICT), Southern Alberta Institute of Technology – SAIT Polytechnic, Canada. ICT/Cybersecurity specialist with more than 20 years of experience.

**Beatriz Sabino-Moxo** is a full-time Professor-Researcher at the University of La Cañada, member of the academic body Computational Applied Technology. She has published several scientific articles, book chapters and books, she also works as a reviewer of scientific articles in journals of national and international impact. Since 2013 it has been recognized as a Desirable Profile accredited by the SES. His areas of interest are: Digital Image Processing, Medical Informatics, Augmented Reality and Bioinformatics.

**Miguel Sánchez-Acevedo** is a Professor-Researcher of the University of La Cañada since 2015. Director of the Academic Body Computational Technology Application and Director of the undergraduate program in Computing of the University of La Cañada. Master of Science with Specialty in Electrical Engineering from CINVESTAV Unidad Guadalajara. Phd student in Information Technology and Electronic Business at the Popular Autonomous University of the State of Puebla. He has participated as a speaker in national and international conferences. He has written several articles and three book chapters. His areas of interest are: Collective Intelligence, 3D Modeling, Sensor Networks, Machine Learning, Distributed Computing and Big Data Analysis.

**Adhi Setyo Santoso** is the Director of SetSail BizAccel, the President University Business Accelerator and a Lecturer of the Business Administration Program at the President University. He has published extensively in the area of strategic management, management information system, open innovation, strategic entrepreneurship and digital business. He also serves as the CEO for PT. Biline Aplikasi Digital, a digital agency firm in Indonesia. He has been active as a digital strategy consultant for national and international institutions as well as for major companies in the banking, media, telecommunications and other service industries. He uses his both academic and practical background to advise and mentor technology-based startups. He received his PhD in Strategic Management with a focus on community-based open innovation from Universitas Indonesia..

**Manoj A. Thomas** is an Associate Professor in the Discipline of Business Information Systems at the University of Sydney Business School. His research interests are primarily in areas of emerging technologies, data science, and social computing. He conducts research in diverse settings involving uncertain environments, non-traditional users, and unconventional application of technological solutions. His work has been published and presented internationally. His ICT outreach and research is global, taking him to Asia, Africa, Caribbean, Europe, Oceania, South America, and the United States.

**Roisin Vize** is a lecturer in digital marketing in the College of Business, TU Dublin. Her research interests revolve around web retail services and innovation adoption, business-to-business relationships in an online context including B2B networks (two-sided & multi-sided) and digital trust. Her research has been published in leading academic journals, book chapters and in international research conference proceedings. She supervises research at undergraduate, post graduate and PhD levels.

**Jan C. Weyerer** is a researcher at the Chair for Information and Communication Management at the German University of Administrative Sciences Speyer, Germany. He frequently contributes with studies and reviews to international peer-reviewed journals and academic book projects. His main research interests are artificial intelligence, big data, electronic business, Internet of things, business model management and digital government.

**Emre Yildirim** is currently Lecturer at Sakarya University/Turkey. He received the PhD degreein Production Management and Marketing from Sakarya University. His current research interests include e-commerce, social media, customization, consumer behavior, etc.

**T. Yuri M. Zagloel** (University of Indonesia, 601-800 World University Ranking, 1st Indonesia University Ranking, Jakarta, Indonesia) is currently active as Head of Industrial Engineering Post Graduate Department, University of Indonesia. He has taught several courses in Industrial Engineering in Indonesia, Australia, Taiwan, and Malaysia, including Introduction to Industrial Engineering, Engineering Ethics, and Total Quality Management. His research interests' are in Quality Management and Production System field. Not being restrained by his activities, he is also involved in laboratory activities by becoming a Head of Production System Laboratory. He has achieved Distinguished Educator Award IEOM 6th International Conference, Kuala Lumpur, 2016.

# Index

## V



## W

## Purchase Print, E-Book, or Print + E-Book

IGI Global's reference books can now be purchased from three unique pricing formats: Print Only, E-Book Only, or Print + E-Book. Shipping fees may apply.

**www.igi-global.com**

# Recommended Reference Books

ISBN: 978-1-5225-6201-6
© 2019; 341 pp.
List Price: $345

ISBN: 978-1-5225-7262-6
© 2019; 360 pp.
List Price: $215

ISBN: 978-1-5225-8003-4
© 2019; 216 pp.
List Price: $205

ISBN: 978-1-5225-5763-0
© 2019; 304 pp.
List Price: $205

ISBN: 978-1-5225-6980-0
© 2019; 325 pp.
List Price: $235

ISBN: 978-1-5225-7808-6
© 2019; 397 pp.
List Price: $215

**Looking for free content, product updates, news, and special offers?**
Join IGI Global's mailing list today and start enjoying exclusive perks sent only to IGI Global members. Add your name to the list at **www.igi-global.com/newsletters.**

Publisher of Peer-Reviewed, Timely, and Innovative Academic Research

IGI Global DISSEMINATOR OF KNOWLEDGE

**www.igi-global.com** Sign up at www.igi-global.com/newsletters facebook.com/igiglobal twitter.com/igiglobal

www.ingramcontent.com/pod-product-compliance
Lightning Source LLC
LaVergne TN
LVHW080604300926
849162LV00005B/1044